LANDSCAPE CITIZENSHIPS

Landscape Citizenships, featuring work by academics from North America, Europe, and the Middle East, extends the growing body of thought and research in landscape democracy and landscape justice. Landscape, as a milieu of situated everyday practice in which people make places and places make people in an inextricable relation, is proving a powerful concept for conceiving of politics and citizenships as lived, dialogic, and emplaced.

Grounded in discourses of ecological, environmental, watershed, and bioregional citizenships, this edited collection evaluates belonging through the idea of landscape as land*ship*, which describes substantive, mutually constitutive relations between people and place. With a strong international focus across 14 chapters, it delves into key topics such as marginalization, indigeneity, globalization, politics, and the environment, before finishing with an epilogue written by Kenneth R. Olwig.

This volume will appeal to scholars and activists working in citizenship studies, migration, landscape studies, landscape architecture, ecocriticism, and the many disciplines which converge around these topics, from design to geography, anthropology, politics, and much more.

Tim Waterman is an Associate Professor of Landscape Architecture History and Theory at the Bartlett School of Architecture, UCL.

Jane Wolff is an Associate Professor at the University of Toronto's Daniels Faculty of Architecture, Landscape, and Design.

Ed Wall is an Associate Professor of Cities and Landscapes and Head of Landscape Architecture and Urbanism at the University of Greenwich.

LANDSCAPE CITIZENSHIPS

Edited by Tim Waterman, Jane Wolff,
and Ed Wall

Routledge
Taylor & Francis Group

LONDON AND NEW YORK

First published 2021
by Routledge
2 Park Square, Milton Park, Abingdon, Oxon OX14 4RN

and by Routledge
52 Vanderbilt Avenue, New York, NY 10017

Routledge is an imprint of the Taylor & Francis Group, an informa business

British Library Cataloguing-in-Publication Data
A catalogue record for this book is available from the British Library

Library of Congress Cataloging-in-Publication Data
A catalog record has been requested for this book

ISBN: 978-0-367-47883-4 (hbk)
ISBN: 978-0-367-47882-7 (pbk)
ISBN: 978-1-003-03716-3 (ebk)

Typeset in Bembo
by Newgen Publishing UK

CONTENTS

FIGURES

CONTRIBUTORS

Anna S. Antonova works at the Rachel Carson Center for Environment and Society at LMU Munich as director of environmental humanities development. Her research brings together humanities and social science approaches to examine social, environmental, and policy change in the contemporary European context, particularly in coastal landscapes. She was previously a Marie Skłodowska-Curie research fellow as part of the ENHANCE network at the University of Leeds, where she wrote her dissertation on the conflicting narratives about environment and society emerging from the Yorkshire North Sea and Bulgarian Black Sea coastlines.

James Bird is from both the Nēhiyawak and Dënesųłiné Nations and affiliated with the Northwest Territory Metis Nation. After a career as a carpenter and cabinet maker, he earned a bachelor's degree in Indigenous Studies and Canadian History at the University of Toronto. A member of the steering committee that prepared the university's response to Canada's Truth and Reconciliation Commission's Calls to Action, he also belongs to the Royal Architecture Institute of Canada's Indigenous Task Force. Two years ago, Mr. Bird began studies in the master's degree programme in architecture at the University of Toronto's Daniels Faculty of Architecture, Landscape, and Design.

Danika Cooper is an Assistant Professor of Landscape Architecture and Environmental Planning at the University of California, Berkeley, where the core of her research centres on the geopolitics of scarcity, alternative water ontologies, and designs for resiliency in the world's arid regions. Her work incorporates historiographical research methods, data visualizations, and theories of urban infrastructure to evaluate and design for environmental and social justice. Specifically, Cooper is focused on emphasizing alternatives to the prevailing nineteenth-century conceptions that the arid lands should be overturned through technocratic solutions and neoliberal politics.

Paula Crutchlow is a performance maker and cultural geographer who collaborates as a writer, director, and dramaturg on national and international trans-disciplinary art and public realm research projects. A founding member of arts collective Blind Ditch, Paula has created unexpected happenings in everyday spaces across the South West of England, engaging publics in art practice as both thinking citizens and active spectators. As a scholar and educator she was an Associate Lecturer in Theatre at Dartington College of Arts, Devon 2001-2010, and an ESRC funded Doctoral Researcher in Cultural Geography at University of Exeter 2014–2019. The Museum of Contemporary Commodities, Paula's art geography research performance in collaboration with cultural geographer Ian Cook, was a touring exhibition and ethnography exploring the future heritage of commodity culture through lively, digital 'hacktivism', and is the subject of her 2020–2021 ESRC funded postdoctoral Fellowship.

Catherine De Almeida is an Assistant Professor of Landscape Architecture at the University of Washington. She is a certified remote drone pilot, and received her BARCH from Pratt Institute and MLA from Harvard University Graduate School of Design. Trained as a landscape architect and building architect, her research and teaching investigate the materiality and performance of waste landscapes through exploratory methods in design research and practice. Her work, landscape lifecycles, is a framework for investigating the performance, visibility, citizenships, emotions, and injustices of waste materials and landscapes. Her scholarship has been supported by numerous grants, and recognized in national and international publications and media outlets, including *The PLAN Journal, Landscape Research Record, Scenario Journal, Journal of Landscape Architecture,* and *Journal of Architectural Education.*

John Drever. Operating at the intersection of acoustics, audiology, urban design, sound art, soundscape studies, and experimental music, Drever's practice represents an ongoing inquiry into the perception, design and practice of everyday environmental sound. Drever is Professor of Acoustic Ecology and Sound Art at Goldsmiths, University of London, where he co-leads the Unit for Sound Practice Research (SPR). He is Goldsmiths' academic lead for the PhD consortium, CHASE. Drever is an avid collaborator and has devised work in many different configurations and contexts. Commissions range from the Groupe de Recherches Musicales, France (1999), WDR Studio Akustische Kunst, Germany (2011), Shiga National Museum, Japan (2012). He is a Visiting Research Fellow at Seian University of Art and Design, Japan. In the summer of 2017 he was a Guest Professor in The Department of Digital Design and Information Studies, Aarhus University, Denmark. He is a founding member of arts collective Blind Ditch.

Chris Hunt is a creative technologist, trading as Controlled Frenzy, a creative prototyping, research, and development consultancy. Chris works with clients to develop engaging audience-focussed technology prototypes, products, and installations to explore their data and ideas so they can grow and adapt to new ways of working, communicating, and doing business. Chris also delivers talks, workshops, and support on creative technology to learners of all ages. In 2019 he became a trustee of Plymouth Music Zone, was selected to be a Creative Technologist Resident of the Studio Wayne McGregor Questlab Network, and was awarded a South West Creative Technology Arts Producer Fellowship. In 2020, Chris became a fellow of the RSA.

Pete Jiadong Qiang is currently a PhD candidate in arts and computational technology at Goldsmiths and trained in architecture at the Architectural Association School of Architecture. Pete's work focuses on the specific investigation of the bridges and interstices between pictorial, architectural, and game spaces. Pete's works range from architectural drawings, paintings, moving images to augmented reality (AR) drawings, virtual reality (VR) paintings, and video games. It forms an idiosyncratic cosmotechnics of research methodology intra-acting the physical and virtual spaces with ACGN (Anime, Comic, Game, and Novel) and fandom contexts. Pete Jiadong Qiang is often referred to in architectural Maximalism.

Joern Langhorst is an Associate Professor of Landscape Architecture at the University of Colorado Denver. His research and teaching explores the processes, forces and actors that make and unmake place, space, and landscape. How humans 'make sense' of both the places they occupy and their own identity, and the role of immediate and mediated experience is fundamental to his exploration of how landscape operates as both cultural construct and cultural agent and has significant impact beyond its material performances. He is drawn to places of incisive and radical change, such as post-colonial, post-industrial, and post-disaster cities. His approaches involve multiple perspectives and disciplines, and establish a methodology he calls 'landscape forensics'. He examines how concepts such as resilience and sustainability are conceived and implemented, arguing for a 'right to landscape'. Langhorst scrutinizes the role of emergent technologies, alternative processes and the relationships between traditional and new actors and agents, and foregrounds contestation and conflict as unavoidable processes central to landscape and place change.

Ange Loft is an interdisciplinary performing artist and initiator from Kahnawake Kanienkehaka Territory, working in Toronto. She is an ardent collaborator, consultant, facilitator, and mentor working in storyweaving, arts-based research, wearable sculpture, and Haudenosaunee history. Ange works within the community, art, and education sectors as a speaker, co-creator, and advisory member. She's also a vocalist with the Juno- and Polaris-nominated band YAMANTAKA// SONIC TITAN.

Ewa Majewska is a feminist philosopher and activist, affiliated with the Institute of Cultural Inquiry (ICI) Berlin. She taught at the University of Warsaw and the Jagiellonian University, she was also a visiting fellow at the University of California,

Berkeley; ICI Berlin and IWM in Vienna. She published four books and some 50 articles and essays, in journals, magazines, and collected volumes, including: *e-flux, Third Text, Journal of Utopian Studies,* and *Jacobin.* Her current research is in Hegel's philosophy, focusing on the dialectics of the weak; feminist critical theory and antifascist cultures. Her next book, *Feminist Antifascism. Counterpublics of the Common*, will be published in 2021. She lives in Warsaw.

Jala Makhzoumi is Adjunct Professor of Landscape Architecture, American University of Beirut and co-founder and President of the Lebanese Landscape Association. In her practice and teaching, she explores place and culture responsive design solutions that are ecologically sensitive and community driven. Her areas of expertise include sustainable urban greening, landscape heritage conservation, and postwar, post-disaster recovery. Jala has published many articles and book chapters. Her books include *Ecological Landscape Design and Planning*, with Gloria Pungetti, *The Right to Landscape*, co-edited with Shelley Egoz and Gloria Pungetti, and *Horizon 101*, a reflective collection of paintings and prose on landscape and identity. She is recipient of the European Council of Landscape Architecture Schools Lifetime Achievement Award, 2019, for her outstanding contribution to landscape architecture education and practice.

Volkhardt Müller is a multidisciplinary artist, collaborator, and core member of Blind Ditch artist collective. As a German resident in the UK since 2002, Müller has developed a trans-cultural perspective of landscape practices, and the constituent factors of land experience and production. Though site specific and participatory approaches, his work interrogates rural–urban narratives, habits, and horizons of land experience and use, spaces of dwelling and transit, landscape myths and iconographies, as well as the pervasive politics of land ownership in Britain.

Mary M. Nelan is an Assistant Professor in the Department of Emergency Management and Disaster Science at the University of North Texas. Dr. Nelan earned her PhD in Sociology at the University of Delaware, and was a research assistant at the Disaster Research Center. Her research interests focus on the field of Disaster Science, specifically investigating volunteers and donations that enter disaster zones in the days, weeks, and months following an event. In addition to her research into disaster volunteers, Dr. Nelan volunteered following five events between 2010 and 2013 both in the United States and internationally in Haiti. These experiences inform her research and drive her passion to better understand the aftermath of disasters.

Kenneth R. Olwig is a Professor Emeritus of Landscape Planning at the Swedish University of Agricultural Sciences (SLU) in Sweden. He has also worked in the geography departments of five other universities internationally, and as a research fellow at universities and advanced research centres in Denmark, Norway, and California. His interests focus on a combination of aesthetic, legal, literary, and cultural geographical approaches to landscape and place, and the relationship between society and nature. These issues are the topics of two monographs: *Nature's Ideological Landscape* (1984; reprint in press) and *Landscape, Nature and the Body Politic* (2002). These, along with his recent collection of essays *The Meanings of Landscape* (2019) and his many articles, are a key part of this volume's underpinnings in the developing and associated realms of landscape justice, landscape democracy, and landscape citizenships.

Eglé Pačkauskaité is BA Architecture graduate and currently an independent scholar. Eglé co-runs a design collective duo Studio Moon, which explores speculative architectures, archives, landscapes, and atmospheres through theory and drawing at all scales. Studio Moon formed as a result of a collaborative mapping exercise, which set the scene for the projects that followed, all of which value the importance of psychogeography, with architecture being part of the landscape it belongs to and therefore possessing the intelligence of the place before being imagined and drawn. Being a Lithuanian citizen and UK resident, Eglé has always questioned the meaning of citizenship by exploring ideas of belonging, orientation (in space), and place-making.

Steven Palmer is Senior Lecturer in Physical Geography at the University of Exeter. His research focuses upon remote sensing of the cryosphere, glacier hydrology, and ice dynamics and ice sheet basal conditions, with an especial focus in recent years upon Greenland. Steven thus uses a variety of airborne and satellite techniques to investigate changes in ice geometry and motion, in order to understand the dynamic response of glaciers to recent climate forcing. Steven undertook his PhD research at the University of Edinburgh. Following this, he conducted fieldwork on the Larsen-C ice shelf in Antarctica as part of a project investigating Antarctic Peninsula ice shelf instability. Before joining the University of Exeter in 2013, Steven participated in a NERC-funded airborne geophysical survey of Greenland outlet glaciers, during which time he was based at the Scott Polar Research Institute, University of Cambridge.

Maria Gabriella Trovato is an Assistant Professor in the LDEM Department at the American University of Beirut. She has a PhD (2003) in Landscape Architecture from the Universities of Reggio Calabria and Naples. Maria Gabriella's most recent researches focuses on landscape in emergency, ADMIGOV-Advancing Migration Governance funded by the EU Horizon programme, MEDSCAPES project funded by the ENPI/CBCMED, Landscape Atlas for Lebanon, and on FLRM (Forest and Landscape Restoration Mechanism) project funded by FAO and MOA. As the chair of the IFLA group Landscape Architecture Without Borders (LAWB), she is working on landscape in emergency research with particular focus on Syrian informal settlements (ISs) in Lebanon. She has worked in several countries including Italy, Morocco, Tunisia, and Canada, and was a lead partner in EU-research programmes.

Ed Wall is an Associate Professor of Cities and Landscapes and Head of Landscape Architecture and Urbanism at the University of Greenwich. He is a Visiting Professor at Politecnico di Milano, and in 2017, was City of Vienna Visiting Professor: Urban culture, public space, and the future–urban equity and the global agenda (TU Vienna). Ed trained in Manchester (MMU) and New York (CCNY) before completing a PhD at the Cities Programme of the London School of Economics. Ed's research explores the intersection of environmental justice, urban equity, and design practice. He was awarded a research grant by the Graham Foundation in 2020 to explore global forms of architectural practice. He has written widely, most recently guest-editing *The Landscapists: Redefining Landscape Relations* (AD/Wiley 2020) and co-editing, with Tim Waterman, *Landscape and Agency: Critical Essays* (Routledge 2017). His design and research work has been published in The Guardian, Architects' Journal, Building Design, Abitare, and Arch Daily and exhibited at the Van Alen Institute, Royal Academy, Building Centre, Garden Museum, and Des Moines Art Center.

Tim Waterman is an Associate Professor of Landscape Architecture History and Theory at the Bartlett School of Architecture, UCL. His research addresses imaginaries: moral, political, social, ecological, radical, and utopian. This forms the basis for explorations of power and democracy and their shaping of public space and public life; taste, etiquette, customs, belief and ritual; and foodways in community and civic life and landscape. He is Chair of the Landscape Research Group, Vice-President of the European Council of Landscape Architecture Schools, and a Non-Executive Director of the digital arts collective Furtherfield. He is the author of *Fundamentals of Landscape Architecture*, now in its second edition and translated into several languages, and, with Ed Wall, *Urban Design*. He has recently edited two collections: *Landscape and Agency: Critical Essays* with Ed Wall and the *Routledge Handbook of Landscape and Food* with Joshua Zeunert. A collection of his essays is forthcoming from Routledge.

Jane Wolff is an Associate Professor at the University of Toronto's Daniels Faculty of Architecture, Landscape, and Design. Her design research investigates the complicated landscapes that emerge from interactions between natural processes and cultural interventions; its goal is to articulate terms that make these difficult (and often contested) places legible to the wide range of audiences with a stake in the future. Her subjects have ranged from the western Netherlands and the California Delta to post-Katrina New Orleans, the shoreline of San Francisco Bay, and the metropolitan landscape of Toronto.

John Wylie is Professor of Cultural Geography at the University of Exeter, UK, and one of the editors of the journal *Cultural Geographies* (Sage). He first became interested in questions of landscape and experience as a PhD student, and has pursued this ever since, through studies of walking, haunting, and drawing amongst other topics. In recent years, he has written on the politics and ethics of landscape belonging, and has tried to explore the intersections of landscape and creative expression through collaborations with visual artists, performers, and other creative practitioners.

Burcu Yigit Turan is senior lecturer in the subject field 'planning in cultural environments' at the Swedish University of Agricultural Sciences, Department of Urban and Rural Development, Division of Landscape Architecture, Uppsala. Her researches combine the disciplines of urban planning and landscape architecture with semiotics, political theory, cultural geography, and cultural criticism.

PREFACE

Landscape Citizenships challenges assumptions that *landscape* and *citizenship* belong in different intellectual arenas. In geography, landscape architecture, and urban planning, landscapes are defined through relationships between people and land: *landscape* means that a place has been transformed by human action and its consequences, expected or unexpected. In the social sciences, *citizenship* is defined through relationships between people and states: notions of *citizenship* highlight political belongings, rights, and identities. *Landscape Citizenships* brings together those meanings to examine conditions where people belong to a place—and to other people connected to that place—in ways that transcend and question structures of political belonging. Defined by the essays that follow, landscape citizenships include material, social, spatial associations; they encompass neighbourhoods we live in and territories we pass through. Landscape citizenships provide frameworks to explore contests over belonging, rights to landscapes, and constructions of identity in relation to particular sites and circumstances.

Landscape Citizenships began as a symposium in London in 2018, in the shadows of the Brexit vote, the Trump presidency, and heightened concerns about migration into Europe. The book has been developed and finalized in the midst of 2020's global pandemic. As it goes to press, widespread public conflict over displacement, rights, asylum, nationalism, and patriotism has only intensified, and debates about belonging have expanded to address colonial histories, Indigenous rights, and systemic racism. News stories, political agendas, and academic discourse have intersected with the lives of millions of people who struggle for rights and recognition in the streets, parks, neighbourhoods, cities, and countries where they live.

Landscape citizenships has affinities with other discourses focused on the relationships that landscapes make manifest in order to understand people's work with, lives in, and concerns for places. Notions of ecological, watershed, and bioregional citizenships have evaluated human belonging in terms of more-than-human systems and processes. Landscape justice and landscape democracy have highlighted social relationships in relation to environmental conditions. Landscape urbanism has proposed landscapes as the primary structure of city form and experience. All of these discourses offer the chance to investigate landscape relationships across scales, social contexts, political agendas, and technologies; all provide lenses to look back in time and to look forward.

From planning the symposium to editing this book, we have been influenced by Kenneth Olwig's work on landscape, particularly by his insistence that belonging to places emerges from sustained practices of inhabitation. The significance of time—of long periods of engagement in the formation of landscape citizenships—focuses on landscapes' relationships to specific temporal dimensions and thresholds. In the book's chapters, different scales of time illuminate varied (and sometimes conflicting) understandings of landscape citizenship. As Anna Antonova asks in 'Narrating Landscape Citizenship on the Coast', how long do individuals or communities need to spend in a place to claim or establish belonging? The answer varies across the conditions and issues each chapter explores, but consistent themes emerge: tensions about who belongs; contests over who has rights; and the possibility of forming identities that are simultaneously fluid, layered, and powerful.

Jane Wolff and Ed Wall, November 2020

ACKNOWLEDGEMENTS

This book emerged from the symposium *Landscape Citizenships*, convened in November 2018 at London's Conway Hall (home to the South Place Ethical Society, now the Conway Hall Ethical Society). This historic venue's independence provided an open platform (and an inspiration) to bring together old colleagues and new friends. The research they shared spoke to landscape citizenships across a broad spectrum of disciplines and viewpoints, from environmental art to land law and from marginalization to political empowerment. We thank the day's speakers: Anna S. Antonova, Dane Carlson, Danika Cooper, Catherine De Almeida, Rachael DeWitt, Joern Langhorst, Jala Makhzoumi, Christopher Marcinkoski, Zannah Mæ Matson, Mary M. Nelan, Maria Gabriella Trovato, and John Wylie. We are grateful to Kenneth Olwig for his stimulating summation of the conversations and to Shelley Egoz for her many contributions, from initial conversations about the symposium theme to chairing one of the sessions. The symposium was made possible through a research grant from the Landscape Research Group and funding from Advanced Landscape and Urbanism at the University of Greenwich. We would like to thank Suzanne Louail, Karen Ward, and Fran Reeves for their support in organizing the symposium. From the symposium to the book we have been consistently supported by the Bartlett School of Architecture, University College London, University of Toronto, University of Greenwich, Landscape Research Group, and NMBU Centre for Landscape Democracy.

The book includes essays by some of the symposium's participants and by a few additional authors. To the symposium participants, thank you for sticking with us. To the additional authors, thank you for joining us. We are grateful to all of you for your enthusiasm and commitment. Our editors at Routledge, Grace Harrison, Julia Pollacco, and Rosie Anderson, have been equally keen. And this book would not have been possible without the meticulous work of Joshua Bauman and Amelia

Hartin to bring the final manuscript together. Thank you for your dedication, energy, patience, and good humour.

Tim Waterman would like to thank Tom Moylan, who pushed him to pursue the idea of landscape citizenships during conversations at utopian studies conferences and in correspondence. He is also grateful for his conversations and correspondence with Ruth Levitas, James E. Block, Allen S. Weiss, Jala Makhzoumi, and Shelley Egoz (and an inspirational evening drinking by candlelight at her conference with Don Mitchell, Ken Olwig, Joern Langhorst, Richard Alomar, Andrew Butler, and Paula Horrigan listening to Stan Getz's 'Focus' and Charles Mingus's 'The Black Saint and the Sinner Lady' echo around the stone courtyard of Oscarsborg fortress). He would also like to thank his family, first and foremost his mother, whose passion for the Transcendentalists has contributed to the formation of his ideas. Then of course to Jason Tong whose love and support made it easier to get through the process of book creation. Then, in the spirit of 'Avuncular Architectures', his wonderful nephews Alasdair and Eliot.

Jane Wolff thanks Ed Wall and Tim Waterman for the sustained and thought-provoking exchanges that have emerged from their *Landscape and Critical Agency* symposium in 2012. She thanks James Bird and Ange Loft for the many conversations that led to their jointly authored chapter. She thanks Amir Gavriely for everything.

Ed Wall thanks Antonia Dawes for a stimulating collaboration on the *Cartographies and Itineraries* workshop; Suzanne Hall and Fran Tonkiss at the London School of Economics' Cities Programme for their careful guidance on the research of Elephant and Castle Market; Alexis Xiaotong Liu for development of GIS mappings; students and colleagues at University of Greenwich, Politecnico di Milano, and TU Vienna for informing the *Incomplete Cartographies* approach; and Antonella Contin and Sabine Knierbein for making him at home in Milan and Vienna, respectively. He also thanks his families, in particular Kristin and Carly.

INTRODUCTION

Landships

Tim Waterman

Everywhere people are courted by sophisticated media engines—political, corporate, marketing—which seek to tie their identities to brands, nations, races, sports, monarchs, deities, and shared desires and animosities. Many people's senses of belonging, likewise, are similarly formed, informed, or influenced. Citizenship, often conceived as something singular, monolithic, and bounded, can be manipulated in the same way, and rather than being seen as something which is practised and verified, is often seen as a birthright or as something which is given or claimed. Andrew Dobson writes that citizenship

> is a condition for which one requires qualification, and those who do not qualify in the relevant ways are denied it. In this sense, citizenship is a good that is distributed and, as in any regime of social justice, its distribution is, by definition, discriminatory.
>
> *(2003: 68)*

It is important to frame qualifications for citizenship in ways other than *jus sanguinis* (birthright, ethnicity) and to find more ethical ways to discriminate. Environmental belletrist Gary Snyder, in advancing the idea of 'watershed' or 'bioregional' citizenships, seeks to tie belonging to substantive landscapes, asking that we do not take land and practices of dwelling for granted, and that relations to land should be part of the definition of citizenship (1995: 223). Kenneth Olwig also frames similar ideas, grounded in a discourse of landscape justice and landscape democracy. Land, he writes, "is something to which people *belong*, as to a commonwealth" (2005: 20).

What landscape and citizenship have in common is first, perhaps surprisingly, a suffix. As Olwig explains in his book *The Meanings of Landscape*, the suffix *-scape* signifies on one hand, *shape*, and the sense of creating and carving, while it is also "cognate with the suffix *-ship*, which gives the concrete a more abstract character in the sense of a condition as in *friendship* or *scholarship*", or the sense of a condition more as a quality or state, "as with the case of *township* or *fellowship*" (2019: 130). Combined, the senses of *-scape*, *-ship*, and shape compose sets of substantive relations which have form, perhaps topography, and which have both concrete and abstract meanings which interact symbolically, representationally, and practically in everyday life. A word commonly employed in landscape practices is stewardship, which is commonly conceived as a practice *upon* or *for* rather than *with*, as the suffix should rightly suggest. There is a productive resonance here with Donna Haraway's notion of *sympoiesis*, of 'making with' as a crucial mode of being and thinking (2016: 33). This can inform and enrich the idea of stewardship, (and other -ships) as a quality, state, or condition that is mutual and shared. Were it not such a startling neologism, we might have named this book simply *Landships*, and indeed one of the contributors, Catherine De Almeida, has taken up the term to describe landscape as it is commonly conceived in scholarship today, as a relationship between people and place—topography, culture, work, play, ecology, climate, language, politics—that is mutually constitutive. Note that a plural is necessary here. Landscapes and land*ships* are always particular.

Citizenships, conceived as landships, also ask that the criteria for belonging issue not from a given, but from affinity, experience, and applied landscape knowledge and interactions. People might come to belong to a landscape through work, inhabitation, and showing an understanding of its operation, and to have their citizenship approved and validated by others for the same reason. This book conceives such citizenships as the next necessary evolution of the developing fields of landscape justice and landscape democracy, and the idea of the right to landscape, so compellingly advanced in the work of scholars Shelley Egoz, Gloria Pungetti, and Jala Makhzoumi (2011), two of whose work—by no accident—appears in this volume. Indeed the genesis of the ideas presented in this book can be traced back to 2015 to the 'Defining Landscape Democracy' conference held in an island fortress in the fjord south of Oslo. The book which resulted from that conference is an important companion to this one (Egoz et al., 2018).

It is also no accident that many contributors to this book are landscape architects. In recent decades the work of landscape architecture has been continually enriched by the burgeoning and increasingly post-disciplinary field of landscape studies, which deals with landscape relationships—urban, rural, and everything in between—across the widest set of scales from the local to the continental. Landscape architects have increasingly become involved not only with the traditional disposition of plants and geometry, but with the design of landscapes as the design of relationships and an engagement with processes and forces both organic and inorganic, and social, political, and cultural. Thus, the key technique of design is now not the deployment of colour, form, and materials, but the ability to work through complex scenarios which envision processes of dwelling. Scenarios are also crucial tools for moral, ethical, and political judgement, and it is this dimension of design which is increasingly leading landscape architects to ask political questions and to insist that the political most often arises from, and comes back to dwell, in substantive landscapes.

This is not to say that landscape is not still a tool for the representation of earthly wealth and power. Landscape has an optimistic side and a dark side: bottom-up struggles and top-down forces, and it is this opposition that helps to make it such a useful term. On the one hand it is a relation between people and place, and on the other it is a fixed, reified, commodified representation of the state, of wealth, and/or of overpower. The first provides a model of citizenships as lived, practised, negotiated, researched, and shared, while the other exists as an image to be transmitted and which binds people together around an abstract set of ideals. Substantive landscapes to which people belong are by definition practised places, and citizenships within them are affirmed by the relational practices of everyday life. Symbolic landscapes signify citizenships in a different way: these are given or claimed, often as birthright, and are often collapsed into racial or ethnic identities, often in sinister ways in blood-and-soil forms of nationalism. The essays in this book agitate these notions, seeking, as does for example Joern Langhorst, to embed citizenships in lived landscapes, or as does Burcu Yigit Turan, to question the nature of citizenships as given, who has the right to a claim, and through what mechanisms they can be empowered to make that claim.

This book rests on the idea that equality, justice, and democracy are social practices that *take place* and that citizenships, thus, of necessity are also enacted in places as forms of conduct which require solidarity: "mutuality, accountability, and the recognition of common interests as the basis for relationships among diverse communities" (Mohanty, 2003: 7). Equality, justice, democracy, freedoms, and citizenships are mutually assured, practiced, and verified through everyday life in substantive landscapes and through the work of the political and moral imagination, focused on places while transcending borders.

Bibliography

Dobson, Andrew (2003). *Citizenship and the Environment*. Oxford, UK: Oxford University Press.

Egoz, Shelley, Deni Ruggeri, and Karsten Jørgensen, eds. (2018). *Defining Landscape Democracy: A Path to Spatial Justice*. Cheltenham, UK: Edward Elgar Publishing.

Egoz, Shelley, Jala Makhzoumi, and Gloria Pungetti, eds. (2011). *The Right to Landscape: Contesting Landscape and Human Rights*. Farnham, UK, Surrey, UK, and Burlington, VT: Ashgate.

Haraway, Donna J. (2016). *Staying with the Trouble: Making Kin in the Chthulucene*. Durham, NC and London: Duke University Press.

Mohanty, Chandra Talpade (2003). *Feminism Without Borders: Decolonizing Theory, Practicing Solidarity*. Durham, NC and London: Duke University Press.

Olwig, Kenneth R. (2005). 'Representation and Alienation in the Political Land-*scape*'. *Cultural Geographies*, 12(1), pp 19–40.

Olwig, Kenneth R. (2019) The Meanings of Landscape: Essays on Place, Space, Environment and Justice. London and New York: Routledge.

Snyder, Gary (1995 [1992]). 'Coming into the Watershed'. In: Snyder, Gary ed. *A Space in Place: Ethics, Aesthetics, and Watersheds: New and Selected Prose*. Washington, DC: Counterpoint, pp 219–235.

1

LANDSCAPE CITIZENSHIPS

A conversation among treaty people

James Bird, Ange Loft, and Jane Wolff

Why begin this mostly theoretical book with a conversation about tangible experiences and observations of a particular landscape? From the start, landscape citizenships are personal: they arise from a sense of connectedness between people and places—and between people and other people who care about the same place.

Landscape citizenship in Canada is neither abstract nor uncomplicated. Our country is famous for welcoming immigrants and refugees, but its settler colonial culture has disenfranchised and dispossessed the Indigenous peoples who have been here since time immemorial. Canada's foundations rest on treaties, nation–to–nation agreements between Indigenous peoples and European colonizers to live as allies on this land.[1] Every Canadian is a party to those treaties: a treaty person. And every settler Canadian bears responsibility for the fact that the Canadian government has not honoured its treaty obligations.

Being a treaty person means knowing that landscape citizenship, like all questions related to land, exists in a web of relationships with Indigenous compatriots. It requires conversation. And so, during the first summer of a pandemic that has shed a harsh light on longstanding crises of inequity and injustice in the city and country I call home, I asked my friends, colleagues, and guides James Bird and Ange Loft to talk with me about how we might understand landscape citizenship together. A few words about the three of us:

James is from both the Nēhiyawak and Dënesųłiné Nations and affiliated with the Northwest Territory Metis Nation. After a career as a carpenter and cabinet maker, he earned a bachelor's degree in Indigenous Studies and Canadian History at the University of Toronto. A member of the steering committee that prepared the university's response to Canada's Truth and Reconciliation Commission's Calls to Action (Steering Committee for the University of Toronto Response to the Truth and Reconciliation Commission of Canada, 2017), he also belongs to the Royal Architecture Institute of Canada's Indigenous Task Force. Two years ago, James began studies in the master's degree program in architecture at the Daniels Faculty of Architecture, Landscape, and Design. His wisdom, experience and generosity have made him a teacher to many people at our school, including me.[2]

Ange, an interdisciplinary performing artist and initiator from Kahnawake Kanienkehaka Territory, works in ways that defy classification: her endeavours include arts-based research, oral history, outdoor performance, community art design, wearable sculpture, and project planning. As Associate Artistic Director of Jumblies Theatre, she initiated and directed *Talking Treaties*, a collaboration with historian Victoria Freeman in which film, spectacle, dance, large-scale puppetry, sculpture, and humour are used to examine the controversial treaty over land that now comprises most of Toronto. She created the *Toronto Indigenous Context Brief* for the Toronto Biennial of Art's Advisory Councils for 2019 and 2021, and she is a vocalist with the band YAMANTAKA//SONIC TITAN, which has been nominated for the Juno and Polaris prizes for Canadian music. Ange and I met in connection with the Toronto Biennial. From our first conversation, she has been an inspiration and mentor.[3]

And I am an associate professor at the University of Toronto's Daniels Faculty of Architecture, Landscape, and Design. My education in documentary filmmaking and landscape architecture has led me to work that investigates and articulates language for complicated places subject to change. The methods use drawing and writing to tell documentary stories that represent different ways of understanding the landscape; the goal is to offer terms for discussion that can be broadly shared among the range of audiences with a stake in the future. At present, my work concerns the everyday landscapes of Toronto, where dynamic processes and phenomena are hidden in plain sight.

James, Ange, and I are all Torontonians, and like many people here, we came from someplace else. Toronto is the largest city in Canada. Established in the late eighteenth century as a colonial outpost on the shore of Lake Ontario, it expanded across (and transformed) a glaciated landscape dissected by rivers and streams. In recent decades the city has become the centre of a metropolis, and arrivals from across the country and around the world have made it extraordinarily diverse. The municipal government estimates that there are 70,000 Indigenous people living here. That makes Toronto home to the largest Indigenous population in the province of Ontario and the fourth-largest in the country (Indigenous Affairs Office, n.d.). Common stereotypes imagine Indigenous cultures on reserves and in the past; many people don't know that the city is home to such a vital, substantial Indigenous community.

James and Ange had not met until this conversation. Instinct told me that their points of view would resonate, and our triangle gave rise to wide-ranging questions, subjects, and insights about the importance of rootedness; the excitement of movement; the problem of exclusion; the need for ritual and ceremony; the presence of the sacred; the liveliness of language; the knowledge in names; and the power of kindness and care to engender meaningful relationships among people and with places. Together, these topics sketch an outline for landscape citizenship. What follows is a record of our exchange, held over Zoom and edited and condensed for clarity. *JW*

Jane: When my colleagues Tim Waterman and Ed Wall asked me to work on a book about landscape citizenship, I began to ask myself what that means for a treaty person. I know that I'm only a little part of the picture in the place where I live (and where I feel like I belong). Rather than writing from my point of view, I wanted to have a conversation with you—like the conversations we've had over the last couple of years about so many things relating to our lives here and together. I didn't send questions in advance because I'd like to start with what you think and see where that takes us. I hope that's okay.

Ange: Yes, but I want to hear what you think first.

Jane: Because of this project, I began reading about ideas that connect the meaning of landscapes to people's processes of participation in a place over time.[4] For me that was interesting: I'm a member of Jewish diasporic culture that didn't have a particular attachment to place for thousands of years. I'm an immigrant citizen of Canada and a settler.

The question of citizenship seems very freighted to me, and I've begun to think more about *belonging*. I feel at home here because I have relationships that make me feel at home. Some are to people who have kindly welcomed me here—like you!—and some are about my interactions with the landscape—like the way water from Lake Ontario goes into my body and sustains me and then comes out of my body and goes back into the lake. Just walking around the city has caused me to understand a set of very old processes in a new way. I've never lived in a landscape that was so clearly shaped by glaciation. And until I came here, I really didn't have awareness of myself as a treaty person, even though I came from settler culture in the United States. I think about my belonging here as having to do with generosity—by people and by the place itself.

Ange: Belonging has been tied a lot to thoughts about place and homelessness. We're being displaced. We're also being priced out of our neighbourhood. I've lived with a bunch of roommates in the same neighbourhood for many years, and now I'm moving to a place that's going to be different. If I ever choose to move back, I'll be paying much more. To be able to belong to a place actually involves the financial ability to have a home. It's impossible for me to own a home in Toronto, and it always will be.

I think about how very limited my access to outdoor spaces is because I don't have any land and because I don't have any space to go. My community is six hours away. Even in my own community, I don't have any land access. I'm actually very restricted from feeling like this place is my own, even though I've been in this city for 11 years. Having your own house makes it feel like you have a space—and makes it feel like this land can own you. As you say, your water moves through the same pipes every day.

We're growing squash now, but it won't be mine because I'm leaving before the end of the harvest. I'm moving to Scarborough for about a year. I'm moving in between two tributaries of the Don River,[5] and it's going to be a 15-minute bike ride to get to any river edge from this particular spot. That's going to be different. Right now my closest access to a river would be biking towards the Humber,[6] which would take me about half an hour. Or I could bike down to the waterfront, where I don't want to go. I never want to go to the downtown core these days.

Even if you put down roots in Toronto, they're always incredibly precarious. As a community artist, the ability to say this place owns you and that you make art for and of this place is completely destabilized by not having a house and not having a place to call your own and not having a place to put your things. It's going to make a huge difference to leave my small Toronto West/Dufferin community. It's going to be a whole other world. It's hard for me to think about belonging to a place when it's financially unfeasible.

James: When I think of landscape citizenship the first thing that comes to mind is the Dënesųłiné[7] ways of knowing the land and our relationship to land. We [Dene people] believe that our language is from the land and that the land is the language. This interwoven relationship has a metaphysical aspect built into language.

When I went up north this last week, I brought my medicines[8] and my eagle feather that we used in Sun Dance. It was a nice connection to the land because I slept on the land. I had some ceremony, which really helped me connect to the land in my Dene ways of knowing the land.

It's interesting what's happening with all of us during this pandemic and how we're forced now to really think about who we all interact with space, and how our space (at least my space) has become very sacred. I've examined every corner of my space, being stuck in it, in the beginning. I had to re-imagine my little apartment [in] downtown Toronto: my home office, my studio, my everything. I do ceremony at home. Many of the Sun Dances were cancelled this summer; the PowWow at Six Nations was cancelled; the one in Albuquerque that is very popular in the summers, the Festival of Nations, was cancelled. All the big dances. I've had to reinvent my own ceremony with land.

When we talk of this landscape citizenship, I only speak from my own worldview and my own cosmology. For me, the use of land is so based in language. I noticed when I went to Bon Echo, in Algonquin Territory, that I was speaking more of my language. I was speaking to the land. There's a ceremony when we're getting ready to bury somebody: we begin to speak to the land where we're going to put that person. I felt that connection again. It was a healing week to be out.

This is a chance for all of us to heal through the land because the land is giving us an opportunity. The Dene and the Hopi have a prophecy about this time, and the Navajo as well. In essence—in English—the prophecy talks about Mother Earth taking back her sovereignty. This time is about that. We're at a pivotal point in our journey on Earth, not just Native people but everybody. It's everybody's concern and everybody's responsibility to really look at our relationship to the land and the spirit of the land. I am letting the landscape take precedence and listening to what we need to do. When [the pandemic] started, the elders met in northern Alberta, and they talked about the prophecies and about what's going on. A lot of them were enthusiastic and excited about this time because it's no longer about us.

When I saw the title *Landscape Citizenship*, the first thing that came to my mind was speaking land. Land-speaking. The language of the land. That's my take.

Ange: I'm never home in the summer. I'm always on tour, or working on a show. I do outdoor theatre and travel with my band. It's weird for me to be here—to be home so much. I've been driving from place to place and going so fast for years. I'm really destabilized when I'm in cars or planes or under water too deep. I like to be on the ground level. I don't like to be in apartment complexes that are too tall, or things that are too far off the ground. All my family are iron workers, so there's no reason for that, except for the opposite being born: that fear got passed down.

I do prefer working on projects in one spot for a while. After spending enough time in a certain spot, following a specific track, and taking a specific path, I can finally start to settle down. But then I like to find a new thing to throw myself into. I definitely feel comfortable on the road and in situations of high stress. How does that connect to my landscape citizenship? I find that, as an artist and as somebody who makes things, I go hunting for that feeling of making complications for myself and feeling really excited.

James: Ange, do you think that's where your inspiration comes from?

Ange: It's been hard to stay still. I only go to three places in the whole city now! I get my food from the same Portuguese deli, I get my fruits from the same Economy Fruit store, and that's it. That's all I do now. My landscape is literally just the path I walk between my house, to those stores. I would usually be off in a different town or across the world. Our networks are so wide—it's weird to feel like they don't exist.

James: Ange, I feel the same way. My little neighbourhood now is a village within the city. It's a bit like that on the reserve, too. How vulnerable I am within this system. We're so dependent on electricity and running water. If these things went out, what would survival be like in this little community that I've formed? I could survive in the bush because I was raised on a trapline,[9] but in the city it's very different. A time to carefully consider my own footprint on the landscape.

Ange: Yeah, it's really odd. I've been trying to learn to make everything at our house. All I do is look up Depression-era teachings—find out, "How do people make soap?" I need to learn how to make rope. I haven't started yet, but I want to! I'm going to take a thousand steps back … I don't know what's going to happen when we don't have the Internet, so I can't find out how to make my new white corn flour cookies. White corn flour—that's my new thing. You can go to South American stores and find something very close to Iroquois-style white corn. So my new experiment is, shop only from the tiny grocery stores within my triangle and pick up bags of dried hominy, cancha corn, ground white corn masa, all the different styles of corn and make the best Iroquois stuff I can, wherever the corn came from.

I think a lot about adaptability, about people who've had to move significant distances from their homes, international people who live here in this area. Everyone else had to adapt their stuff …

James: My friend in Forest Hill, who lives in privilege and opulence, has forced himself to dig up a corner of his lot and plant a garden. During the Second World War, there were victory gardens where people would plant food. He's finding himself in a very weird place with his privilege, and he's coming out of his comfort box because he doesn't know how long he's going to have his privilege for, he is now questioning his place within a privileged system. In fact, what we're all sharing here is kind of like—landscape citizenship stories.

Ange: So a good friend of mine used to live two doors down, and she had a fence between her house and my house. On her fence, she planted a few raspberry bushes but then she had to move, priced out like everyone else in the area. And now the raspberries are huge, and I can't pick them because they're on the other side of the fence! I see all of this fruit from my backyard, and it's dying. It's not mine, and it's on the other side of the fence, but *is* it mine? My friend planted it, and she said I was allowed to go get it. I get a kick out of those places. Even in my own community, I got yelled at for picking raspberries on the way to my mom's house. It was outside of a store I used to work at, that'd been closed and grown over with berry bushes. Someone yelled: "Don't take those raspberries, those are my kids' raspberries!" And there were so many raspberries and I could reach them from the street. I really wanted to yell back, "I used to work here!"

When does ownership of a place stop, and when do you just give up on gates and fences and lines, and pick the berries? Thinking about Indigenous sovereignty and land practice, I'm like, "I'm just taking the berries, I don't care!" They're going to rot if you don't take them. I've been playing by the Western rules for a long time, and I've seen that they haven't given me anything. So now I want to learn to plant things and make seed bombs … I bought a bunch of heritage, heirloom seeds, and couldn't use them all in my garden. I could theoretically just drop them all over my neighbourhood.

I love when people use the municipal land areas for planting, even though it's totally illegal. When I first got here, I used to walk along the train tracks instead of the roads. Along some of the tracks, there were full gardens. People are growing food, just growing it there, anywhere with dirt. I know that we don't want to plant underneath wires or over things we're not supposed to, but I've been looking at all of the unused ground in the city lately, and at all of the untapped greenspaces … Now I'm growing vertically: squash and all these old little melons and things like that. So I could feed my house and two more houses next to me for a good week at the end of the summer. What would happen if we all did that, if we all put our extras out on the front porch? What would happen if we were *allowed* to do that and it wasn't littering? There are so many municipal regulations on urban farming. Who has space to grow food? Who is allowed to have chickens, allowed to have different kinds of self-sustaining things—not places that have lower rents. I just so happened to have some yard space in a shared house. Most people I know have nowhere to grow.

I've experienced more subtle racism in garden centres than anywhere else, and it's happened at multiple garden centres over my life. One time someone wouldn't sell me and my mom a bag of dirt (potting soil) in Chateauguay, for not speaking French. I accept that these places are places of privilege, places where you can get earth and you can make things grow—and you have to pay 50 dollars to make that happen. The people that go there can afford to have yards. I've been shocked at the difference having money, having that land, and having that municipal freedom in your area makes for becoming self-sustaining. If I want land access, I'm going to have to pay for it first. If we ever wanted freedom for artists of colour to be able to grow food outside, to get away from the city and go onto a piece of land, we're going to make that place—because nothing like that exists right now. If we go for it, we're going to have to leave the city.

Jane: I'm living about ten or twelve minutes' walk from Wychwood Park,[10] where Taddle Creek was dammed to make a pond. I usually go there in the morning to see the trees and turtles and water—there's so much life! And I've thought a lot lately about how strange it is that the creek became a pond because of the will of the people who owned the property around it. There was a time when I would have felt shy to walk around that neighbourhood every morning—like it wasn't for me. But there's something about the pandemic has made me think, "Well, I'm just going to go there anyway!" It's not harmful to anybody, and it helps me so much to follow the creek—to listen to the creek. I start to hear it in the catch basins as I'm walking there. I see it because I see the grade of the streets changing.[11] It's made me ask myself how this living space can be part of something that's called a private street. It feels strange to me.

Ange: I've been trapped with only what I have, so I've been doing all kinds of experiments. I had an installation at MOCA last year made with PVC pipes inside of net sculptures, and my garden is made out of those. I'm using up old art pieces because at this point in my life, the physical things I make—I make big things, spectacle stuff—it's all tools. In one day everything got flipped to being tools: materials and tools. The art doesn't matter in this context. I could put them back in the art pieces, later, but so much has changed.

James: I've been talking with Douglas Cardinal[12] about this project we might do with rock. He forms a lot of his architecture on the organic. Often before he will work on a building, he'll do sweat lodge[13] or he'll do ceremony on the land, and he says to me how important and how almost obvious that is now.

His architecture hasn't changed because of the pandemic. I think he truly would say, "This is what I've been doing for fifty years as an architect". He goes and spends time on the land before he builds a building—he believes in the metaphysical connection with land and what he's about to do. And he believes he gets that inspiration: he believes in the metaphysical part. The spiritual is thinking that's part of the 'box' for him—that's not *outside* the box. So his architecture is part of that spirituality, that connection to land.

Ange: With the *Talking Treaties* project I've been leading, I've been trying to tackle some of those questions for myself, around the sacred in relation to land and place. I've been hesitant around speaking about sacredness. It turns some people off, so I play symbol games instead to try to get around it. We learn some history, come up with images, reinterpret other people's pictures, create texts and performances. I've been using theatre games and art research to find the right symbols to talk about these big concepts.

It's strange doing this from outside my community and without keeping up my connections to people I could have learned these things from [in] my own town.

FIGURE 1.1 Following Taddle Creek. It runs beneath the street but comes to the surface at Wychwood Park (credit: Amir Gavriely)

I kicked myself out of being a traditional person because I thought I was too queer and decided to live in the city instead. Now, I'm finally taking time to re-learn some of those foundational practices I've been avoiding for a while … Doing the Ohèn:ton Karihwatéhkwen[14] in your own words, putting your own perceptions into it and making those connections. And going through and introducing yourself

FIGURE 1.2 "A Foreign Source of Extraordinary Power" by Ange Loft for MOCA (2019)

FIGURE 1.3 Hand moving homes around Toronto's lost rivers; "By These Presents: 'Purchasing' Toronto" (credit: Jumblies Theatre and Arts)

to your plants. I did this in an activity with Jill Carter,[15] where we tracked out personal 'Kin-Stillatory' relations and introduced ourselves to our plants. It's the oldest idea ever! I watched an interview with Rick Hill,[16] where he talked about another old idea. We say that the next generations are "The faces yet to rise from the earth", and he wondered what that meant. So he was talking about digging a

FIGURE 1.4 The confirmation of the purchase; "By These Presents: 'Purchasing' Toronto" (credit: Jumblies Theatre and Arts)

hole with his hands and finally seeing the little spurts of water that are coming from underneath the ground, all of this life starting. I'm 37 and I'm finally making these connections. And I think that's because of spending the past 20 years in the city and never putting my hands in the ground that whole time. I've been talking a big talk but not walking it. It's been humbling to put my energy into those things which we say are foundational—but I never put the work into making it a foundation for myself. I guess I may become a different artist if I pay attention to these things. Or slow down.

James: What were your ideas, Jane, about landscape citizenship, and what was your motivation for this?

Jane: Well, a couple of years ago, when my colleagues Tim Waterman and Ed Wall asked me to work with them on a symposium about landscape citizenship, I said yes. The people who submitted papers came from Europe, the Middle East, Canada, and the United States, and the day's discussions centred on affinities—on ways in which people felt they belonged (or didn't belong) to different landscapes. And because of the symposium, I became acquainted with Olwig's and Ingold's writing about how people build meaningful connections to landscapes through practices and over time.

I had been thinking for years about what it means to belong because of the experience of being an immigrant. I chose to leave the country I came from because I felt that I *didn't* belong, even though it was the place that made me (and the place where my great-grandparents and their families became citizens after being persecuted in Europe). So I began to think about who I was in this place, and who I had been in other places I'd lived. And I began to think about the development of love for a place—not something we talk about in my world of rational work at a university!

Why did I come to feel at home here? I came alone, without knowing anyone, and to my amazement, I began to feel like I *did* belong. It was because of people's kindness and also because of the chance to observe the landscape and to see how my own life was part of a set of living relationships. But the structures that allowed me to become a political citizen—and a landscape citizen—are exactly the structures that have led to the disenfranchisement of Indigenous people. And so I have to find ways to acknowledge that we share this place.

I wanted to talk to the two of you about what all of this means because your work and your friendship and your willingness to share who you are has changed my idea about who I am here—and that's something for which I feel so grateful.

James: You know, the whole conversation—also some of your art, Ange—reminds me of something. When I made a conscious decision to heal from Residential School,[17] part of it was going back to old teachings of landscape and land, and how I would allow the land to help heal me through ceremony—but not only that, how I would force myself to build relationships with settler people because that was part of the healing. And I heard a Holocaust survivor say the same thing, that part of his healing was meeting German people again and helping to build relationships forward. Part of my healing was building these new relationships so I could feel comfortable in a society I had no choice but to live in. And I think that landscape has been a huge part of this healing. Land and landscape, they don't have agendas. They're pretty loving environments. When I walk amongst the pine, these big needles don't care if I'm black, Jewish, First Nations, whatever, and so there's this incredible presence of natural love. And there's so much of it in Canada. It's pretty much virgin.

I just wanted the landscape to be part of the healing, and it sounds like what happened for you, Jane. Discovering yourself, you had to get out of where you were. I left the Northwest Territories 30 years ago, one, because I felt really threatened because of my sexuality (even though we have ancient teachings about being two-spirited and gay) and the other, I didn't feel comfortable even though it was my homeland, and I came here to discover myself. And through that process much more has happened.

It's interesting to talk about this stuff. You talked about the garden stores … A friend of mine, an Elder, a gardener, said that every garden centre should have a land acknowledgement above the entrance: this dirt we have to buy now; these plants that are growing in this soil. And I think that would be a powerful move. I know it would piss off a lot of people, but it would force us to think about the land. It would ignite conversation that is silent, which is sometimes the most important conversation, and the biggest conversation. Wouldn't that be an interesting thing, a movement for land acknowledgements at the garden centres?

Ange: Even if everyplace put a list: "This is actually an invasive plant. Don't plant this". Or a re-naturalizing zone at every garden centre.

James: Plants often have a Greek or Latin name. How about the Haudenosaunee name, or the Algonquin name, or the Anishinaabe name? One of the things in decolonizing landscape architecture would be to acknowledge these plant names in the Native language. [18] We do a land acknowledgement, but how about if we acknowledged the plants, too? I'm not Haudenosaunee, and I can't speak for them, but if you talk to a Haudenosaunee Elder, each plant has its own story. Each plant is a storyteller.

Ange: We're talking about the logic of introducing yourself by your name, and learning the plant's name in whatever language they're from, because things like to be called by their name.

Jane: James, one of the first conversations I had with you had to do with trees. If I remember, you said that in your language, the words for trees have to do with the life of the tree …

James: Well, the spirit of the tree. Each type of tree has a different spirit. When I built that Truth and Reconciliation grove, an Elder told me to put a tree in the centre of the memorial park—not a stone monument but a living tree.

It was going to be a sacred maple from the Haudenosaunee, and how well we did with reconciliation would be reflected in how well the tree grew, and if songbirds would nest in there. These would be certain signposts as to how we're really doing.

I had a whole tree teaching with a West Coast Elder taking me to an old growth forest in a ceremony where I was introduced to a 1,000-year-old tree as a living being, as a place maker where it stood. This tree knows its territory and has knowledge of a 1,000 years of stories of standing there. The Dene people have our own tree teaching for the trees that are native to my homeland; the Haudenosaunee have their sacred tree teachings of the pine and the maple.

FIGURE 1.5 Architectural model of Reconciliation Grove (2017); model and photograph by James Bird

Jane: So many of the things you've both talked about have to do with the recognition and acknowledgement of what's around us—maybe that's a way to begin to think about what it means to belong.

Ange: We did an interview with Lee Maracle[19] quite a while ago for *Talking Treaties*, and she talked about treaties with the trees, too, from the West Coast. But then, when talking about the *Dish with One Spoon*,[20] she pretty much said, "If you want to make any agreement with me going forward, that's going to be the first thing that you start with for this area, so start there". And it's true, if all of those practices of caring for the dish are in place, then I'm comfortable. Take only what you need, make sure there's some left, keep it clean. If those are in place, I'm good. Those are the things that make me the most comfortable, actually. You can build on and apply the *Dish* to so many aspects of your life. You can use it to divide up the last three hotdogs when you have a family of ten people: how are you going to deal with those three hot dogs? Take your share.

If a place is clean … I could be in a bombed-out building in Western Europe sitting on a clean stone windowsill, it can feel comfortable. If there's enough to go around, where I can feed myself, where I can help other people get access to the same things that I have—a place where there's something left for the future.

Longevity's a really big thing with me, with projects and locations and places and activities happening in a place. I don't want to take part in projects that drop into a community, do something grand, and then disappear. I want to know that people will be able to learn skills and carry things on afterward. I want to work within the trajectory of a project and not feel like I'm dropping myself in out of nowhere. I like stepping into stuff that's already rolling. I like working where somebody's already got an idea going and where other people really know the area.

James: As Ange says that, Jane, I think of artists and architecture or landscape, and I think it's becoming landscape's time. One of the things that I think architecture really suffers from [is] its horrific ego in built form. Architects build monuments to themselves. We get way off track, way out of whack with reality. When I came to architecture, one of the prime motivators was to answer the question of horrific housing conditions on my reserve, or in Attawapiskat: third-world housing. That's always been my focus, and it will always be my focus. But now landscape has an opportunity to be investigated in ways it's never [been] before.

One of the things I like about Ange's art: it's so interactive that you can't help but look at the connectivity. I think it's very therapeutic, and I wish architects would go along with that. I'd love to work with you.

Ange: You know what: I've been asked by multiple non-Native architecture firms to partner on their Indigenous proposals, and it has always happened to me between one and three days of when they need to put their proposal in. There's something wrong with how architecture is approaching Indigenous people. Definitely.

James: Very wrong indeed.

Ange: And I haven't said yes to any of them, because clearly there is no relationship. I don't know anything about them, I look up their firms, and I find out the history of some of the other people that have been brought in to be part of these municipal bids …

James: A lot of times, it's so they can get one of the boxes checked off, to make the proposal look good, look presentable. We've formed a consortium now through the Royal Architectural Institute of Canada called the ITF, the Indigenous Task Force on Architecture. We're going to start building relationships with Indigenous art minds—artists, sculptors, I do believe their insight is so important to built form.

Ange: I was looking at your reconciliation structure as a performance space. When I saw the model, I thought, "Oh wow, I can see exactly how to activate the space—how and where to put things, and what should happen in which corners, what break-off places seem like they're for what types of activities".

When I'm facilitating groups, I do this thing I call 'scooping a room', making sure all of the people on the outsides are engaged and that they know what can happen in the middle—but they don't have to come in. A lot of these community arts and performance practices translate to architecture really easily—think about breakout rooms and hiding spots. Decision making is often done in a big circle, where you can see each other, and be transparent. This is again, connected to the *Dish* and that idea of introducing yourself through your kin networks. I'm going to be using some of the ideas in the *Dish* to make some dance work this year. As part of it, I'll be hosting a few dance workshops [with] grown and senior women of colour, because, when I'm talking in university settings, returning students are always paying the most attention. Who can take up these teachings and wrap them into their lives? I think that grown women are the key to this happening. They've been navigating the social sphere; they've been moving around the city for so long. They know their paths, they know what is what, and they know what is important here. And their kids know those same paths, they're the stewards of this city. So those people are the key to making any kind of information change. Now, imagine if all those paths grew food …

James: That really sounds like an interesting project, Ange.

When I was introduced to the Elder from the west coast, I began to build a relationship with him, this was the reason the project worked. It's the importance of building relationships; both with the land and the knowledge keepers of the land. I think that for successful Indigenous architecture to work, building long-term relationships with people like yourself is integral to the design. And that's what Douglas does. For him it's all about relationship-building—and it's an obligation to his architecture. This reminds me of the wonderful book by Shawn Wilson called *Research is Ceremony: Indigenous research methods* (2008). He shares a different research paradigm, an approach like what I think *landscape citizenship* could inspire in design.

Jane: A last question: because we're all people who draw, would you like to make a little drawing for the book?

James: A drawing of this conversation today? I like that! What words can't do but imagination and spirit see.

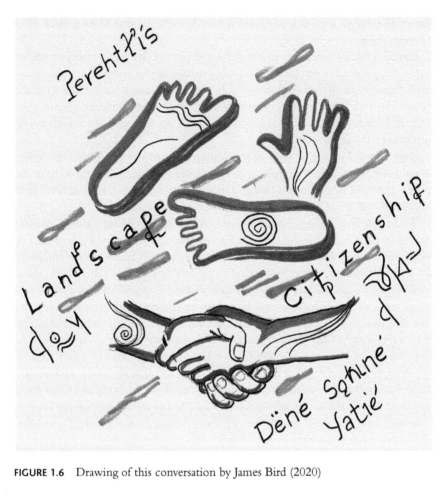

FIGURE 1.6 Drawing of this conversation by James Bird (2020)

FIGURE 1.7 Drawing of this conversation by Ange Loft (2020)

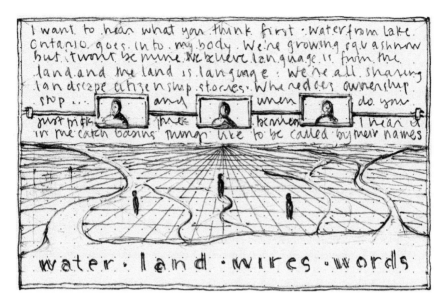

I want to hear what you think first. water from Lake Ontario goes into my body. We're growing squash now but it won't be mine. We believe language is from the land and the land is language. We're all sharing landscape citizenship stories. Where does ownership stop... and when do you just... in the catch basins things like to be called by their names

water · land · wires · words

FIGURE 1.8   Drawing of this conversation by Jane Wolff (2020)

Notes

1 For an introduction to Canada's treaties, see "We are all Treaty People", www.oise. utoronto.ca/abed101/we-are-all-treaty-people/ and "Canada by Treaty", canadabytreaty. cargo.site/Stack-2. (Accessed 5 October 2020). For settler citizens' responsibilities toward reconciliation, see Truth and Reconciliation Commission of Canada (2015).
2 Biographical information compiled from "Truth and reconciliation at U of T". www. utoronto.ca/news/truth-and-reconciliation-u-t; "U of T to Take Action on Truth and Reconciliation Commission". www.magazine.utoronto.ca/campus/u-of-t-to-take-action-on-truth-and-reconciliation-commission-jonathan-hamilton-diabo-james-bird-megan-easton/; and "Indigenous U of T students represent Canada at the Venice Biennale of Architecture". www.utoronto.ca/news/indigenous-u-t-students-represent-canada-venice-biennale-architecture.
3 Biographical information compiled from "Ange Loft: interdisciplinary performing artist and initiator". angeloft.ca/; "Ange Loft—*Canadian Art*". canadianart.ca/author/ange-loft/; "Jumblies Theatre: Ange Loft". jumbliestheatre.org/jumblies/about/staff/ange-loft; jumbliestheatre.org/jumblies/talking-treaties-history-project; talkingtreaties.ca/.
4 I was particularly influenced by these two essays: Tim Ingold's "The Temporality of the Landscape" (1993); and Kenneth Olwig's "Representation and alienation in the political land-*scape*" (2005).
5 The Don River runs through the east side of Toronto. Scarborough is the city's eastern-most district.
6 The Humber River runs through Toronto's west side.
7 Dënesųłiné language is the language and nationhood of a northern Indigenous peoples of Northern Alberta and Saskatchewan and into the Northwest Territories of Canada.
8 Indigenous medicines refer to plant-based medicines—sage, punk (a type of fungus), sweetgrass, and tobacco.

9 *The Canadian Oxford Dictionary* (Barber, 2004) defines 'trapline' as "a series of traps set outdoors for catching animals".

10 Wychwood Park is a private enclave in Toronto. It was built as an artists' colony at the turn of the twentieth century. The street and grounds are owned by the residents, not by the city, and the houses there are extremely expensive.

11 The city of Toronto is dissected by rivers and creeks. Many have been put into underground pipes and culverts since the late nineteenth century.

12 Canadian architect Douglas Cardinal is a proponent of Organic Architecture, which his firm's website defines as

> a holistic enterprise where all members of the architectural process conjunctively create a legacy for the future … each building takes a life of its own as a living, organic being shaped in close partnership with the client and all stake holders.
>
> *(Douglas Cardinal Architect, n.d.)*

13 The Sweat Lodge is a ceremonial lodge where hot rocks are placed and water applied to the rocks—its uses are for prayer and offerings.

14 Ohèn:ton Karihwatéhkwen is a Haudenosaunee practice commonly referred to as the "Opening Address" or the "Thanksgiving Address". A more literal translation, in the words of Elder Tom Porter, is "what we say before we do anything important".

15 Jill Carter (n.d.) is an Anishinaabe-Ashkenazi scholar and theatre practitioner and an assistant professor at the University of Toronto. A collaborator on *Talking Treaties*, she also works as a researcher and tour guide with First Story Toronto and as a facilitator of Land Acknowledgement workshops for theatremakers across Toronto.

16 Rick W. Hill Sr. (n.d.) is Tuscarora of the Beaver clan. He is a professor of American History, an artist, a photographer, and a leading authority on contemporary Native American art, images, and multimedia representations.

17 The Residential School system in Canada was a programme of cultural genocide that lasted over a hundred years. It was the colonial government's attempt at total assimilation of Indigenous peoples into European social models, predominantly those of English society.

18 Native refers to the local Indigenous names.

19 Lee Maracle, (2019) of Salish and Cree ancestry, is a member of the Sto:loh Nation. One of the first Indigenous novelists to be published in Canada, she is an award-winning author and teacher, a lifelong activist, and an expert on First Nations culture and history.

20 The *Dish with One Spoon* is an Indigenous Nation to Nation agreement, extended from the Haudenosaunee to allied nations, including the Anishnawbe in 1701. It describes how land can be shared to the benefit of all its inhabitants. The treaty's name is a reminder that everyone lives from the same land; parties to the treaty agree to share resources so that they can live together in peace and cooperation.

Bibliography

Barber, Katherine (ed.) (2004). 'Trapline', in *The Canadian Oxford Dictionary*. [online] Oxford, UK: Oxford University Press. Available at www.oxfordreference.com (Accessed 5 October 2020).

Carter, Jill (n.d.). Available at www.provost.utoronto.ca/profile/jill-carter/ (Accessed 5 October 2020).

Douglas Cardinal Architect (n.d.). 'Philosophy'. Available at www.djcarchitect.com/philosophy (Accessed 5 October 2020).

Glover, Fred (2020). *A Dish with One Spoon*. Available at www.thecanadianencyclopedia.ca/en/article/a-dish-with-one-spoon (Accessed 7 October 2020).

Hill, Rick W., Sr. (n.d.). Available at www.pbs.org/warrior/content/historian/hill.html (Accessed 5 October 2020).

Indigenous Affairs Office (n.d.). 'Indigenous people of Toronto'. Available at www.toronto.ca/city-government/accessibility-human-rights/indigenous-affairs-office/torontos-indigenous-peoples/ (Accessed 7 October 2020).

Ingold, Tim (1993). 'The temporality of the landscape'. *World Archaeology*, 25(2), pp. 152–174.

Maracle, Lee (2019). Interviewed by Shelagh Rogers for *The Next Chapter*, 12 October. Available at www.cbc.ca/radio/thenextchapter/full-episode-july-20-2020-1.5316490/lee-maracle-reflects-on-her-legacy-as-one-of-canada-s-most-influential-indigenous-writers-1.5317159 (Accessed 5 October 2020).

Olwig, Kenneth R. (2005). 'Representation and alienation in the political land-*scape*'. *Cultural Geographies*, 12, pp. 19–40.

Steering Committee for the University of Toronto Response to the Truth and Reconciliation Commission of Canada (2017). *Answering the Call: Wecheehetowin*. Available at www.provost.utoronto.ca/wp-content/uploads/sites/155/2018/05/Final-Report-TRC.pdf (Accessed 7 October 2020).

Truth and Reconciliation Commission of Canada (2015). *Truth and Reconciliation Commission of Canada: Calls to Action*. Available at www.trc.ca/assets/pdf/Calls_to_Action_English2.pdf (Accessed 5 October 2020).

Wilson, Shawn (2008). *Research is Ceremony: Indigenous Research Methods*. Black Point, Nova Scotia: Fernwood Publishing.

2

UNEARTHING CITIZENSHIPS IN WASTE LANDSCAPES

Catherine De Almeida

In a suburban neighbourhood of St. Louis, Missouri, the smell of burning trash fills the air as an underground landfill fire creeps ever closer to buried radioactive waste. A thousand miles away in Toronto, Canada, a truck dumps construction debris beside an urban lake, where floating seeds will collect and take root next spring. We have been culturally conditioned to perceive both of these waste landscapes with disgust, and therefore to fear and disassociate our social structures from them. While these landscapes conjure strong, pervasive reactions, they also offer rich opportunities for cultivating highly specific place relationships.

Spatial marginalization generates disassociation. Devalued byproducts, when concentrated in dispossessed communities and on peripheral lands, become invisible in the collective consciousness of most urban dwellers. Communities consigned to residing in places of waste concentration are forced to live with the daily experiential impacts of others' waste. These landscapes are complex entanglements of biophysical, technological, political, economic, and sociocultural processes reinforced by the structural fabrication of social, racial, and material oppression. Twenty-first century waste landscapes are byproducts of multi-scalar webs of material extraction, movement, and disposal. Who belongs to these enigmatic places, who is responsible for them, and how and why do these place attachments manifest? Does enhancing the legibility of waste inscribe more productive and intentional citizenships?

In this chapter, I argue that citizenships in waste landscapes are constituted through active, operative, and transformative measures that reinforce a range of voluntary and involuntary dwelling. I first examine the role of 'disgust' in forming social and emotional attitudes toward waste conditions. I use this lens to investigate case studies of waste landscapes—three toxic and two non-toxic—and their complex involuntary and voluntary citizenships. The cases enable us to reconsider waste landscapes as contemporary commons, and suggest more deliberate, constructive, and inclusive ways forward as voluntary and conscious citizens.

Waste, disgust, and distancing

As our relationships with waste have evolved over centuries, we have been cultur-ally conditioned to see waste landscapes as not only useless, but also repellent.[1] Fear drives their removal from our social spheres, and therefore any sense of belonging to them. Social and emotional experiences form the basis for citizenships, and the lands to which people do or do not belong (Olwig, 2005: 20). Relationships with place are abstract emotional connections between people and lands—or *landships* (ibid.). Landships exist as hybrid conditions within the mind and around the body when we occupy physical lands; they elicit emotional reactions that influence behavioural responses.

The fallacy of dualistic western objectivity has separated the rational mind from irrational emotional and bodily experiences, but they cannot truly be uncoupled (Waterman, 2018: 517; Toulmin, 1990). Complex relationships between percep-tion, attitudes, and behaviours drive emotional reactions to our environments, including desire and disgust. Like taste for food, "our social and aesthetic tastes are determined in relation to one another and to particular times and settings, they must be evaluated always as both particular and general" (Waterman, 2018: 519). Relational culture drives our social and aesthetic tastes, which are manipulated and shifted through our interpersonal experiences. These tastes are simultaneously indi-vidual and collective cultural norms.

'Disgust' as a term that emerged from aesthetic taste depends more on appearance than reality, as architectural historian Vittoria Di Palma (2017) describes. Playing a role in aesthetic judgement, "[disgust] focuses exclusively on how [a] repellent object *appears to us* rather than on what it *is*" (Di Palma, 2017: 19). Originating from the Latin word 'gustus', or the physical experience of mouth-taste, it is more visceral than aesthetic taste. In a 1980s experiment by the psychologist Paul Rozin, in which fudge made to appear as faeces was presented to volunteers, "the reaction of disgust was both immediate and universal", and irreversible (Di Palma, 2017: 19). Although volunteers rationally knew they were presented with fudge, the sense of disgust overpowered reason—body over mind.

Aesthetic perceptions and emotional attitudes of waste project anti-value[2] onto *all* excess material form lacking identity,[3] regardless of the true danger they pre-sent. In the United States (US), even non-toxic, potentially valuable leftovers from useful products and processes (Dillon, 2014: 207) are viewed as undesirable impur-ities to be avoided (Lynch, 1981: 1). Fear and disgust as emotional reactions to waste exist to protect our bodies against potentially dangerous toxic substances and landscapes—aversion can be a vital form of self-preservation. Like the Rozin experiment, though, not all seemingly disgusting waste conditions warrant disasso-ciation and abandonment.

These cultural constructions also apply to spatial conditions that lack perceived value and embody emptiness, or *wastelands* (Di Palma, 2014). Historically, western attitudes toward wastelands applied even to ecologically rich forests, marshes, moun-tains, and other wildlands—any land absent of white colonization and cultivation

(Di Palma, 2014: 22). Today's waste landscapes, by contrast, have been manufactured and overused by humans: they are the undesirable byproducts of consumption. Elizabeth Meyer describes these 'disturbed sites' as "direct manifestations of the unacknowledged and largely unseen consequences of technological processes and industrial manufacturing" (2007: 62). They are hybrid conditions—technologically engineered sites existing within and affected by biophysical systems.

The sanitation infrastructures of our modern cities make waste invisible by removing it out of place, detaching its repercussions from its production. Waste landscapes may therefore be understood as sites of ambiguous landships: simultaneously belonging to everyone and no one through private creation and public accountability (Engler, 2004: 14). Waste production and management are part of the commons—our shared public spheres—constituted through a set of practices derived from creating and confronting it. While this work and labour take many forms, waste is often commodified by private enterprises designated to handle it.

The maintenance art of Mierle Laderman Ukeles reconnects us to the social and environmental networks and infrastructures that work to manage waste. By making these interconnections visible through everyday actions, we are able to experience our waste in the places it goes and the humans labouring to handle it. Her works, such as *Touch Sanitation* (1978)[4] and *Flow City* (1983–1995),[5] highlight waste and garbage as a pervasive social process that connects us all, "making visible labour which is usually invisible to those who rely on it but do not do it themselves" (Feldman, 2009: 49). As contemporary art curator Helen Molesworth (2000) argues, Ukeles's work "opens public space to the pressures of what it excludes or renders invisible", driven by aversion and disgust (2000: 82). Although we do not always see or immediately experience waste, as creators and participants, it is a part of us. Increasing its visibility enables us to establish more meaningful relationships. Like other systems that are frequently employed to oppress and privilege certain groups, we are complicit in waste production, distribution, and concentration, even as unwitting participants.

Waste landships are multi-scalar. We belong to what we create, individually and collectively through common practices, even if we are unaware of the ways our waste manifests in the environment or how it impacts other communities. Different groups of citizens experience waste asymmetrically, depending on source, type, proximity, and longevity. Waste landships exist across spectrums of democracy, equity, power, participation, agency, awareness, complacency, and unintentionality. A range of experiences in a range of places can create meaning (Manzo, 2005: 67), and landships are malleable gradients reinforced by ecological and social feedback loops. In the case studies that follow, I frame involuntary and voluntary waste landships as spectrums of agency, asymmetrical experiences, and intentionality relative to toxic and non-toxic waste conditions.

Involuntary citizenships in toxic waste landscapes

Disgust towards toxicity and its adverse effects on human and environmental health became palpable in the 1960s following publication of *Silent Spring* (1962). Written by American marine biologist and conservationist Rachel Carson while her health was deteriorating from cancer, her book discusses widespread use of the chemical compound dichlorodiphenyltrichloroethane (DDT). Created and used by the synthetic pesticide industry, this toxin had detrimental effects on human and non-human populations. DDT was only one of numerous instances of toxic byproducts impacting the environment. The Cuyahoga River in Cleveland, Ohio, caught fire 13 times over a 100-year period. The last instance was on 22 June 1969, when sparks from a passing train landed on oily slicks and debris floating on the surface (Ohio History Central, n.d.). It was a dramatic and potent event that put America's environmental toxicity on full display, and demonstrated to consumers the interconnections between consumption, production, and pollution (Meyer, 2007: 63). Carson's book, the Cuyahoga River fire, and countless other environmental disasters around this time inspired a grassroots environmental movement that led to the signing of the National Environmental Policy Act into law on 1 January 1970, and ultimately, the establishment of the Environmental Protection Agency (EPA) in December 1970 (Paull, 2013: 2). A decade after her book was released, the EPA banned usage of DDT in the US (ibid.). Pollution-producing industries responded to these policies by relying on waste management infrastructures to remove pollution and toxicity from daily consumer experiences.

The Love Canal and Superfund

While removed from the lived experience of many, waste ends up somewhere. Concern over contaminated brownfield sites hit a tipping point in the late 1970s following the highly publicized incident at Love Canal (De Sousa, 2006: 155; De Sousa, 2008). Located in Niagara Falls, New York, Love Canal set in motion US legislation to confront contaminated sites that posed public health risks. The canal, originally dug for power generation, was repurposed as a dumpsite: first for municipal waste in the 1920s, and then for industrial chemicals in the 1940s. Improperly sited, managed, and regulated, the landfill was inadequately capped with earth and clay, making its hazardous waste invisible, but not benign. The site was sold to the City for one dollar in 1953, and a suburban working-class neighbourhood of 100 homes and a school were built directly on top of it.

Colourful, foul-smelling waters seeped into residents' basements when it rained, and over decades many fell ill. As it became clear to the community that the source of their afflictions was the very land where they had raised their families, these involuntary citizens of a toxic landscape were left with no choice but to organize. With no clear villain to blame and a government in denial, this landship became a matter of life and death. Local residents wrote letters and led protests voicing their knowledge, experiences, and concerns in order to bring visibility to the site. For years they received no governmental response. Finally, the site's toxicity literally and metaphorically rose to the surface when a record rainfall eroded the soil cap and exposed the waste. Only then did President Jimmy Carter declare an unprecedented federal health emergency on 7 August 1978, ordering the relocation of the approximately 700 families that lived on site (Blum, 2008: 25). The residents' involuntary waste landship was inscribed through a labour of protest and activism. They resisted a toxic condition that shaped their daily experiences and sense of place, and fought to protect themselves and their loved ones. Residents of Love Canal challenged the injustice of hidden toxicity, undemocratically assigned to them, through material, artistic, intellectual, and linguistic practices (Means, 2011: 32–33).

This helped mark a significant turning point in the environmental justice movement, leading to the passage of the Comprehensive Environmental Response and Liabilities Act (CERCLA) in 1980. Also known as 'Superfund', CERCLA set aside remediation funds for state governments to hold those accountable for environmental contamination to pay for cleanup (De Sousa, 2006: 155). However, this process of determining who has been responsible for polluting landscapes and subsequent responsibilities to remediate has progressed slowly through obscure bureaucratic processes. The first site on the CERCLA National Priorities List, the Love Canal was only officially de-listed in 2004, nearly 25 years later (De Sousa, 2008: 9–10). The federal taxes that fund CERCLA are disbursed through an opaque, undemocratic process, leaving citizens with little agency.

Widespread invisibility of waste and toxicity not only muddies the distinction between toxic and benign wastes but also makes hazardous substances more dangerous. Toxicity hidden from view can evade a collective sense of urgency to address hazards. Visible toxicity, in contrast, has the capacity to invite a healthy attitude of disgust toward truly dangerous waste, and makes apparent that different types of waste deserve distinct management strategies. Hiding waste restricts its potential as a visible indicator of potential problems. Toxic waste is most legible when infrastructures designed to contain it fail. Its reveal appears to be a sign of crisis, but in truth the crisis was already ongoing. Involuntary landships with toxic waste sites are often bred out of the necessity. Those most directly affected by these sites become responsible for laying these conditions bare before disaster strikes. Over 40 years after Love Canal, a smouldering underground landfill fire in St. Louis, Missouri, has been poisoning nearby residents for nearly a decade, as it inches ever closer to buried radioactive waste.

The radioactive Bridgeton-West Lake Superfund Landfill

The air stinks of "petroleum fumes, skunk spray, electrical fire, and dead bodies" (Johnson, 2017). Wind patterns and toxic air are monitored by residents using carbon monoxide detectors outside their homes (Cammisa, 2018). They and their children are ill with cancers and respiratory diseases from exposure to migrating radioactive contaminants in the soil, water, and air. The Spanish Village subdivision, a working-class neighbourhood of St. Louis, Missouri, is adjacent to the Bridgeton-West Lake Landfill, the most high-profile and complex Superfund site in the US (Gray, 2018). Trapped by debt and disinvestment, these residents are involuntary citizens whose futures are attached to the fate of this toxic waste landscape.

When the stench intensified in fall 2012, resident Dawn Chapman began asking questions and combing through reports about the landfills in her backyard. She contacted a representative from the EPA regional office with more questions and concerns, who assured her there was no danger. But she continued digging through technical documents about the fire and learned that the two contiguous landfills—West Lake Landfill that included illegally dumped nuclear waste and Bridgeton Landfill that contained a fire started from a chemical reaction—had been listed on the CERCLA National Priorities List in 1990. The fire was 1,800 feet (550 metres) across, 150 feet (45 metres) deep, and less than 1,000 feet (300 metres) southeast of the buried nuclear waste (Johnson, 2017). On the brink of unfathomable disaster, this site and its residents are entangled in a complicated history of weapons production, federal-corporate agreements and liabilities, and complacency.

During the Manhattan project, as the US raced to be the first nation to create a powerfully destructive bomb, the West Lake Landfill became one of over 300 landscape relics of these efforts (Johnson, 2017). Mallinckrodt Chemical Works in downtown St. Louis was responsible for all the uranium refinement for the world's first atomic bombs, including those detonated in Hiroshima and Nagasaki, Japan, during World War II (ibid.). Between 1942 and 1957, Mallinckrodt refined uranium (St. Louis Remediation Task Force, 1996, ES-4) under a contract with the US Atomic Energy Commission (AEC). The linkages between this uranium refinement and the modern crisis at the Bridgeton-West Lake Landfill span decades of mismanagement and over half a dozen public and private actors, obscuring and diluting the prolonged widespread impacts of these wastes.

By the mid-1940s, Mallinckrodt ran out of space for storing radioactive wastes onsite and began shipping them to the St. Louis Airport Site. These materials were stored uncontained, unattended, in bulk, until the mid-1960s. In 1962, the wastes were sold at public auction and, before going bankrupt in 1966, the company moved them to a private storage facility a mile away (Johnson, 2017). The following year, Cotter Corporation of Colorado purchased the wastes, dried the material to ship to their facility in Colorado, and left behind an 8,700 tonnes (7,900 tonnes) pile of leached barium sulphate—a residue of uranium ore processing. In 1973, in an effort to clean up the site, B&K Construction, a subcontractor for Cotter, mixed the leached barium sulphate with 48,000 tonnes (43,500 tonnes) of toxic soil (Barker, 2015). This soil was illegally dumped in the West Lake Landfill, a former limestone quarry turned municipal waste repository, as a daily cover for three months (Gray, 2018). Cotter claimed responsibility and reported it to the AEC, and while they found that the waste disposal violated federal standards, no strong penalties were enforced (Barker, 2015). Laidlaw Waste Systems owned the landfill until 1996, when Allied Waste purchased the property. In 2008, Republic Services acquired the site in a merger and created the registered subsidiary Bridgeton Landfill LLC. It is not widely understood whether knowledge of the wastes and agreements for responsibility were provided during this convoluted property transfer process. However, since the late 1980s and the site's CERCLA listing in 1990, the EPA completed studies and reports of the site acknowledging the risks but neglecting to act to fully address them. Although the US Superfund Program has been established for almost 40 years, residents have continued to live with environmental injustice and inaction for over 30 years, highlighting the reactive rather than pre-emptive approach of CERCLA.

In early 2013, Chapman participated in an air-sampling event with other neighbours and environmental activists outside the fenced site. As dials on their monitors jumped and beeped, these citizen scientists documented what they suspected—the air in their environment was toxic (Johnson, 2017). Chapman began working with other neighbours to uncover the history of toxicities they had inherited, grasping policies, technical jargon, and transfers of land ownership. They began pressuring local, state, and national representatives, who would sometimes ask who they were. "We're just moms", Chapman and her neighbour Karen Nickel would say, "we're just citizens concerned about the health and safety of our kids and our community" (ibid.). With another neighbour, Beth Strohmeyer, they formed the Just Moms STL advocacy group.

Since 2013, Just Moms STL has held monthly meetings to raise public awareness of the site. They have three goals: a safe and permanent solution to the radioactive waste in West Lake Landfill, voluntary buyout for all residents within 1 mile (1.6 kilometres) of the landfill, and property assurance for those within 5 miles (8 kilometres) (Just Moms STL, 2012). The EPA's initial inability to acknowledge the threat of fallout, should fire reach the radioactive waste, caused frustration. Just Moms STL organized third-party studies of contamination in and around the site to disprove studies conducted by the EPA that said they were safe. They put pressure on government agencies to release reports, and helped facilitate conversations between the community, EPA, Republic Services, and other stakeholders. They travelled to Washington, DC, to protest and meet with EPA administrators on the Superfund process. Just Moms STL led the process towards action, contrasting the inaction of government and private companies that participated in the making of this toxic waste landscape.

There are different spatial and temporal scales of belonging to the Bridgeton-West Lake Superfund landfill, spanning gradients of landships and responsibility. Public and private entities are intertwined in this landscape, from the US federal government (EPA, AEC [now the Nuclear Regulatory Commission], Department of Energy), to state, regional, and local governments (the Missouri Department of Natural Resources, St. Louis County, and City of St. Louis), to private entities (both current and former), to the thousands of impacted residents. US citizens funded the secret Manhattan Project and celebrated the detonation of atomic bombs in Japan (Johnson, 2017). Mallinckrodt's labourers who processed uranium ore for over 15 years, or who mixed the radioactive waste with contaminated soil on behalf of Cotter, all participated in transforming this landscape. The federal government, meanwhile, tends to deny the dangers of wasted landscapes for as long as possible.

By necessity, the involuntary citizens of this polluted landscape bear the cost of voluntary actions in the past. Even in the face of sickness and death, they have assumed responsibility to make this waste visible before it explodes. They have worked tirelessly to learn, organize, disseminate, question, and protest, grasping obscure policies and language to hold the government and private entities accountable. The

West Lake Landfill is a present-day Love Canal gone nuclear. Nearly a decade after residents began demanding change, the EPA announced plans in September 2018 to excavate approximately 70% of the radioactive contaminants before the fire reaches the area. But it will take four and a half years to complete with a $205 million price tag, paid out by a handful of public (US DOE) and private (Republic Services and Exelon Corp., whose subsidiary, ComEd, formerly owned Cotter) entities (Gray, 2018). The excavated contamination will be sent out-of-state to a licensed facility in Utah, Idaho, Michigan, or Colorado to address "the concerns of people who live around the site" (Gray, 2018). In other words, contaminants will be moved to other disposal sites near different human and non-human communities. As the waste is moved from one place to the next, the extent and multiplicity of landships will correspondingly expand.

Waste landships as distributed networks: New York City's 'poop train'

The commodification and transportation of waste fuels national and global distribution networks, diluting responsibility and expanding landships. New York City's (NYC) 14 wastewater treatment plants treat over 1.3 billion gallons (5,000 cubic metres) of wastewater and produce between 1,000 and 1,400 wet tonnes (900 and 1,300 tonnes) of biosolids every day (NYCDEP, 2016: 3). For decades, this waste was dumped 106 miles (170 kilometres) out in the ocean, but in 1988, Congress passed the Ocean Dumping Ban Act to stop such practices. This drove NYC to develop a beneficial use programme for biosolids in 1992: 50% as fertilizer for farms in Colorado and Texas and 50% to a pelletization plant in the Bronx (ibid.: 6). After the 2008 financial crisis, landfilling became less expensive, and NYC opted to sign contracts with private waste management companies for landfill disposal (ibid.: 7–8). This fragile capital-driven network led to the now infamous 'poop train'.

Alabama's lack of regulations, inexpensive land, and permissive zoning laws make it easy for landfill operators to open landfills in rural and poor communities and make large profits (Martin and Reeves, 2018). The 'poop train' is one of many instances in which waste commodities produced in one state are shipped to another due to differences in local and state regulations. Since 2017, NYC has shipped its biosolids 1,000 miles (1,600 kilometres) to the Big Sky Environmental landfill in Adamsville, Alabama (Street, 2018). When the train arrived at the town of West Jefferson, the loads were transferred to trucks to haul the biosolids ten miles away (Victor, 2018). After receiving complaints from residents, in January 2018 officials representing West Jefferson filed an injunction about the odours and Big Sky's mishandling of the waste, grinding the train to a halt in March in the poor rural community of Parrish (Victor, 2018), which lacks the zoning regulations to block the train cars (Martin and Reeves, 2018). "Parrish had become purgatory in a little-known disposal pipeline that imports materials no one wants, for profit— big city waste from the North that heads to landfills in the rural South" (Street, 2018). The odours seeping from the 42-car train engulfed the two-square-mile town, preventing residents from going outdoors and causing public outcry. Disgust, experienced through smell and taste, exposed the active collapse of the tenuous system of waste disposal in southern states. Over two months later, after Parrish's town council threatened to file a lawsuit, Big Sky loaded their trucks from the rail yard to transport ten million pounds (4.5 million kilograms) of biosolids 20 miles away to their landfill, agreeing to discontinue receiving biosolids from NYC (ibid.). The vocalization of residents brought visibility to this waste condition. In doing so, residents became involuntary citizens of Big Sky's private landfill in order to create a shift in their immediate environment.

The varied waste landships of NYC's biosolids begin, continue, and end along a vast multi-modal network. Any NYC resident who flushes a toilet is an unwitting citizen of not only the Big Sky landfill and the countless other landfills where their biosolids are dispersed, but also every town along the various train and truck routes that transport it. In contrast, citizens of Parrish and other small rural towns through which the 'poop train' passes and stops have an involuntary landship with the landfill that transcends territories of waste accumulation. The 'poop train' may offer a window into the future of the West Lake Landfill's radioactive waste, planned for removal and transport elsewhere.

These involuntary landships demonstrate the complexities of entangled parties and responsibilities in relationship to toxic waste landscapes. Residents living on top of or adjacent to these sites are agents of change—working to understand why their environments are suddenly poisoning them in order to push for accountability and action. Involuntary landships influence the trajectory of a toxic landscape. If not for the involuntary citizens of Love Canal, the West Lake Landfill, the 'poop train', and countless others, these landscapes would continue on their catastrophic trajectories. Their unwavering labour of survival unearths truths of waste injustice and oppression which those in power tried to bury. Waste is not discrete from the environment. Disgust revealed through sight, smell, and taste call our attention to environmental changes caused by mismanaged waste: our bodies index contaminant data more than our minds can comprehend. Citizens play critical roles in enacting social and environmental change, working to create visibility and legibility for toxic waste landscapes. Through this individual and collective participatory work, waste landscapes have the potential to become commons.

Wastelands as commons

If they were treated as resources rather than liabilities, waste and wastelands could become new kinds of commons. Commons are more than *things* and *resources;* they are "*a resource + a community + a set of social protocols*" (Bollier, 2014: 15). Practices and values establish social protocols that influence how work is performed and commons are managed (Linebaugh, 2014: 13). Labouring activities in the commons remake landscapes and establish attachments. As spaces produced through labour, landscapes are "*nothing but* work", and through this work, landships are strengthened (Mitchell, 2017: 189).

In seventeenth-century literature, 'waste' and 'common' were used to characterize notions of use, not site ecologies (Di Palma, 2014: 25). In eighteenth-century England,[6] waste was a useful and valuable aspect of common life, giving everyone access to common resources (Neeson, 1993: 173). For those who lacked formal rights to land, "fuel, food, and materials taken from common waste … gave them a variety of useful products, and the raw materials to make more" (ibid.: 158–159). The use of common waste carried social meaning—a sense of belonging, localness, and freedom (ibid.: 177–179).

Landships in wastelands are established through practices—in eighteenth-century England, through gleaning, foraging, and poaching (Di Palma, 2014: 33); and in cases of involuntary landships such as the West Lake Landfill, through material, artistic, intellectual, and linguistic practices of protest and activism. Such practices in and of place fortify and verify meaning and attachment to place. Perceptions and attitudes of waste and wasteland in the early commons were not grounded in 'disgust', but in resource use, community building, and local economy. How did attitudes about waste evolve towards disgust and repulsion?

Social and emotional constructions of waste have changed dramatically over the past 200 years, responding to changing sociocultural contexts of needs, knowledge, and technology. Stemming from hygienic concerns and a developing understanding of sanitation, the 'proper' management of waste came to be a measure of class and status, "a primary ingredient of comfort, a factor of health, an aesthetic element, and a major component contributing to efficiency and leisure" (Engler, 1997: 62). Today, waste is becoming everyone's issue once again—reconsidered as "a resource, a symbol of environmental crisis, and its possible cure" (ibid.). Reconceiving waste as a resource of the commons may be one of our greatest challenges in the twenty-first century, possibly making waste the antidote to waste. Changing perceptions and attitudes towards waste can generate new landships in waste landscapes and transform human–waste relationships from passive generation to actively productive. Commons can arise almost anywhere, and be highly generative in unlikely circumstances (Bollier, 2014).

Waste landscape as commons—the Leslie Street Spit

The shores are made of brick and concrete rubble whose edges have been smoothed by waves, like beach glass. Visitor-created sculptures of stacked construction debris, including bent and twisting steel rebar, seem to defy gravity as they are strewn throughout this landscape of demolition fragments. Songbirds rare to the region hop and sing in stands of trees. The Leslie Street Spit in Toronto, Canada, is an active construction debris-dumping site that is simultaneously a marginal wasteland, beloved city park, and ecological sanctuary.

Belonging to the public domain, the Spit as a waste heap was originally invisible from the collective consciousness of city residents—citizens of a landscape they unknowingly relied on. In the early to mid-1900s, many cities' waterways were marginal, industrial ports. Originally constructed as a breakwater for a deep-water port, the Spit is an artefact of material byproducts that have accumulated since 1959. Composed of dredged sediment from the Toronto harbour and construction debris from disassembling and reconstructing the city over decades, it extends the Lake Ontario shoreline by 3 miles (5 kilometres) and adds over 2 square miles (500 hectares) of land (Foster and Sandberg, 2014: 1046). Over decades, other actors have become rooted in this substrate of waste: "seeds and plant matter dispersed by wind, birds, water, and deposited material" instigated primary ecological succession on the artificially manufactured peninsula (Schopf and Foster, 2014: 1086). The Leslie Street Spit represents both present and historical definitions of wasteland—at once dumping grounds and urban wilds. Wildflower meadows, cottonwood forests, coastal marshes, cobble beaches of brick, and sand dunes have become a public park, sharing space (but not time) with active waste dumping logistics (Yokohari and Amati, 2005: 55). As a "polyfunctional infrastructure" (Belanger, 2009: 82), the Spit hosts active dumping of dredged sediment and construction debris from 9:00 am to 4:00 pm and park access 4:00 pm to 9:00 pm during the week, with full park access on the weekends (TRCA, 2018). The "accidental ecology ... that took over this large wasteland" can be accessed on foot from downtown Toronto (Belanger, 2009: 82). This world-renowned bird sanctuary and spontaneously emerging eco-logical reserve (Foster and Sandberg, 2014: 1046) is celebrated by Torontonians as a symbol of wilderness in the city (Schopf and Foster, 2014: 1087), home to over 390 plant species and 290 animal species (Yokohari and Amati, 2005: 55).

After the first 10–20 years of dumping, ecological succession had produced thriving wildlife communities while environmental attitudes were changing in the 1970s. The "hideous industrial by-product" had begun to attract threatened bird species (Foster and Sandberg, 2004: 91). This phenomenon led to a collective per-ceptual shift reclassifying the landscape as value-full instead of value-less and pro-ducing "a widely cherished feature of the Toronto landscape that is fiercely guarded from development by a network of community-based and nonprofit organizations" (ibid.: 91). The Spit is a hybrid waste landscape of dumping and land manufac-turing (Belanger, 2009: 83), made visible through its appropriation by human and non-human assemblages. As value has been assigned to this landscape, it has become legible both as a waste landscape and as a common resource, engendering a shared experience of beauty in contradictions.

Humans and wildlife co-produced and continue to remake this waste land-scape commons through acts of dumping, nesting, feeding, observation, and advo-cacy. Landscapes themselves are not necessarily significant to people; rather, it is "what can be called 'experience-in-place' that creates meaning" (Manzo, 2005: 74). These 'place attachments' are dynamic and socially produced, reinforced and made visible through performative lived experiences (Manzo and Devine-Wright, 2019: 136, 139).

How did residents find beauty in disgust; how was value assigned to a value-less landscape? As Di Palma explains, "disgust [as] an emotion harbours a contra-dictory duality, a mixture of repulsion and allure ... we are both repelled and transfixed" (2017: 19). Environmental psychologist Lynne Manzo connects this to place meaning and attachment, describing strong affections for particular places as topophilia, and aversion as topophobia (2005: 70). Place meaning can develop "from an array of emotions and experiences, both positive and negative" (ibid.: 67). Misunderstanding can create fear and misperception. Landship follows from beauty, ownership, love, respect, and compassion—in order to find beauty in disgust, one must grow to understand it through participation. Voluntary landships evolve not out of necessity, but out of desire and curiosity, inciting the selective adoption of waste landscapes as commons.

The changing landscape of the Leslie Street Spit attracted citizens to encounter this waste landscape in a new light, resulting in a new commons and a new voluntary landship. When certain waste conditions are framed as desirable, value is assigned, creating a feedback loop. Frequented by families, birdwatchers, cyclists, hikers, photographers, and artists for recreation, pleasure, inspiration, and respite, commoning activities at the Leslie Street Spit increase its value over time. The more first-time and daily visitors participate in these practices, the more its collective ownership as a commons is verified. Local knowledge and understanding can make such wastelands legible as a part of the commons, and this legibility engenders an experience of beauty. While voluntary landships propel the adoption of waste landscapes as commons, a collectively beloved, novel place may also fall subject to commodification and privatization.

From commons to commodity: the Blue Lagoon[7]

At the Blue Lagoon, a human ecology and economy developed out of materials and spatial conditions originally perceived as waste (De Almeida, 2018). An unintended byproduct of renewable geothermal energy generation, this waste landscape was named one of 25 wonders of the world by National Geographic in 2012 (Blue Lagoon Iceland, 2019). The lagoon, famous for its milky sky-blue colour, formed by accident when the Svartsengi Geothermal Power Station dumped geothermal waste effluent in the surrounding lava field. As the geothermal effluent was exposed to air, minerals in the water hardened, forming deposits in the bedrock's pores and rendering the ground impermeable, creating the lagoon. At first, the National Energy Authority declared this destruction of an 800-year-old geologically significant lava field an environmental disaster.

In Iceland, it is common practice to bathe in geothermal waters wherever they are found. Nearly every city and town has public geothermal pools, hot tubs, saunas, and steam rooms. There are both well-known and secret geothermal bathing areas across Iceland's landscape. As nights get longer in the winter, daily geothermal bathing becomes an important ritual for health, well-being, and maintaining social connections. Ingrained in Icelandic culture, geothermal water is used for nearly every activity, from agriculture and aquaculture to energy production and recreation. After the lagoon adjacent to Svartsengi formed, local residents began bathing in the warm waters, adopting the wasteland as a commons and giving rise to an unexpected recreational oasis (De Almeida, 2019). Nearly a decade later in 1987, its popularity led to the construction of a modest bathhouse in order to improve public access (Blue Lagoon Ltd, 2016). The Blue Lagoon Limited company was formed in 1992, and eventually constructed a complex of buildings housing a spa, an R&D centre, and a psoriasis clinic in 1999, enclosing this commons (Gudmundsdóttir, Brynjólfsdóttir and Albertsson, 2010: 3; Albertsson, 2008: 5).

The Blue Lagoon is now the top tourist destination in Iceland, hosting over 1.3 million visitors in 2017 (Arnardottir, 2018). Blue Lagoon Limited produces and sells high-priced skin care products from the unique microbial and algal ecosystem of its mineral-rich water, such as silica mud, algae, and mineral masks (Gudmundsdóttir, Brynjólfsdóttir and Albertsson, 2010). Once a recreational commons for people who lived nearby, the lagoon has become a commodity, a package of experiences and products. Its developers cater to the tastes of tourists from cultures that perceive all industrial waste as undesirable and potentially harmful (De Almeida, 2019). Although the waste-exchanging relationships between power plant and spa industry continue to evolve, this foreign negative perception has resulted in marketing strategies that make these processes less legible to visitors in an attempt to choreograph a high-end, 'natural' spa experience (ibid.). The site's design and continuous reconfiguration increasingly conceals the underlying waste reuse frameworks that created the landscape and its unique conditions. Tall lava rock berms have been installed to create visual barriers, while ad campaigns and site signage celebrate the site's 'natural' beauty. International tourists are kept unaware of the Blue Lagoon's fascinating origins as well as the ongoing innovative development of shared waste resource streams within a hybrid industrial-recreational landscape. The juxtaposition of industry and recreation, alluring to its early visitors, has been disrupted.

Waste legibility can be an asset shared by active industrial operations, a novel ecological community, and recreational uses, as demonstrated by the Leslie Street Spit. In the case of the Blue Lagoon, the privatization of this wasteland commons negated a democratic space by creating a high-end, unaffordable, inaccessible spa industry. This shifted the ways people interact with this waste landscape from communal to individual—public to private. The Blue Lagoon is now a landscape of consumption. No longer a landship commons, it is a privatized, globalized tourist commodity. Rather than embracing its fascinating paradoxical qualities, the origin story of this waste landscape has been suppressed in order to cater to globalized tastes and the expectation that a spa and industrial activities do not belong together (De Almeida, 2019).

Cultivating voluntary citizenships of waste landscapes

We all unwittingly participate in the making of waste landscapes. We each belong to many, some locally and others globally, with varying degrees of separation. They are all a part of us, our communities, and our bodies, but the responsibility for them is distributed unequally. We can choose to ignore them, or we can choose to question why they exist and uncover their latent opportunities for connecting, healing, material sourcing, and reconsidering our roles in their creation and new futures.

Waste landships are individually and collectively practiced, but they are not fixed. Landships emerge, evolve, vary, and at times are indeterminate. In some cases, people have agency to voluntarily develop landships, and in others, people are involuntarily subjected to toxic lands that slowly and violently kill them and their families. Rather than being given or claimed, equality (and inequality) is practiced and verified in both immediate and distant lands (Rancière, 1991: 137). Landships are practiced, formed through acts of becoming, driven by the ever-shifting sociocultural, political, and economic contexts in which we experience, create, and change landscapes. Active practices make waste visible and continually verify landships, thereby reinforcing emotional connections and place attachments. Whether those relationships are voluntary or involuntary, and whether the waste is benign or toxic, informed citizens use their bodies and minds to imagine and practice new realities that challenge or strengthen belonging. In all these cases, landships affirm that experience *is* knowledge.

The case studies I describe only reveal a few waste landships in a world with millions of similar places. Countless lives, landscapes, and places are sacrificed for the convenience of removing from our public spheres what society deems valueless. While each is distinctive in place, materials, and community, what drives their creation is ubiquitous—marginalization, disassociation, and the creation of things and places destined for failure, disposal, and erasure. Visibility forces communal action when waste is dangerous, and when it is not a threat, visibility makes possible its adoption into the commons as a collective resource. Each waste landscape condition is unique and requires specificity and precision in their transformation. Some of these landscapes instil fear, while others spark curiosity. Some are assigned to us without our knowledge or consent, while others invite exploration and adoption as our own. How can we apply these lessons to future sites? Instead of reacting to dangerous or serendipitous accidents, how can we pre-emptively work with them?

Rather than avoiding their challenging conditions, waste landscapes are filled with opportunity. Understanding the nuances of waste landscapes, and questioning the emotions they inspire, enables us to reconsider what it means to belong to them. What would happen if waste was destined to be remembered, or rethought as a condition filled with opportunity? Waste materials and landscapes are resources that are not solely raw materials, but things and places with rich histories of labour, production, processes, and life. Such histories, combined with adaptation, can unlock uncharted possibilities of novel conditions that break away from conventional practices that further marginalize and disguise. They provide new, unpredictable possibilities that have yet to be imagined. Active, productive, and transformative relationships with waste as a common ground can be cultivated by enhancing legibility, fostering a sense of responsibility, and empowering citizens to become agents of change.

Acknowledgements

I first presented this research at the *Landscape Citizenships Symposium* organized by Tim Waterman, Ed Wall, and Jane Wolff. I am grateful for their and the Advisory Group's constructive feedback throughout this process, helping to shape the exciting trajectories this piece and my work have taken. I would also like to thank the University of Nebraska-Lincoln for supporting the initial development of this piece. Finally, I thank Amelia Jensen for her insightful comments, ongoing conversations, and enduring support for me and the development of this work.

Notes

1 Vittoria Di Palma explores the role of 'disgust' as an aesthetic term in the way in which we perceive waste landscape in her critical essay 'In the Mood for Landscape'. Di Palma states,

> Disgust is an emotion that operates powerfully in the formulation of a culture's ordering systems: it establishes and maintains hierarchies: it is fundamental to the construction of a moral code. Disgust can therefore help to shed light on the systems through which different kinds of landscapes are valued, and the reasons why ethical or moral arguments so often appear in the context of discussions regarding derelict or polluted sites.
>
> *(2017: 18)*

2 Gay Hawkins and Stephen Muecke describe waste as a construction of value in their introduction, stating, "waste isn't just the uselessness that sustains utility, or the place where only the symbolic is in play; it has a complex role in formations of value" (2003: x).
3 As Dietmar Offenhuber describes in his recent book, "the process of becoming waste implies a loss of information … discarded objects lose their value by losing their characteristic properties and becoming part of an undifferentiated mass … individual components are rendered invisible" (2017: 2).
4 In *Touch Sanitation* (1978), Mierle Laderman Ukeles spent 11 months shaking the hands of 8,500 sanitation workers across New York City's five-borough system, saying "Thank you for keeping New York City alive" to each one (Jackson, 2011: 99).
5 Ukeles' *Flow City* (1983–1995) was an ambitious public art project that made garbage visible as a dynamic process by providing visitors with a place and lenses to view and understand the processes and places our waste goes (Feldman, 2009: 52–53).
6 See J.M. Neeson's chapter 'The uses of waste' (1993: 158–184).
7 For more on the Blue Lagoon as a landscape of waste reuse, see De Almeida (2018). For more on the commodification of the Blue Lagoon, see De Almeida (2019).

Bibliography

Albertsson, Albert (2008). 'Glimpse of the history of Hitaveita Sudurnesja Ltd. – the Resource Park concept'. *United Nations University 30th Anniversary Workshop*, 26–27 August. Reykjavik: United Nations University, Geothermal Training Programme. Available at: orkustofnun.is/gogn/unu-gtp-30-ann/UNU-GTP-30-22.pdf (Accessed 12 July 2020).

Arnardottir, Hronn (2018). Interviewed by Catherine De Almeida. 'R&D Specialist at the Blue Lagoon R&D Center'.

Barker, Jacob, St. Louis Post-Dispatch (2015) *Pointing Fingers: Exelon says feds knew radio-active waste was being dumped at landfill*. Available at: stltoday.com/business/local/pointing-fingers-exelon-says-feds-knew-radioactive-waste-was-being-dumped-at-landfill/article_d41d1052-2d80-53e9-afc0-6d531c5cdbdf.html (Accessed 19 July 2020).

Belanger, Pierre (2009). 'Landscape as infrastructure'. *Landscape Journal*, 28, pp 79–95.

Blue Lagoon Iceland (2019). *Is the Blue Lagoon a Wonder of the World?* Available at: bluelagoon.com/stories/is-the-blue-lagoon-a-wonder-of-the-world (Accessed 15 July 2020).

Blue Lagoon Ltd. (2016). *Blue Lagoon Press Kit*. Grindavík: Blue Lagoon Ltd., pp 1–5.

Blum, Elizabeth (2008). *Love Canal Revisited: Race, Class, and Gender in Environmental Activism*. Lawrence, KS: University Press of Kansas.

Bollier, David (2014). *Think Like a Commoner*. Gabriola Island, BC, CA: New Society Publishers.

Cammisa, Rebecca (2018). *Atomic Homefront* [Online]. HBO.

Carson, Rachel (1962). *Silent Spring*. Boston, MA: Houghton Mifflin.

De Almeida, Catherine (2018). 'Performative byproducts: The emergence of waste reuse strategies at the Blue Lagoon'. *Journal of Landscape Architecture*, 31(3), pp 64–77.

De Almeida, Catherine (2019). 'The Blue Lagoon: from waste commons to landscape commodity'. *Scenario Journal 07: Power*. Available at: scenariojournal.com/article/blue-lagoon/ (Accessed 12 July 2020).

De Sousa, Christopher (2004). 'The greening of brownfields in American cities'. *Journal of Environmental Planning and Management*, 47(4), pp 579–600.

De Sousa, Christopher (2006). 'Green futures for industrial brownfields'. In: Rutherford H. Platt, ed. *The Humane Metropolis: People and Nature in the 21st Century*. Boston, MA: University of Massachusetts Press, pp 154–168.

De Sousa, Christopher (2008). *Brownfields Redevelopment and the Quest for Sustainability*. Oxford, UK: Elsevier.

Dillon, Lindsey (2014). 'Race, waste, and space: Brownfield redevelopment and environmental justice at the Hunters Point Shipyard'. *Antipode*, 46(5), pp 1205–1221.

Di Palma, Vittoria (2014). *Wasteland: A History*. New Haven, CT: Yale University Press.

Di Palma, Vittoria (2017). 'In the mood for landscape'. In: Christophe Girot and Dora Imhof, eds. *Thinking the Contemporary Landscape*. New York, NY: Princeton Architectural Press, pp 15–29.

Engler, Mira (1997). 'Repulsive matter: Landscape of waste in the American middle-class residential domain'. *Landscape Journal*, 16(1), pp 60–79.

Engler, Mira (2004). *Designing America's Waste Landscapes*. Baltimore, MD: Johns Hopkins University Press.

Feldman, Mark (2009). 'Inside the sanitation system: Mierle Ukeles, urban ecology and the social circulation of garbage'. *Iowa Journal of Cultural Studies*, 10(1), pp 42–56.

Foster, Jennifer and L. Anders Sandberg (2004). 'Friends or foe? Invasive species and public green space in Toronto'. *Geographical Review*, 94(2), pp 178–198.

Foster, Jennifer and L. Anders Sandberg (2014). 'Post-industrial urban greenspace: Justice, quality of life and environmental aesthetics in rapidly changing urban environments'. *Local Environment,* 19(10), pp 1043–1048.

Gray, Bryce (2018). *EPA Reaches Cleanup Decision for Radioactive West Lake Landfill Superfund Site*. Available at: stltoday.com/business/local/epa-reaches-cleanup-decision-for-radioactive-west-lake-landfill-superfund/article_70796e6f-d975-5122-8670-16a67154b442.html (Accessed 15 October 2018).

Gudmundsdóttir, Magnea, Ása Brynjólfsdóttir, and Albert Albertsson (2010). 'The history of the Blue Lagoon in Svartsengi'. *Annual International World Geothermal Congress*, 25–29 April. Bali. Available at: geothermal-energy.org/pdf/IGAstandard/WGC/2010/3314.pdf (Accessed 12 July 2020).

Hawkins, Gay and Stephen Muecke, eds. (2003). *Culture and Waste: The Creation and Destruction of Value*. Lanham, MD: Rowman & Littlefield Publishers, Inc.

Jackson, Shannon (2011). *Social Works: Performing Art, Supporting Publics*. New York, NY: Routledge.

Johnson, Lacy M. (2017). 'The Fallout'. *Guernica* [Online]. Available at: guernicamag.com/the-fallout/ (Accessed 19 July 2020).

Just Moms STL | West Lake Landfill (2012). *Our Mission*. Available at: stlradwastelegacy.com/about-us/our-mission/ (Accessed 19 July 2020).

Linebaugh, Peter (2008). *The Magna Carta Manifesto: Liberties and Commons for All*. Oakland, CA: University of California Press.

Linebaugh, Peter (2014). *Stop Thief!: The Commons, Enclosures, and Resistance*. Oakland, CA: PM Press.

Lynch, Kevin (1981). *Wasting Away*. San Francisco, CA: Sierra Club.

Manzo, Lynne (2005), 'For better or worse: Exploring multiple dimensions of place meaning'. *Journal of Environmental Psychology*, 25, pp 67–86.

Manzo, Lynne and Patrick Devine-Wright (2019). 'Place Attachment'. In: Linda Steg and Judith IM de Groot, eds. *Environmental Psychology: An Introduction*, 2nd ed. Hoboken, NJ: John Wiley & Sons Ltd, pp 135–143.

Martin, Jeff and Jay Reeves (2018). '"Poop Train" full of NYC sewage raises stink in Alabama town'. *Associated Press* [Online]. Available at: apnews.com/28f5b0563b1441b6ac5d29f7dfd7f5f7 (Accessed 20 July 2020).

Means, Alex (2011). 'Jacques Rancière, education, and the art of citizenship'. *The Review of Education, Pedagogy, and Cultural Studies*, 33(1), pp 28–47.

Meyer, Elizabeth (2007). 'Uncertain parks: Disturbed sites, citizens, and risk society'. In: Julia Czerniak and George Hargreaves, eds. *Large Parks*. New York, NY: Princeton Architectural Press, pp 58–85.

Mitchell, Don (2017). 'Afterword: Landscape's agency'. In Ed Wall and Tim Waterman, eds. *Landscape and Agency: Critical Essays*. New York, NY: Routledge, pp 188–192.

Molesworth, Helen (2000). 'House work and art work'. *October*, 92, pp 71–97.

Murphy, Raymond (1994). *Rationality and Nature: A Sociological Inquiry into a Changing Relationship*. San Francisco, CA: Westview Press.

Neeson, Jeanette M. (1993). *Commoners: Common Right, Enclosure, and Social Change in England, 1700–1820*. Cambridge, UK: Cambridge University Press.

New York City Department of Environmental Protection (2016). *NYC DEP Biosolids Program: A Review and Update* [Online]. Available at: ned-beecher.squarespace.com/s/Ellis-NYCDEPBiosolidsProgramReview-12Oct2016.pdf (Accessed 12 July 2020).

Offenhuber, Dietmar (2017). *Waste Is Information: Infrastructure, Legibility, and Governance*. Cambridge, MA: The MIT Press.

Ohio History Central (n.p.). *Cuyahoga River Fire* [Online]. Available at: ohiohistorycentral.org/w/Cuyahoga_River_Fire (Accessed 13 July 2020).

Olwig, Kenneth (2005). 'Representation and alienation in the political land-*scape*'. *Cultural Geographies*, 12, pp 19–40.

Paull, John (2013). 'The Rachel Carson letters and the making of *Silent Spring*'. *SAGE Open*, 3, pp 1–12.

Rancière, Jacques (1991). *The Ignorant Schoolmaster: Five Lessons in Intellectual Emancipation*. Trans. Kristin Ross. Stanford, CA: Stanford University Press.

Relph, Edward (1985). 'Geographical experiences and being-in-the-world: The phenomenological origins of geography'. In: David Seamon and Robert Mugerauer, eds. *Dwelling, Place & Environment: Towards a Phenomenology of Person and World*. New York, NY: Columbia University Press, pp 15–31.

Reykjanes UNESCO Global Geopark (2015). *About Us* [Online]. Available at: reykjanesgeopark.is/en/about-us (Accessed 15 July 2020).

Schopf, Hiedi and Jennifer Foster (2014). 'Buried localities: Archaeological exploration of a Toronto dump and wilderness refuge'. *Local Environment*, 19(10), pp 1086–1109.

St. Louis County, MO (2014). *West Lake Landfill Shelter in Place/Evacuation Plan.* (Accessed 19 July 2020).

St. Louis Remediation Task Force (1996). *Report.* Available at: coldwatercreekfacts.com/media/reports/13_1996-stlouis-remediation-task-force-report.pdf (Accessed 19 July 2020).

Street, Erin Shaw (2018). 'A "poop train" from New York befouled a small Alabama town, until the town fought back'. *The Washington Post* [Online]. Available at: washingtonpost.com/news/post-nation/wp/2018/04/20/a-poop-train-from-new-york-befouled-a-small-alabama-town-until-the-town-fought-back/?noredirect=on (Accessed 15 October 2018).

Toronto and Region Conservation Authority (TRCA) (2018). *Visiting Tommy Thompson Park* [Online]. Available at: tommythompsonpark.ca/visitor-information/ (Accessed 15 October 2018).

Toulmin, Stephen (1990). *Cosmopolis: The Hidden Agenda of Modernity.* Chicago, IL: University of Chicago Press.

Ukeles, Mierle Laderman (1990). 'Sanitation manifesto!' *The Art*, 2(1), pp 84–85.

Ukeles, Mierle Laderman (1996). 'Flow city'. *Grand Street*, 57, pp 199–213.

Victor, Daniel (2018). 'Free of New York's stinky sludge train, an Alabama town is still steaming'. *New York Times* [Online]. Available at: nytimes.com/2018/04/19/nyregion/poop-train-alabama.html (Accessed 15 October 2018).

Waterman, Tim (2018). 'Taste, foodways, and everyday life'. In Joshua Zeunert and Tim Waterman, eds. *Routledge Handbook of Landscape and Food.* New York, NY: Routledge, 517–530.

Yokohari, Makoto and Marco Amati (2005). 'Nature in the city, city in the nature: Case studies of the restoration of urban nature in Tokyo, Japan and Toronto, Canada'. *Landscape and Ecological Engineering,* 1(1), pp 53–59.

3

NARRATING LANDSCAPE CITIZENSHIP ON THE COAST

Conflicting views from the Bulgarian Black Sea and Yorkshire North Sea shores

Anna S. Antonova

Marshall Berman once wrote that modernity is "a struggle to make ourselves at home in a constantly changing world" (Berman, 1982: 6). In that sense, global society has never been more modern. Belonging and citizenship alike can be challenging forms of existence in a globalizing, increasingly unequal, increasingly environmentally damaged world. In the era of the Anthropocene, as human beings make sense of the planetary temporal and spatial scales of their impact on the planet (Crutzen and Stoermer, 2000), in some ways we 'belong' more than ever; yet this far-reaching agency as a mode of belonging is not easy to recognize, and indeed often alienating. Defining citizenships is similarly complex. In a global world, we may be more connected to each other than ever, but this has not eradicated old injustices and has produced new ones.

In this context of global environmental and social strife, the task of conceptualizing 'landscape citizenship' presents many challenges. In this chapter, I discuss two important tensions with landscape citizenship arising in two key dimensions: first, time—or the perception that longevity of embeddedness may be necessary before landscape citizenship can be achieved—and second, spatial scale—or the question of how far and to whom landscape citizenship of any landscape can be extended. I think specifically about time and space for philosophical and practical reasons. Philosophically, the onset of the Anthropocene, as Dipesh Chakrabarty has argued, necessitates reflection across different spatial and temporal scales both to recognize human impact as planetary and to question existing conceptualizations of historic time (2009). Pragmatically, how do individuals and communities conceptualize their belonging in any landscape? Narratives, as a key mechanism to human meaning-making, are vital to that process; and, as Bakhtin has theorized, narration itself is always simultaneously spatial and temporal (1981). Thus, thinking about time and space is important for landscape citizenship both in the philosophic sense—is the concept fit for the intellectual challenges of negotiating human belonging in the

Anthropocene?—and pragmatically, in terms of assessing how well landscape citizenship speaks to the ways in which communities narratively construct their own belonging in a landscape.

I derive the observations about time and space I make in this chapter thanks to insights from 2017–2018 fieldwork I conducted on two specific coastlines, the Bulgarian Black Sea and the Yorkshire North Sea. For this chapter, each of these two case studies helps me describe and define one of the two tensions with landscape citizenship. The first tension, time, emerges through my discussion of narratives from the Bulgarian Black Sea coast and the second tension, space, through narratives from the Yorkshire North Sea coast. For each tension, I draw on the specificity of that landscape's maritimity. By intersecting narratives of inheritance and belonging on the Bulgarian Black Sea coast, I reflect on the complication of adding *temporality* to the notion of landscape citizenship; I ask how Olwig's sense of land-ship (2005) may change if we look at the longer timelines of narrative attention necessitated by the Anthropocene. Next, by examining narratives from the Yorkshire North Sea coast about property in the sea, I outline the problematic tension that emerges when we think on the notion of belonging across wider spatial scales. I consider the challenge of formulating belonging and by extension citizenship in any way without simultaneously creating exclusions. The ontological potential of maritime space that these two coastlines share helps me bring these insights together at the end of this chapter.

My fieldwork on both coastlines formed part of a larger project that examined the links between narratives about environment and society and the varying transformations and contestations each community experienced within its landscape. Narratives were the core analytical focus of my work. Their relevance in helping me navigate between the analytical frameworks of environmental humanities, political ecology, and critical policy studies came especially from their complex social function: narratives can be simultaneously political and affective, direct, and metaphorical. Accordingly, I explored the extent to which narratives' ability to express individual and shared affective experiences can also reveal important aspects of how communities experience, navigate, and indeed engender political and environmental change. To pursue this question, I developed an interdisciplinary narrative analysis approach that allowed me to draw on multiple types of materials (Antonova, forthcoming). In 2017 and 2018, I collected interviews with people representing different contemporary perspectives on the Bulgarian and Yorkshire shorelines—from environmental activists through government officials to members of local community. In parallel, I also analysed a range of textual primary sources, including literature, historic materials, policy and legal documents, and media pieces, from the 1950s onward.

My work on each shoreline focused on the material and metaphorical expressions of concrete environmental, political, and societal transformations: in Bulgaria, these were broadly related to the shoreline's rapid tourism-related (over)construction following the country's post-1989 transition, while in Yorkshire, the transformations related to the increased enclosure of maritime space in recent decades due to new uses of the sea, like wind farms or marine protected areas. In both contexts, participants also thought and spoke about the varying measures of belonging to their shoreline. For this chapter, I draw specifically on this theme from the narratives I collected. I outline two important tensions that the theme of belonging raised for me with respect to the concept of landscape citizenship: the one to do with temporality and belonging as an inheritance; and the other with the different spatial scales associated with belonging—and the exclusions they entail.

Together, the two case studies enable me to decentre the idea through the specific context of the coast: an interface where multiple ontologies meet and where epistemological constructs can therefore be destabilized. To place these notions of landscape citizenship and belonging on the coast is to tease out their implications against the wider questions of temporal and spatial scale posed by the Anthropocene. The coast presents its own ontological problems: as maritime anthropologists like Christer Westerdahl and David Berg Tuddenham have argued, this landscape activates reflections across varying spatial, temporal, and epistemological dimensions (Tuddenham, 2010; Westerdahl, 1992, 2007). Moreover, the coast's difficult ontology also helps foreground the tensions associated with devising landscape citizenship for the Anthropocene. Here, Astrida Neimanis's suggestion in *Bodies of Water* that we imagine the Anthropocene not through the lithosphere but through the hydrosphere—the "fascia that lubricates and connects the Earth's lithosphere to biosphere and atmosphere, those most popular players in the Anthropocene drama"—has relevance (2017: 160). To Neimanis, to view the Anthropocene through an aqueous perspective—or more specifically through the ecofeminist emphasis on flows—is of necessity to reflect on the differences and connections incorporated in the collective human agency: the same 'we' that Chakrabarty finds, in 'Climates of History', so problematic (2009; Neimanis, 2017: 161–174). Neimanis draws on the central ecofeminist postulation that human lives and their varying modes of existence are all supported by a series of relationships and flows. She proposes a "watery embodiment" that represents the flows and connections of our bodily experiences: our leaks into the sea and into cisterns of underground reservoirs, the flows from and into other human and nonhuman bodies ("a kissable lover, a blood transfused stranger, a nursing infant") (Neimanis, 2017: 2). For the context of a maritime Anthropocene, these links are important. In Neimanis's words: "Bodies of water, as figuration, invite us to amplify a relational aqueous embodiment that we already incorporate, and trans-corporate" (ibid.: 169). Or, in other words, thinking with an aqueous (in this case maritime) perspective helps foreground the relationships that already exist between human beings, as well as between humans and their landscapes, in ways that pay attention to the relations of power and their consequences for the disempowered. It complicates usefully the problem of belonging. Applying this perspective to some of the narrative material I collected on the Bulgarian Black Sea and Yorkshire North Sea shore, I examine both how landscape citizenship plays out against the ontological tensions of these coasts in the aqueous Anthropocene.

Abstract or enacted landscapes and citizenships

To unify landscape and citizenship in one term is to recognize that representation is always political. The scholarship that first helped elucidate the landscape as an analytical lens concerned itself much more with its material or cognitive—rather than social and political—aspects of representing landscapes. This is apparent in the now-classic debate between Cosgrove and Daniels's (1988) vision, in which the landscape appears as a collection of cultural images, and Tim Ingold's (1993) postulation, in which the landscape is instead the sum of material, lived experiences layered over time. For Cosgrove and Daniels (1988), the act of representation itself is vital; for Ingold, representation is but the result of "engaging perceptually with an environment that is itself pregnant with the past" (1993: 153). These early accounts do not engage extensively with the possible tensions between different types of communities and their engagements with the landscape. Queries about (dis)possession, justice, and power remain, at best, background to both Cosgrove and Daniels's concern with representational processes and Ingold's interest in individual and collective incorporations within physical space.

By 1998, when Cosgrove published the second edition of his *Social Formation and Symbolic Landscape* (1984), this criticism had been voiced—and, as Cosgrove himself acknowledged, with valid reason (ibid.: xvii–xxiv)—by various critics. Later scholarship engaged with the political much more closely, especially through its legal expression in the form of the European Landscape Convention (Council of Europe, 2000). In particular, Kenneth Olwig's notion of 'land-ship' introduced a helpful response to Cosgrove and Daniels' and Ingold's debate (Olwig, 2005). Olwig foregrounds landscape representation for its tangible—that is, political and social—consequences (ibid.).

Alienation, to Olwig, is a vital concern, and he elucidates its representational and organizational aspects. In Olwig's interpretation, thinking of 'landscape' or in his term 'land-ship' adds expressions of social and political relations to the material reality of the 'land' (2005). Olwig emphasizes this fusion of material and social entity: "The physical land, thus–*scaped*, becomes an expression of the *nature, state*, or *constitution* of the land as the embodiment of a *res publica* or commonwealth" (ibid.: 23). Olwig contrasts this socially and materially constructed state of landscape relations to the alienated and abstract version of land and landscape that came to underpin private property management after the Enlightenment (2005). The alienation of common property from its political and material role in people's daily lives, he explains, has occurred in linguistic representation, institutionalized through the practices of law, as well as through artistic representation, institutionalized through cartography (ibid.).

Bruno Latour and Raymond Williams point to the abstractification and alienation of 'nature' as a defining aspect of modernity (Latour, 1993; Williams, 1980); Olwig elucidates some very concrete ways in which this might have occurred. Representation plays a vital role in this account: "[l]andscape as scenery thus provides a means of masking and envisioning the land as a national territorial unity, its boundaries (rather than its parliament) defining its state" (2005: 25). Thus, landscape's *abstract* representation helps execute a transformation of the value ascribed to natural space. It shifts from a collectively negotiated resource to something entirely abstract, delineated, and no longer held in common.

Olwig argues that the process through which land enclosure delivers "the embodiment of a *res publica*" inevitably reward some while dispossessing others. He writes how in the initial processes of land enclosure during the Enlightenment,

> land that might previously have been held in common was alienated from the commoners ... The transformation of land into private property was also alienating in the psychological and social sense, particularly for the poor, whose rights in common land disappeared when the land was enclosed. This not only reduced their resource base, leaving them much more dependent upon the property owners, it also estranged them from their sense of having a place in the land as a polity.
>
> *(2005: 28)*

Olwig's insights about property and cartographic representation of the land help elucidate the mechanisms through which the landscape is highly political. If, as Raymond Williams (1980) argues, the creation of 'nature' and 'the human' as separate abstractions underpins the intellectual landscape of modernity in the West, then Olwig and Williams together invite us to reconsider human history since the Enlightenment. Both Williams and Olwig demonstrate how Western modernity adopted its specific political representation of landscapes as 'other' or alienable in order to ground its intellectual and economic accomplishments.

As we look to resolve this entanglement by trying to imagine alternative political representations—specifically through the concept of landscape citizenship—we come against an inherent difficulty. In "The Climate of History: Four Theses", Dipesh Chakrabarty (2009) reminds us that the conceptual separation between human and geological timelines has historically supported the various narratives—like humanism, rationalism, or freedom—that underpin Western modernity and its most promising improvements. "The mansion of modern freedoms", he writes, "stands on an ever-expanding base of fossil-fuel use. Most of our freedoms so far have been energy-intensive" (ibid.: 208). Thinking through new kinds of human–nature relations in the Anthropocene must account for what would be lost, and what could be reimagined, when we problematize human agency. Here, Chakrabarty (2009) expresses a concern that the universalizing narrative—in this case of humanity as one (dominant) species—cuts against the necessary work of differentiating between the various different human experiences. Universalist thought is necessary in the Anthropocene; at the same time, thinking about humanity in universals is often at best politically naïve and at worst an undoing of the lessons of postcolonialism, feminism, and class critique. This leads to a quandary: how can we tell the universal story of human agency on a planetary scale "while retaining what is of obvious value in our postcolonial suspicion of the universal? The crisis of climate change calls for thinking simultaneously on both registers, to mix together the immiscible chronologies of capital and species history" (ibid.: 219–220). The concept of landscape citizenship faces the same conceptual difficulty. How can landscape citizenship both attend to the universal problems of human belonging in the Anthropocene *and* account for the plural ways of belonging that should rightly co-exist?

These difficulties are reflected when applying landscape citizenship on the coast. On the one hand, as DeLoughrey (2010) has argued, the oceans have simultaneously enabled and embodied the flows of modernity, militarization, and waste along with their unequal consequences. Flows of capital and conceptual thinking moving through the oceans liberated some even as (and often because) they enslaved others (DeLoughrey, 2007). This difference must be recognized in contemporary notions of landscape citizenship on the coast. On the other hand, blue environmental humanities perspectives like Neimanis's "watery embodiment" show how the oceans highlight the flows and connections between individual human bodies and "a more-than-human hydrocommons" (2017: 2). Transposing landscape citizenship to a maritime setting therefore replicates both the difference and the universality affiliated with representation and landscape citizenship. In the following sections, I demonstrate this in more depth by relaying narratives from the Bulgarian Black Sea and Yorkshire North Sea coastlines and demonstrating the tensions that they raise with belonging and the construction of landscape citizenship. Reflecting Chakrabarty's efforts to reconcile the universalist with the particular, I understand 'landscape citizenship' in the singular as the concept that can hold the multiple possibilities of how landscape citizenships are encountered. In the following two sections, through the two case studies, I debate this perspective on landscape citizenship in two key dimensions: time and space.

Tension one: inheriting the coast and the temporality of citizenship

Discussing questions of environmental preservation and societal consciousness on the Bulgarian Black Sea coast, several participants in my research lamented the transience of local people, whose sense of belonging in the coastal landscape they saw as interrupted by historic events multiple times and in multiple ways. For these speakers—environmental activists, NGO employees, local policy officers—this sense of temporal displacement contributed to the coastline's ongoing over-development by eroding locals' economic and affective links to the sea and thereby undermining their stewardship of the landscape. When voicing these opinions, participants referred to the complicated history of this stretch of the Black Sea coastline since the Ottoman Empire's decline (the 1800s and early 1900s). Until the early 1800s, this shoreline's population was predominantly of Greek cultural ethnicity (Shterionov, 1996). Political events surrounding the Russian–Turkish war of 1828–1829 resulted in roughly two-thirds of this coastal population leaving the region, causing a significant demographic shift in the region that now forms the southern Bulgarian coast (ibid.). Most of these emigrants never returned. They took away with them not only their maritime economic expertise—particularly where it came to ship-building—but also their political and societal activism, thereby changing the region's social fabric in its entirety (ibid.).

Around the early 1900s, the southern coastal region became repopulated through land grants offered by the re-established Bulgarian state. Most of the new arrivals were refugees of Bulgarian cultural ethnicity, displaced from the inland during either the Ilinden–Preobrazhenie uprising of 1903 or the 1912–1913 Balkan War (Raychevski, 2008; Shterionov, 1996). These newly established inhabitants of the coast were primarily agricultural folk, for whom, as local manager Kaloyan told me, "the sea was perfectly foreign" (Interview, my translation). The sea was foreign to them, he explained, because they had not inherited the knowledge of continuous maritime practices. Crossing Ingold (1993) with Neimanis (2017), one might argue that they did not figure their sense of self and community through embodied practices connected to the sea, the coast, and its flows. As a result of the sea being 'foreign', Kaloyan told me:

> [T]he people related to the coastline itself, those who exercise some control over their councils, which in turn direct local development, have no lasting generational link to this precise territory. They came as refugees here, in two or three generations they became settled, but I think that, on a very deep level, as an archetype, these are not the typical sea people who could go back generations, generations, generations [sic] in time and *defend, with some honour, with some dignity, their attachment to and their roots in precisely this place.*
>
> *(Interview, my translation and emphasis)*

Much could be read from Kaloyan's insistence, in the block quote above, that transience precludes not simply care but also the dignified, honourable defence of a whole mode of existence—let alone the environment that supports it. It may be the kind of existence in space theorized more formally by Olwig's (2005) linguistic and philosophical writing on 'land-ship': an intricate relationship between nature, state, and constitution, wherein the notion of citizenship itself connotes idealized, as well as institutional, qualities. Thus, the interlinked concepts of nature, state, and constitution pertain to social relations in the moral, and not only in the physical or in the legal sense (ibid.). One dwells, one becomes embedded in the environment through a long-lived, generationally reiterated experience; and in turn one defends, and defends *with*, a full set of rights, but also dignity and honour.

A close reading of Ingold's (1993) essay "The temporality of landscape" tells us that temporality underwrites the processes through which a community embeds itself in a landscape. Temporality, in Ingold's view, relates to a perspective on history and time in which events are not strung out "like beads on a thread" but instead form a pattern, essentially one of causality: the past, through the present, leads to the future (ibid.: 157). In this formulation, "temporality and historicity are not opposed but rather merge in the experiences of those who, in their activities, carry forward the process of social life" (ibid.). And these activities or experiences together, embedded in the landscape, then form the taskscape (ibid.). In Kaloyan's account, the cognitive, imaginative aspects of a community's experience take on a similar role. For Kaloyan, the continuity in temporality proves so integral because it helps shape practices of thought. The inheritance the community has not yet received is as much one of narrative practices as of material ones. To be able to develop any kind of sustainable existence in the landscape, Kaloyan told me, one "needs to know what will be *in the heads of one's neighbours* for the next twenty, thirty, forty years" (Interview, my translation and emphasis). Because it carries and reiterates traditions of mutual understanding and trust, temporality is important in the organization of material practices, and in the performance of social functions. Kaloyan saw *longevity* as what bridged the community's past with its future (Ingold, 1993) as well as what helped to combine the institutional with the moral aspects of being a citizen in that community and in that place (Olwig, 2005).

In Kaloyan's account, a further destabilization in the lines of inheritance comes from political turbulence over the last century. Temporal disruption becomes in Kaloyan's account a mechanism for the kind of alienation Olwig describes. Olwig (2005) writes of the representational processes that transform social value into abstract ideal: geometrically delineated property lines. A similar trajectory appears in the narratives of the property transformations on the Black Sea coast: land was nationalized at the beginning of socialism in the 1940s, and property restitutions at the beginning of the democratic process in the 1990s exacerbated rather than helped resolve local issues of longevity and continuity. By way of illustration, another participant, Andrei, described the contemporary transformation of a salt lake ecosystem into an unsustainable tourism-oriented landscape. Construction, and the opportunity to earn money from selling property to tourists, was the ultimate developmental fantasy in this context, Andrei explained. In the first half of the twentieth century, this area had been the largest producer of salt in the country and the local economy was further supported by balneological tourism. These industries were almost completely destroyed in the course of Bulgaria's political transformations: first the onset of the socialist period and suspension of private ownership, and then the post-socialist transition and subsequent property restitutions. Disruptive as they were, these political events, to Andrei, were not *in themselves* the cause of local social or ecological trauma. More worrying by far was the negation of those locally sustainable, entrepreneurial economic models that *had* existed and *could* still exist here: salt production, health, and nature tourism. To Andrei, who works as a local environmental NGO officer, the difficulty was rooted precisely in longevity: "society in Bulgaria simply is not developed enough to that level" (Interview, my translation). Local heirs to property, in his account, were not interested in salt production because they had lost their knowledge and therefore connection to that industry. The promise of capitalist profit (and by extension the dependence on external capital) replaced their connection:

a large part of the town here, including the municipality, the school, all of it was built over salt flats that were destroyed, filled in. The last to be destroyed and filled in … transformed into a neighbourhood of apartments for sale, chiefly to Russians or the former Soviet republics.

(Andrei, Interview, my translation)

The reintroduction of economic value to lands, when these lands had already been alienated and emptied of their communal or heritage meaning, did not lead to the return of sustainable models or stewardship that might have existed prior to socialism. Quite the contrary: the transformation of inherited property added to the sense of rupture in the social fabric of the coast. It has spurred the perverse, profit-driven logic Olwig describes.

Together, these accounts elucidate the tension temporality brings to landscape citizenship. If belonging and connection are contingent on longevity, as these accounts insist they are, then how do we reconcile landscape citizenship with either global connection or mobility? Given the evidence that political rupture interrupts lines of inheritance and affective belonging, we must ask if landscape citizenship is not oxymoronic. Citizenship, as Hayward argues (2006), must exist within a polity that allows for negotiations and contestations as well as for political change. Yet belonging to a landscape, if Ingold, Olwig, and the participants I spoke to are to be believed, depends on a political stability that cannot always be guaranteed. How are these two propositions to be reconciled? Further, if the temporal dimension is vital to landscape citizenship, how do we reconcile these very *human* temporalities with the necessity to think on geologic or lithic scales in the Anthropocene? Are we not in danger (Massey, 2006) of ascribing human constructs of solidity and permanence to the material features of the landscape that the rocks and strata themselves—in their natural fluctuations—do not objectively support? These difficulties remain pertinent in the narratives I collected on the Yorkshire North Sea coast.

Tension two: spatial dimensions of belonging and landscape citizenship

If acquiring landscape citizenship or care for the landscape in Ingoldian or Olwigian terms relies on continuous practices and experiences through time, then landscape citizenship excludes those who do not fit the same time frames. This seems to disqualify refugees or immigrants, whose belonging is of shorter duration. But to do so would be unjust, as well as a slippery slope. In Kaloyan's account above, the refugees to whom the sea is still 'foreign' have been settled on its shores for more than a century. So how long is long enough for a community to establish its belonging to that landscape?

If time is problematic, what about space? Could a spatial delineation situate belonging in an embodied way with strong affective links, bodily, emotive means of relating to the landscape that speak to what we might instinctively understand as landscape citizenship? At the same time, landscape citizenship must go beyond the local level and consider bigger scales. Landscape citizenship encounters Chakrabarty's (2009) challenge of balancing the universal with the diverse. Neimanis's (2017) thinking on watery embodiments, emphasizing the links between individuals and all that sustains them, could resolve how we might define landscape citizenship in spatial terms.

I turn here to narratives pertaining to fisheries and environmental conservation I encountered in my 2017 fieldwork on the Yorkshire North Sea coast. In many ways fishing provides an apt opportunity to debate human forms of belonging in and with the nonhuman landscape. Fishing is a symbiosis of human and nonhuman agency, as the health of the stocks (and the ecosystem) is closely interlinked with the economic and cultural fate of the communities that depend on it. At the same time, especially in the UK, it is an industry that places the community's sense of self in a broad, global, oceanic context. Fishing has a long history on the Yorkshire coast; as one participant, Connor, told me that "we've lived and worked in and around [the sea] for thousands of years … human fact is part of that environment, has been since Neolithic times" (Interview). The longevity that Kaloyan and Andrei missed on the Bulgarian Black Sea coast exists here in terms that would be recognizable to Ingold: material experiences, especially of weather, risk, danger, or damage to their bodies, create people's sense of belonging. Writing about Hull's long-distance trawling community, for instance, historian Alec Gill (2003) has pointed out that the hardship and danger associated with the industry was pivotal for the community's sense of self. In the UK more generally, commercial fishing remains the most hazardous peacetime occupation (Roberts, 2010). And for some participants, danger was something transferred into the body through the fisherman's labour. When I spoke to James and Grace, a fishing family, they often discussed the connection between weather, work, body, and danger, as below:

> James: When it's fine [out] it's alright [to trawl]. It's a hard job; it is a hard job. There's certainly harder jobs than ours, there certainly is. But the rougher [the weather] is, as it gets rougher it gets harder and progressively harder and if it's really rough it's very, very difficult. That's when it starts to become dangerous if you like.
>
> *(James and Grace, Interview)*

James's comment speaks to a strong sense of the individual's physical entanglement with the environment through the prism of danger. The body is a site of risk where the material tasks and dangers associated with the coast and the weather become intertwined with the fisherman's life experiences; the technology he or she utilizes extends the body. We may read here Ingold's (1993) postulation that the material practices of the day to day are what shape the landscape and the societies on it.

Reading through Astrida Neimanis's reflections on "watery embodiment" (2017), James's words offer another insight. Her criticism of most Anthropocene narratives is in their inattention to "the ways in which circuits of matter and differentiated flows of human power are entangled" (Neimanis, 2017: 166). To

James and Grace these flows seem clear. The fisherman's body plays host to the dangers and fluctuations in the environment. Much as Neimanis might suggest, fishing means engagement with flows and continuities: familial, bodily, and techno-logical. James and Grace spoke of their roots, as well as of their son, and the likeli-hood of his moving "lock, stock, and barrel" to fish exclusively out of a bigger town once James retires (Interview).

Similarly, Connor described "threads of continuity": fourth or fifth gener-ation fishermen and boats that had kept "the same name for many generations". In these examples, the sense of belonging (and property) is passed on through the generations along with the labour, tasks, risks, and dangers. This historical cognitive connection further extends also to the sensory experience of the coast. Connor asserted that

> Unless you're a fisherman, you've never seen [areas 70 miles off land], unless you've looked out the window of an airplane … What's your stake in that? You have opinions about it, I don't doubt, they're legitimate, valuable opinions, but opinion holders are not the same as stakeholders.
>
> *(ibid.)*

'Seeing' attains value only through tangible experience. Fishermen's connection to the environment becomes visceral via a material relationship with work, tasks, and body. Or, as Grace succinctly put it when speaking of students from Hull visiting the coast: "They'll have seen the sea but it's not the same as going to the seaside and seeing [sic] the beach" (James and Grace, Interview). 'Seeing' the sea must be affiliated with special knowledge, a participatory engagement that would need to be explicated and communicated to the outside. Fishing—especially small-scale fishing—provides an example of landscape citizenship on the Yorkshire coast.

Other participants in the Yorkshire study complicated this definition by pointing to a wider dimension of landscape and environment through reflections on fisheries governance on the Yorkshire North Sea shore. Often introduced to target large-scale commercial fishing, restrictions ended up affecting local small-scale commu-nities disproportionately. "It's one of those", local policy officer William remarked,

> where if you do it top-down, it isn't gonna work, but you can't do it bottom-up either, because it's just not gonna work. Because by the time it gets to there, you've got an issue because you've emptied the seas. Well, from the top down, you've got an issue because by the time you get here, nobody's fishing because you've put them out of business.
>
> *(Interview)*

William's apt comment on the difficulty of reconciling these scales sums up an important tension, a classic concern of fisheries governance, and pertains to the difficulty of delineating landscape citizenship. His words highlight the challenge of maintaining meaningful local existence (which, as James and Grace's comments illustrate, is wider than 'business') while preserving the wider oceanic environment necessary to guarantee it. Viewed from an Anthropocenic lens, the means of belonging through fishing is enabled by a series of abstractions with respect to the environment. Locke's idea, quintessential to modernity, that property is a right arising from the addition of human labour to a thing (Rose, 1985) shares a spirit with Hugo Grotius—widely considered to be the father of international law—who maintained that fish are public while in the ocean but become private property through the agency of fishermen's nets and hooks (Mansfield, 2004; Steinberg, 2001; Grotius, 1609). Both Locke's and Grotius's logics depend on an understanding of nature, particularly *oceanic* nature, as separate, alien, and abstract (Hugo Grotius, after all, famously postulated the idea of "the freedom of the seas"). For them, maritime space's *otherness* enables the vital construction of meaning in labouring the sea, sustaining loss, or taking risk in one's body and with one's boat.

The execution of this abstraction through cartographic and artistic representations has important societal and geopolitical consequences. The belief that the oceans were literally 'inexhaustible' (as Thomas Huxley famously insisted of fish stocks at the 1882 Fisheries Exhibition in London (1903)) reflected more than Western tendencies to render nature abstract. It also upheld a related geopolitical view that underpinned global world-economy and imperial power projection for centuries (Steinberg, 2001; Crosby, 1972): by rendering oceanic space both empty and abstract—a negative space—Western thought could frame it as suited to projections of human abstractions. The empty oceanic space could be 'filled' with the influence of global naval powers. As geographer Philip Steinberg (2001) has argued, the ambiguity and marginality of oceanic space made it an area where powerful actors could debate and establish their own legal and political ideas before advancing them on land.

Since an entire worldview depended on the abstraction of maritime space, Hull's long-distance trawling community could derive their own sense of belonging and identity from their work in the environment (Byrne, 2015, 2016). The same worldview, however, also left such communities open to being used to justify national political interests. For example, Hull trawlers' sense of pioneering and labour-justified rights to distant Arctic grounds became an important factor in the Cod Wars, a series of fishing conflicts between Iceland and Britain from the 1950s to the 1970s that prompted the renegotiation of international ocean law (Guðmundsson, 2006; Jóhannesson, 2004). British naval vessels were dispatched to protect the long-distance trawlers. Vessels on both sides collided with serious and often dangerous damage (Guðmundsson, 2006: 97). Trawlers were seen by the UK to represent national geopolitical interests because their work in northern waters was in line with "freedom of the seas" principles that had supported British naval supremacy for centuries (Jóhannesson, 2004: 545–550). Officially, however, the UK justified its actions with the "traditional interest and acquired rights in and current dependency on [Icelandic] fisheries" that British fishermen had (United Kingdom v. Iceland, 1974: 8). The projection of British naval power projection during the Cod Wars framed fishermen's landscape belonging as one that superseded any maritime boundaries. This reveals difficult links between the individual or local sense of belonging—that is, the bodily, inherited belonging that landscape citizenship might depend on—and national territorial claims, where the construction of belonging can be utilized by governments justifying a nation's extraction, power projection, and exclusion.

The resolution of the last Cod War in Iceland's favour (and subsequently the third United Nations Convention on the Law of the Sea (UNCLOS, 1982), which established the legal instrument of the exclusive economic zone) responded directly to the growing Anthropocenic recognition that the oceans were *not* remote, untouchable, inexhaustible, or otherwise abstract. This realization has produced legal and political commitments towards preserving the maritime environment in a way that could continue to guarantee maritime livelihoods. The complexity of calibrating landscape citizenship remains just as difficult. As William observed of the contemporary context, "sometimes some of the maritime regulations that are brought in have no effect on the big people but absolutely destroy the livelihoods of the little people" (Interview). Even more vitally, as Connor explained, these issues of scale were exacerbated by the relative invisibility of the community because of its local scale: "It's always the case with fishing, it exists on the edge, and then it's swept over the edge and disappears" (Interview). The Brexit campaigns of 2015–2016, in which voices to 'take back our seas' attained national importance, could also be viewed from this perspective of representation. These campaigns had been helpful for fishing communities, Connor stressed, because they had made them visible beyond the local scale and made people "aware that we exist as an industry … *that it's a community that exists*" (Interview, my emphasis). To Connor the issue came down to representation as access to polity in a community that felt invisible once its idea of belonging diverged from the wider sense of national pride and power projection. In the Yorkshire coastal context, what we might term landscape citizenship held stronger sway when it coincided with national visions and projects of citizenships, as in the context of the Cod Wars. Once that match between state citizenship and landscape citizenship no longer functioned, those whose livelihood and belonging depended on local embeddedness felt that their lives had been rendered irrelevant.

These examples from the Yorkshire North Sea shore raise important queries about the spatial dimension of landscape citizenship. How do we delineate it? Who gets to be a landscape citizen? And how do we ensure that the belonging that underpins the concept is not co-opted by powerful actors to serve national and political ideas of citizenship?

Conclusion

Narratives of belonging from the Bulgarian Black Sea and Yorkshire North Sea's ongoing transformations suggest two sets of questions about the concept of landscape citizenship. Although I have deliberately derived the questions relating to time from the Bulgarian Black Sea and the questions pertaining to different spatial dimensions from the Yorkshire North Sea, both contexts speak to the challenges communities everywhere face in narrating their belonging within a landscape. "How far back and how far forward in time should we go in order to belong?" is as essential to narratives of belonging as the question "Who gets to be a landscape citizen?" Examining these questions in two material, enacted landscapes highlight how vital a role change, flux, and polity play in communities' situated negotiations of the answers to these questions. In case studies I have examined, belonging has been relational—in the sense of relating to others—as well as relative—that is, configured through the dimensions of time and space.

The theme of belonging and what it has to suggest for landscape citizenship is a central one I raised. Belonging matters to individual communities and their links to the landscape. As Kaloyan argued on the Bulgarian Black Sea coast, the sense of belonging over time grounds a community's care and stewardship for its place. As James and Grace illustrated on the Yorkshire North Sea shore, belonging is materially grounded in communities' daily experiences and labour. However, as Connor explained, belonging cannot always safeguard a marginalized community's existence in a landscape. The fluidity of the spatial dimensions across which belonging should be considered—especially for the Anthropocene—highlights the challenge of devising narratives to encompass this complexity fairly. The fluid ontology of the coastline brings these difficulties to the fore. Blue thinking also opens up new narrative structures that can hold these varying aspects in continuous relation. Focusing on how humans "leak and seethe, our borders always vulnerable to rupture and renegotiation" (Neimanis, 2017: 2) could offer ways forward in conceptualizing belonging—and landscape citizenship—as simultaneously material, flexible, and relational.

These reflections offer possible directions for ongoing conversations about landscape citizenship. In this chapter and book, landscape citizenship has been conceived alternatively in the singular—as a philosophical concept framing scholarly responses in the Anthropocene—and in the plural—as multiple embodied states continuously (re-)figured in place through narrative negotiation. These alternative configurations speak to the difficulty and the importance of simultaneously holding together the seeming contradictions between plurality *and* universalist thought. As the concept of landscape citizenship attends to the multiple forms of belonging that each community narrates for its own landscapes, it should also reflect on the universalist challenges associated with its temporal and spatial dimensions. Landscape citizenship should strive to reconcile longevity, embodiment, and emplacement with global connection and mobility. The fluid possibilities suggested by belonging on the coast offer an avenue for both.

Bibliography

Antonova, Anna S (forthcoming). 'Blending environmental humanities and marine social science: An interdisciplinary narrative analysis approach to hybrid scholarship on the coast'. In: Madeleine Gustavsson et al., eds. *Researching People and the Sea: Methodologies and Traditions*. London: Palgrave Macmillan.

Bakhtin, Mikhail M (1981). 'Forms of time and of the chronotope in the novel: Notes toward a historical poetics'. Translated by Caryl Emerson and Michael Holquist. *The Dialogic Imagination: Four Essays.* Austin, TX: University of Texas Press, pp 84–258.

Berman, Marshall (1982). *All That Is Solid Melts into Air: The Experience of Modernity*, 1988 ed. New York, NY: Penguin Books.

Byrne, Jo (2015). 'After the trawl: Memory and afterlife in the wake of Hull's distant-water fishing industry'. *International Journal of Maritime History*, 27, pp 816–822.

Byrne, Jo (2016). 'Hull, fishing, and the life and death of trawlertown: Living the spaces of a trawling port-city'. In: Brad Beaven, Karl Bell, and Robert James, eds. *Port Towns and Urban Cultures: International Histories of the Waterfront, c. 1700-2000.* London: Palgrave Macmillan, pp 243–263.

Chakrabarty, Dipesh (2009). 'The climate of history: Four theses'. *Critical Inquiry*, 35, pp 197–222.

Chase, Susan E (2018). 'Narrative inquiry: Toward theoretical and methodological maturity'. In: Norman K. Dezin and Yvonna S. Lincoln, eds. *The SAGE Handbook of Qualitative Research*, 5th ed. Los Angeles, CA: Sage Publications, pp 946–970.

Cosgrove, Denis (1998 [1984]). *Social Formation and Symbolic Landscape.* Madison, WI: University of Wisconsin Press.

Cosgrove, Denis and Stephen Daniels, eds. (2008 [1988]). *The Iconography of Landscape.* Cambridge, MA: Cambridge University Press.

Council of Europe (2000). *European Landscape Convention. European Treaty Series.* Vol. 176. Florence, Italy: Council of Europe. Available at: conventions.coe.int/Treaty/en/Treaties/Html/176.htm (Accessed 15 August 2020).

Crosby, Alfred W (1972). *The Columbian Exchange: Biological and Cultural Consequences of 1492. Contributions in American studies.* London: Greenwood Publishing Group.

Crutzen, Paul J, and Eugene F Stoermer (2000). 'The Anthropocene'. *Global Change Newsletter*, pp 17–18.

DeLoughrey, Elizabeth (2007). *Routes and Roots: Navigating Caribbean and Pacific Island Literatures.* Honolulu, HI: University of Hawai'i Press.

DeLoughrey, Elizabeth (2010). 'Heavy waters: Waste and Atlantic modernity'. *PMLA*, 125, pp 703–712.

Gill, Alec (2003). *Hull's Fishing Heritage: Aspects of Life in the Hessle Road Fishing Community.* Barnsley, UK: Wharncliffe Books.

Grotius, Hugo (1609). *Mare Liberum.* James Brown Scott, ed. New York, NY: Oxford University Press.

Guðmundsson, Guðmundur J (2006). 'The Cod and the Cold War'. *Scandinavian Journal of History*, 31, pp 97–118.

Hayward, Tim (2006). 'Ecological citizenship: Justice, rights and the virtue of resourcefulness'. *Environmental Politics*, 15, pp 435–446.

Huxley, Thomas H (1903). 'Inaugural address: Fisheries Exhibition, London (1883)'. In: Michael Foster and E. Ray Lankester, eds. *The Scientific Memoirs of Thomas Henry Huxley.* London: Macmillan, pp 80–89.

Ingold, Tim (1993). 'The temporality of the landscape'. *World Archaeology*, 25, pp 152–174.

Jóhannesson, Guðni Thorlacius (2004). 'How 'Cod War' came: The origins of the Anglo-Icelandic fisheries dispute, 1958-61'. *Historical Research*, 77, pp 543–574.

Latour, Bruno (1993). *We Have Never Been Modern.* Translated by Catherine Porter. Cambridge, MA: Harvard University Press.

Mansfield, Becky (2004). 'Rules of privatization: Contradictions in neoliberal regulation of North Pacific fisheries'. *Annals of the Association of American Geographers*, 94, pp 565–584.

Massey, Doreen (2006). 'Landscape as a provocation: Reflections on moving mountains'. *Journal of Material Culture*, 11, pp 33–48.

Neimanis, Astrida (2017). *Bodies of Water: Posthuman Feminist Phenomenology*. London: Bloomsbury.

Olwig, Kenneth R (2005). 'Representation and alienation in the political land-scape'. *Cultural Geographies*, 12, pp 19–40.

Raychevski, Stoyan (2008). *Странджа: Етноложки изследвания (Strandja: Ethnographic Research)*. Sofia, BG: Захарий Стоянов.

Rimmon-Kennan, Shlomith (2006). 'Concepts of narrative'. *COLLeGIUM: Studies across Disciplines in Humanities and Social Sciences*, 1, pp 10–19.

Roberts, Stephen E (2010). 'Britain's most hazardous occupation: Commercial fishing'. *Accident Analysis and Prevention*, 42, pp 44–49.

Rose, Carol M (1985). 'Possession as the origin of property'. *The University of Chicago Law Review*, 52, pp 73–88.

Shterionov, Shtelian (1996). 'Преселнически движения от 30-те години на XIX век и въздействието им върху историческото развитие на Южното българско Черноморие' (Emigration movements of the 1830s and their impact on the historic development of the Southern Bulgarian Black Sea coast). In *Българите в Северното Причерноморие: Изследвания и материали*, Том Пети. Велико Търново. Veliko Tarnovo: Veliko Tarnovo University Press, pp 263–272.

Steinberg, Philip E (2001). *The Social Construction of the Ocean*. Cambridge, UK: Cambridge University Press.

Tuddenham, David Berg (2010). 'Maritime cultural landscapes, maritimity and quasi objects'. *Journal of Maritime Archaeology*, 5, pp 5–16.

Fisheries Jurisdiction (United Kingdom v Iceland) (1974). Judgement of 25 July 1974. The Hague, NL: International Court of Justice. Available at: icj-cij.org/en/case/55 (Accessed 15 August 2020).

United Nations Convention on the Law of the Sea III (UNCLOS) (1982). Montego Bay, JM, 10 December 1982. Available at: un.org/depts/los/convention_agreements/texts/unclos/UNCLOS-TOC.htm (15 August 2020).

Westerdahl, Christer (1992). 'The maritime cultural landscape'. *The International Journal of Nautical Archaeology*, 21, pp 5–14.

Westerdahl, Christer (2007). 'Fish and ships: Towards a theory of maritime culture'. *Deutsches Schiffahrtsarchiv*, 30, pp 191–236.

Williams, Raymond (1980). *Problems in Materialism and Culture*. London: Verso.

4

SUPERKILEN

Coloniality, citizenship, and border politics

Burcu Yigit Turan

This chapter explores the relationship between the design of Superkilen—a public park project located in the socially diverse neighbourhood of Nørrebro, Mjölnerparken, Copenhagen—and Denmark's coloniality, nation-building myths, and border and citizenship politics. The park was collaboratively designed by BIG Architects (Copenhagen, Denmark), Topotek 1 Landscape Architecture (Berlin, Germany), and Superflex Art Consultancy (Copenhagen, Denmark) between 2009 and 2010. The park was opened in June 2012 after a two-year implementation period. The official narrative of the project's contribution to tolerance, social diversity, belonging, integration, social and cultural values, and democracy in one of Denmark's most diverse areas has been taken for granted among both academic and professional circles in the architecture and the design realm, nationally and internationally. This official narrative diverts attention from ongoing social, political, and spatial realities, as the neighbourhood—located northwest of the city centre—has been racialized, criminalized, and gentrified in parallel with upsurges in nationwide anti-immigrant and xenophobic socio-spatial politics, as well as ethnocentric, white supremacist, and exclusionary boundary drawing around the issues of belonging and citizenship.

Superkilen has been propagated as a showcase of success in the architecture and design realms both nationally and internationally. Notwithstanding a few voices of dissent from other disciplines and environments,[1] award juries,[2] architecture and design festivals,[3] prominent architecture and design blogs, and magazines and academic literature in urban design[4] have consistently reproduced the official narrative, helping it to become the dominant public discourse. The production and reproduction of this narrow framework strip the park of its multi-scalar social, spatial, and political context. This chapter deconstructs the designers' narrative of the Superkilen project, contextualizing the physical, conceptual, and social elements of the project and relating it to larger political questions around colonialism, exclusionary nation-building, and citizenship politics in Danish society. To that end, the discussion includes a relational reading of geo/biopolitics across scales and dimensions through the lenses of border theory, decolonial thought, and citizenship studies, which offer a different story of Superkilen. This theoretical framework provides a rendering of how design reanimates colonialist bordering dynamics, which becomes insidiously infused into the urban social experiences of the postcolonial/neocolonial diaspora, keeping them at the margins of society through processes of material, spatial, and social dispossession. Consequently, this piece attempts to unveil the complicity of the project in the 'colonial' and 'capitalist' politics of belonging and citizenship making, which downgrades or suspends citizenship for certain populations. It also reflects on the possibilities for anti-colonial, anti-racist urban planning and design.

I explore the relationality of the design of Superkilen, with its particular political and socio-spatial formation in the context of Denmark, through the following questions: How is the Danish bordering regime reproduced in the design of Superkilen as a public space? What are the specific aspects of design that reanimate the social, political, cultural, and spatial bordering dynamics? What is the relationship between contemporary landscape representations in public space design with culturalist tones and ongoing colonialist urban warfare and exclusionary citizenship politics?

Centring border theory in public space design research

In this piece, I centre border theory as the analytical framework in examining the interaction of park design with the larger social and political context. The aim is to elaborate the production of hierarchal social difference and divisions, as well as socio-spatial, cultural, and political systems of the dehumanization and exclusion of certain groups, and of oppression based on constructed difference, alterity, and inferiority (Nail, 2016; Balibar, 2002; Anzaldúa, 1987). I build on the recent works of scholars such as Ananya Roy, Stefan Kipfer, and Margaret Ramírez, who explore the internal colonization and creation of 'borderlands'. These works not only reveal similarities between the power that patrols neighbourhoods inhabited by people of colour and patrols 'migrant' bodies, and the power that patrols the physical borders of a colonialist and white supremacist nation-state; furthermore, they reveal similarities between racialized colonial capitalism extracting materials and labour from far-off geographies and capitalist extraction by the urban development projects in the urban areas of colonial states, which tend to be inhabited by diasporic minorities.

Today, over 200 million displaced people from the colonized Global South have tried to reach to the Global North (Sassen, 2016), where they become further colonized (Kipfer, 2007). The racist, capitalist formation of space and its boundary-drawing practices have spread from the colonized geographies in which they originated to the cities of colonial states, becoming inscribed in the development of their urban spaces (Ramírez, 2020; Roy, 2017; Kipfer, 2007). The colonial practices of categorizing, racializing, subordination, displacement, capitalist extraction, and material and symbolic dispossession are materialized in 'urban warfare', while people from subordinated groups' citizenships are downgraded (ibid.).

While current border theory within the field of urban theory mainly focuses on housing dispossession, displacement, homelessness, and segregation issues, how colonialist power utilizes public space design in bordering remains insufficiently theorized.[5] Critical scholars writing on public space production mainly focus on the role of public space production in gentrification or urban capitalist extraction processes (Madanipour, 2019; Larson, 2018; Mitchell, 2017), and in exertion of state power, social control, and capitalist biopolitics (Sevilla-Buitrago, 2017; Davis, 1992; Sorkin, 1992). Mouffe (2013) and Spencer (2016) underline the role of cultural industries and urban design and architecture in capitalist accumulation processes, and in creating images and concepts through semiotic play that mask the detrimental impacts of developments, and furthermore pacify, depoliticize, and even seduce the public and naturalize these interventions. However, critical works on public space production generally focus on political economy, and do not particularly look at colonial bordering dynamics that reveal the racialization processes alongside capitalist accumulation and dispossessions. In this chapter, I contribute to the critical theory literature on public space production by adding one more layer, that of border theory, in order to elaborate how public space design has become part of not only capitalist extraction, but also part of a revival of projects of constructing colonialist nationhood, and racial difference and hierarchy, which implicate a politics of exploitation and exclusionary citizenship delineating a case from contemporary Denmark.

Coloniality, citizenship, and border politics in Denmark

Encountering 'Denmark' and 'coloniality' in the same sentence might be puzzling to those outside Scandinavia and annoying to those from Denmark's ethnic majority. Indeed, the contemporary nation-building myths behind Denmark's image and border regime are based on sidetracking the state's colonial past and contemporary colonial/neocolonial existence in favour of an alternative narrative, one that justifies the dominant ethnic core's sense of entitlement and superiority over the state's territory, while also claiming its innocence—or even its humanism and virtues—in the face of increasing injustices at the global scale due to colonialism (Jensen, 2015; Loftsdóttir and Jensen, 2012; Keskinen et al., 2009). These myths are enforced along lines of reasoning such as an 'indigeneity fiction', which positions Danes as an indigenous population, a denial of colonialism that masks Denmark's past and present, a further denial of the role of that colonialism in the rise of the modern Danish state, and its self-presentation as a humanistic superpower that abolished its slave trade, slavery, and colonial endeavour, and established in their place an anti-colonial, environmentally friendly,[6] self-sufficient, democratic, and tolerant welfare-state (ibid.).

Contrary to the popular narratives that trivialize Denmark's role in colonial history, make the colonial complicity 'forgotten', or as if it is 'false', Denmark is a dynamic and powerful colonial state, with an ever-evolving colonial territory including and excluding different geographies in line with its international politics and ever-evolving tactics, and places in Western imperialism (Naum and Nordin, 2013; Jensen, 2008). It had taken a role in the Atlantic slave trade and slavery; in some of its colonies—such as in the Caribbean Islands or West Africa—it had "planted the colonies" and "displanted the natives",[7] and experimented with different socio-spatial, economic, political, and cultural apparatuses in order to control and maximize the exploitation of land and labour (Hall, 1992). The capitalist expansion of Denmark continued after the official abolition of slavery, since "abolition was a matter of fiscal policy and practicality not necessarily subject to the whims of humanitarianisms" (Meader, 2009: 65). From the literature one can understand how through time, sophisticated ideologies and practices of capitalist expansion, racialization, and extraction have been embedded in the global colonial power complex, labour abuse, spatial segregation, bio-socio-spatial control, boundary drawing, and dehumanization (Meader, 2009; Hall, 1992). These practices have deeply impacted today's formation of previously colonized societies in a way to keep the labour and land exploitation alive (Olwig, 1995), while exploitation has become more invisible (Naum and Nordin, 2013). In parallel with these imperial practices in far-off geographies, an immensely rich cultural realm has developed through colonial literature (Volquardsen, 2014), exhibitions (Muasya, 2018; Mealor 2008) and 'racialized affective consumption' (Danbolt, 2017: 110) which produce extremely othering, abstracting, dehumanizing, and exoticizing images of non-European subjects and objects. These pass racist knowledges into the colonial knowledge production system, which function to maintain the European colonial project of the construction of racial difference and hierarchy (Danbolt, 2017, 2018; Mealor, 2008). Furthermore, landscape has not been detached from these social, political, and cultural constructions. Colonial invasion had profoundly reshaped the relationships between plants, people, and place (Mastnak et al., 2014: 367). It created a particular ideology—a mindscape—that carries a deep sense of entitlement in "the literal planting and displanting of peoples, animals, and plants" (ibid.: 367). This mindscape influenced not only the landscapes in colonial countries, but also—and maybe even more so—the planning of public gardens in the cities of colonizing Western Europe (Certomà, 2013: 24).

Despite recently growing interest in research and activism revealing colonial history, the colonial present, and decolonization in Denmark (Danbolt, 2018), nation-building myths based on the denial of coloniality and the fiction of indigeneity still complete and empower each other (Hervik, 2019; Kvaale, 2011). They dominate ideologies in the cultural realm, and among all political parties across the political spectrum (ibid.). They also establish the most popular images inside and outside of Denmark as the myth of 'Danes as indigenous people' (ibid.). Kvaale asserts that xenophobic anti-immigrant discourse mobilized by the right wing in Denmark and publicly shared by a significant segment of the population has been also based on an argument that:

ethnic Danes—in their capacity of assumed descendants of the first human beings residing on the geological deposit material later to be known as Denmark—are seen as the natural stakeholders of the rights from a historically conditioned and ethno-culturally legitimized perspective.

(2011: 235)

The indigeneity of the ethnic majority of a colonial state is indeed a fiction. Three distinct characteristics of indigeneity reveal why colonial states' claims to it are false:

Distinctiveness, in the sense of being different and wanting to be different. This aspect of being indigenous is closely related to the importance given to the group's self-identification as indigenous.

Dispossession of lands, territories and resources, through colonization or other comparable events in the past, [cause] today a denial of human rights or other forms of injustice.

Lands (located in a specific geographic area) as a central element in the history, identity and culture of the group, usually [give] rise to traditional economic activities that depend on the natural resources specific to the area in question.

(Scheinin, 2004: 3)

Olwig (2003) vividly describes how these colonial images and the fiction of ethnic Danes' indigeneity, and racism and xenophobia, are tightly interwoven into the idea of landscape and landscape representation in Denmark. In turn, landscape representation "has given birth to the 'natives' of these areas" (ibid.: 72).

In contrast to the 'indigeneity fiction', or 'nativity fiction' constructed by ethnic majorities in colonial states—actually in the colonial, neocolonial world order—the Indigenous people of elsewhere have been transformed into migrants (Sassen, 2014). However, the historical amnesia—based on denial, detachment, downplaying, or even the 'sugarcoating' of Danish coloniality (Muasya et al., 2018: 68; Palmberg, 2009)—behind the nation-building project is constantly reproduced by the contemporary media and different sites of the public sphere. This results in a denial of responsibility for the causes of migration (Hervik, 2019; Loftsdóttir and Jensen, 2012). Denial of colonial and neocolonial ties; the economic, material, and ethical consequences of those relations; and self-representation as innocent outsiders to coloniality are still very powerful (ibid.). Meanwhile, colonial ideology shapes gendered and racialized power relations, which have continued to shape the understandings the white ethnic core projects onto minorities (Hervik, 2019; Keskinen et al., 2009; Hvenegård-Lassen, 2008). Meanwhile, histories, agencies, and the rooted lives of minorities within state borders, are ignored, and sensational images and stories of migration are perpetuated (Forkert et al., 2020). In this way, minorities become trapped within the category of 'migrant' (ibid.).

As a result, today's citizens are divided between "'real' and 'not-quite-real' Danes, with the latter group consisting of Danes with an immigrant background" (Rytter,

2019: 689). This categorization has the effect of downgrading the citizenships of people with a 'non-Western' diasporic background (Jensen et al., 2017). Diasporic people's subordination is moralized through the 'fiction of indigeneity' of white 'real Danes' (Rytter, 2019; Kavaale, 2011). Minority rights and citizenship are patrolled by the ethnic majority (Jensen, 2015). Identities are forced to stabilize according to an essentialist idea of culture, which is fixed and bounded, thereby denying subjects' adaptive agency through cultural hybridity (Kvaale, 2011; Wren, 2001). In such a context, the conditions that make a person feel 'at home' and as if they 'belong' never come together, even decades or generations after the migratory event (Jensen et al., 2017).

Racism in new forms is deeply interwoven into the formation of society, and is shared across the political spectrum, becoming mobilized to pass anti-immigration laws (Jensen et al., 2017). However, this is hard to reveal because of the powerful image of Denmark as a humanitarian, tolerant, and egalitarian progressive welfare state (ibid.). Paradoxically, the welfare state (including pro-duction of housing and the built environment) has been dismantled through neoliberalization by the similar exclusionary argument that the welfare system has benefited 'others' who do not belong, or who—or whose ancestors—did not contribute to the construction of the welfare state[8] (Agustin and Jørgensen, 2019; Rytter, 2019). Thus, 'indigeneity fiction' and 'denial of coloniality' together con-stitute the pretext for dismantling the welfare state, as well as for anti-immigrant politics.

Nørrebro as borderland

Against such a background, Nørrebro is not just any neighbourhood. Located north-west of the city centre, it has a centuries-long history as a home for immigrants, the working class, and left-wing activist groups exploring their agency, solidarity, and self-determination through different forms of living, community building, place making, and struggles against structural inequalities (see Schmidt, 2011). The neighbourhood has a particular history of left-wing social movements, uprisings against urban development, racism, and police brutality (ibid.). This history is also represented by spaces such as the Youth House, *Ungdomshuset*, that are linked to movements such as the international feminist movement, and *Byggeren*, a self-built public space and playground erected in an empty lot, which stood between 1973 and 1980 before being demolished to make way for a housing development. The demolition was enforced through violent police intervention (DR Nyheder, 2015 [1980]) (see Figure 4.1), first justified by a declaration that the neighbourhood constituted a threat to the state, and culminating in the formal declaration of a state of emergency (Mikkelsen et al., 2018; Stevnsborg, 1986). The neighbourhood's urban space has a character that lends itself to being quickly transformed into a stage, making dissent visible. This character, and the area's increasing ethnic diversity through immigration, are regularly pathologized and attacked in the national media and politics (Schmidt, 2011).

FIGURE 4.1   Police intervention at the sit-in protest against demolishing the self-built park Byggeren in Nørrebro in 1980 (photo by Søren Dyck-Madsen, 29 April 1980)

While being home to different groups and practices, negative images are constantly trafficked by the media, profiling the neighbourhood as a hotbed of conflict with crime, lack of safety, terror, and undemocratic values and practices (Schmidt, 2011).

In parallel to the geopolitical reflex to profile and pathologize 'ethnically diverse neighbourhoods' in other Scandinavian countries (specifically Norway and Sweden), Nørrebro, Mjölnerparken has been officially listed as a 'ghetto' since 2010, and is depicted as one of the worst low-income immigrant ghettos (Barry and Sorensen, 2018). The neighbourhood also became the target of 'anti-ghetto laws' published in 2018, which aimed to dismantle all ghettos by 2030. The determinants for this label are: unemployment rates over 20%, 50% of its population from non-Western countries or are descendants of migrants from non-Western countries, relatively low-income, low education level, and the presence of criminal activity (Regeringen, 2018; Ministry of Transport, Building and Housing, 2017). The measures to dismantle ghettoes target four areas: "redevelopment of infrastructure, redistribution of the population, targeted policing, and early childhood development" (Mohdin, 2018).

The law includes measures such as at least a 60% reduction in public housing in these areas through privatization, demolition and redevelopment of the areas, eviction of tenants if any household member has been convicted of a crime (Danish Transport, Construction and Housing Authority, 2019), and double sentences for crimes that occur in those areas (Mohdin, 2018). Starting from one year old, the children of immigrants from non-Western countries are taken to day care for

25 hours a week, otherwise parents lose child support (Nørgaard, 2018), and imprisonment of parents who send their children to the countries of emigration for periods long enough for 'reverse enculturation' (Mohdin, 2018). In these areas, achieving the safety and successful integration of young people is defined as dependent on their respect for Danish laws and authorities, and their enabling the police to move around effortlessly. The political discourses constantly trafficked by the media construct a meaning around 'non-Western' in order "to imply people who are backward, traditionalist, violent, criminal, lazy, and non-democratic, with cultures that suppress women and children" (Bendixen, 2018). These discourses pay particular attention "every time the word 'migrant', 'foreigner' or 'refugee' is mentioned; the same sentence is sure to several of these words and terms: 'crime', 'terror', 'Islam', 'insecurity', 'economic expenses', 'demands'" (Bendixen, 2018). In recent decades, police visibility, checks, and interventions have increased dramatically. The anti-'anti-ghetto law' activist from *Almen Modstand* (Common Resistance) first heard the news about the anti-ghetto law on the radio through the noise of patrolling police helicopters above her apartment in one of those so-called 'ghettoes' (Tounsi and Northroup, 2019). In political discourses that define citizenship and belonging through colonial logic and neoliberal capitalism, people and public housing areas are merged into each other, racialized, and othered in order to recolonize over and over, while the welfare state is degraded (General Resistance, 2020; Tounsi and Northroup, 2019). These places and people enter into a state of exception, becoming dark areas where citizenships are taken hostage (ibid.). On the other hand, ethnic majority neighbourhoods and populations are never pathologized with segregation and mapped as a 'ghetto', even those with similar or higher crime rates.

Echoing the national politics, the public housing areas have been targeted for different interventions by the municipality and private development consortia, which have mobilized with the neoliberal turn in city governance and housing, and thereby been turned into extraction zones, where the real estate market can expand (Ösgård and Algers, 2019). In such a context, in 2008, the City of Copenhagen partnered with Realdania, the Danish real estate association, which has a monopoly over the production of built environments, including not only urban areas but also rural and peripheral landscapes,[9] with over 450 projects all over Denmark.[10]

Realdania explains its interest in triggering a change in Nørrebro, including a project of levelling the neighbourhood's value on the housing ladder,[11] and transforming the identity of the neighbourhood by introducing two big parks and a cultural centre.[12] Superkilen is one of the park projects that are part of this development plan.

Superkilen emerging as a soft-war border technology

Superkilen is superimposed onto a pre-existing, one-kilometre narrow strip of green in Nørrebro, Mjölnerparken, between two social housing complexes to its north and some retail shops, cafes, a supermarket, and sport centres to its south.[13]

The existing park has lacked investment and been neglected for a long time by the municipality, in contrast to the well-maintained parks of middle- and upper-class white neighbourhoods of Copenhagen (Bloom, 2020). The old park was made up of patches of open space appropriated by the people themselves, in which "all kinds of things unfold" at different times (ibid.). Flea markets, children playing, self-organized art exhibitions, homeless people sleeping, sometimes drug use, and many other activities characterized the space (ibid.). Local activists constantly revive and work in the neighbourhood to produce plans to make the park better for the residents—but they are ignored by the city of Copenhagen (ibid.).

Instead of steady maintenance and light interventions co-produced with residents, the design and implementation of the park project was commissioned to 'big name' artists and architects, who attracted a lot of international and national attention (Bloom, 2020). These were architectural firm BIG, together with landscape architecture firm Topotek 1, and the artist group Superflex, through a closed competition process. Their proposal was supported by Realdania; meanwhile, an alternative proposal by another design firm—which included more natural elements, and which was supported by the dwellers—was rejected (Bloom, 2013).

The designers express that they expect the park to help integrate ghettoized communities by representing *their* backgrounds rather than being an aesthetic exercise in Danish design (Steiner, 2013). Labelled 'citizen participation extreme', participants were asked to recommend objects from the countries they had migrated from, to be installed in the park (Ingels, 2013). The main design concept of Superkilen, as ordered by the 'clients', revolves around the idea of the 'English landscape garden tradition', 'world exhibition', a 'global theme park' (Rein-Cano, 2016). The 'English landscape garden' idea originates from the colonial practice of displaying objects collected from around the world in a park setting (Mastnak et al., 2014). The idea is adopted by landscape architect Martin Rein-Cano (2016) from Topotek 1. According to the founder of BIG architects Bjarke Ingels (2013) and artists from Superflex (2013), "what the space does not have was a sense of community". He explains that, in order to foster a sense of belonging and ownership among disenfranchised neighbourhood dwellers from more than '50 different nationalities', the participants, many of whom have been Danish citizens for decades, were asked to recommend an element from their 'homelands'. Jacob Fenger from Superflex explains the participatory phase as "where those that raise their hand in the classroom [suggest] something to be included in the park" (Superflex, 2016). Superflex travelled with five groups to Palestine, Spain, Thailand, Texas, and Jamaica to bring back the proposed objects and install them in the park together (ibid.). More than 100 objects, including public furniture, signs, play elements, exercise equipment, soil, and plants from around the world are distributed around the park over a highly aestheticized hard surface (see Figure 4.2), which is divided into three parts coded in eye-catching colours: The Red Square, The Black Market, and The Green Park (see Figure 4.3) (Superflex, 2020).

These 'nationalities' were materialized in over 100 representative objects, which were placed in certain areas of the park. The objects selected to supposedly establish

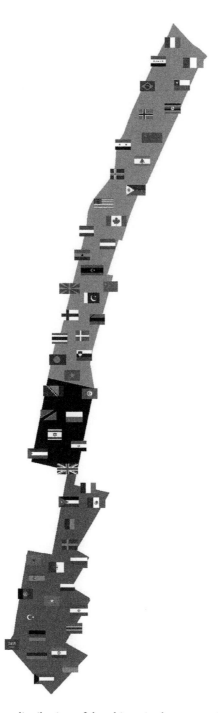

FIGURE 4.2 Superkilen; distribution of the objects in the area symbolized by Flags

FIGURE 4.3 Superkilen; a view from the Black Area

'representation', 'sense of belonging', and 'ownership of the area' (Ingels, 2013) for people with immigration backgrounds. Included objects include a barbecue from South Africa; a picnic table from Yerevan, Armenia; miniature goals from Damascus, Syria; a monkey-puzzle tree from Chile; a Chinese palm; and soil from Palestine (see Figure 4.4). The narratives of the pieces collected from other geographies are revealed to park users through a smartphone application (Superflex, 2020). Categorized under 'other nations', citizens of Denmark with an immigration background are claimed to have 'culture', or are supposed to be able to develop 'a sense of belonging' to and 'ownership of place' (Ingels, 2013) through this caricaturized encapsulation of 'their country' of origin.

The participation process, which was publicized as 'citizen participation extreme', consisted of a ready-made script focused only on the foreign objects that would be spread out on the designed surface. An artist and art critic, who lived in the neighbourhood for five years between 2010 and 2015, observed and took part in grass-roots initiatives. Brett Bloom (2020) defines the participation as a

FIGURE 4.4 Hiba and Alaa, two young women from the neighbourhood, with Rasmus Nielsen from Superflex, are taking soil from Palestine and replacing it at Superkilen

"very curated, very staged" event, where designers' ideas colonize and participants are "used for the right content"—not an open, sharing event. During both the competition and participation processes, the city and Realdania did not let neighbourhood dwellers exert any influence over decisions (Stanfield and Riemsdijk, 2019; Bloom, 2013). In the participatory processes, dissenting groups against the design concept are marginalized; they were not recognized as part of the public that should have a say in shaping the urban space (ibid.). A neighbourhood initiative, which "cares for this space", starts an alternative process of documenting the needs and imaginaries of the residents for the site, but this initiative is also sidelined (Bloom, 2020). The design team worked with a group that gave them exactly what they were looking for (Bloom, 2013, 2020). The resident representative states that residents had less than a 10% input on overall decision making (Stanfield and Riemsdijk, 2019: 1369). At his keynote address, Rein-Cano (2016) asserts that since they thought the neighbourhood dwellers did not have the ideal characteristics to be involved in a participatory design process, or to discuss foundational issues of design, they instead created a well-defined 'fun activity' for them. He explains that, in this way, they "make other cultures more tolerable, nicer" (ibid.), echoing the mixed message of the national politics of culture and diversity.

The prevailing approach of the landscape architectural design is the transformation of the site into a monolithic surface, a *tabula rasa* "crashing self-regulatory things" (Bloom, 2020). Landscape architectural design primarily works on the urban morphology, heavily focusing on bringing this autonomous landscape into the territories of 'urban development' materially. It demolishes elements of partition and relics of rooting and territorialization by the dwellers' activities on the site. It connects this surface to the other parts of the city, tightly stretching the cycling and pedestrian paths across this monolithic surface. It completes the network of flows in the city. This flat, hard surface is primarily open, increasing the possibility of surveillance and policing, echoing the anti-ghetto law that encourages the zone to be kept 'crime free'. Furthermore, this surface is intended as a strong landmark. The designers want the park to become integrated into any Copenhagener's or tourist's cognitive map through its sharp colouring, which is unusual in urban space.[14,15] The colours also function to hide the truth that the park is all naked hardscape that enables it to be surveilled and controlled (see Figure 4.5).

The participatory process and foundational ideas in the design of objects mimic the colonialist culturalist gestures, bringing the site into the territories of 'national landscape' symbolically, while landscape architectural design expands urban development into the site. Founder of BIG, Bjarke Ingels (2013), states that the team did not want to exercise typical Danish design at this site; however, they did exercise typical Danish politics of culture and society.

The portrait of the neighbourhood drawn by the design team echoes the power geometry, the 'ghetto', and 'them' and 'us' narratives have in national politics:

> The neighbourhood is the most ethnically diverse neighbourhood in all of Denmark—a small footprint comprised of more than 60 nationalities—and is the most socially challenged community that includes the closest thing we have to a 'ghetto' in Copenhagen: Mjølnerparken.
>
> *(Ingels, 2013)*

FIGURE 4.5 Superkilen; a view from the Red Area

The objects of the park help to freeze the image and history of inhabitants with diasporic pasts with sensational images and stories of migration experience and connections with 'other' lands. This diverts attention from their histories in Denmark, including existing solidarity, practices, and memories among different individuals in the neighbourhood beyond ascribed ethnic and racial categories. Citizenship, homeland, and belonging are attached to ethnic categories, serving as an implicit reminder that the lands and citizenship of Denmark should only belong to ethnic Danes. The political agency of residents from different backgrounds is denied.

A public sphere under 'citizen participation extreme' is created where subalterns become visible only through the colonialist gaze, as abstracted images of 'other nationalities', coupled with objects detached from their complex realities and subjectivities, and practices rooted in the territories of the state of Denmark. Curating the objects from 'other places', the design simply echoes the colonial exhibitions in the past. Moreover, within some other objects, such as bringing soil from Palestine with descendants of refugees from Palestine, it conveys further symbolic meanings, which contribute to a particular nation-building myth: 'Denmark as anti-colonial humanistic global power'. The messages from urban and national politics are repeated: immigration and people with an immigration backgrounds are problematic. But, there may also be possibilities to consume and enjoy diversity (Schmidt, 2016).

A neighbourhood activist, Uzma—a woman of colour, citizen, and resident of Denmark since she was four—describes the whole project process and design as an operation that cultivates in her mind a deep sense of being rejected as Danish, and as a citizen (Stanfield and Riemsdijk, 2019). For her, the park simply repeats the political message that one with an immigration background cannot be fully Danish, or from Denmark. Uzma remarks:

> I'm not from [another] country; I'm from here … I mean, I've been here my whole life, and other people were born here as well … I think it's curious that we have things [in Superkilen] from China and different places, but I don't really relate to them. Why should I? I mean, I'm from here … It shows that it's an immature thought and it's done in fascination, not in interaction … The situation is like this in Denmark, that if you are Muslim, you can't really be Danish. Because that's what you're told … So you can't say, "I'm from Denmark". But you can say … "I'm from Nørrebro".
>
> *(ibid.: 1369)*

Superkilen has won one prize after another from architectural and design foundations, reinforcing the designers' discourse of integration, social diversity, and democracy. Meanwhile, another public space design—Mimersparken, at another edge of the neighbourhood, whose planning and design process the neighbourhood dwellers meaningfully participated in, and one that is used as much as Superkilen—has attracted no recognition (Stanfield and Riemsdijk, 2019). On the other hand, since the opening of the park, gentrification in the area has accelerated hand-in-hand with 'anti-ghetto laws', causing neighbourhood dwellers to struggle. In the face of evictions and the mass selling of public housing in the area, finally, the neighbourhood dwellers show another example of their solidarity and sense of community. They filed a lawsuit against the Danish government in May 2020 over legislation that authorized dismantling neighbourhoods designated as 'ghettos'. In the lawsuit, the plaintiffs argued that measures that discriminate on the basis of ethnicity and skin colour alienate segments of the population, and risk creating "a lesser class of Danes"; moreover, those "name[d] in the lawsuit are all Danish citizens and include people of various ethnicities and races, including white":

> "This is where my three kids have grown up and this is where I want to grow old", says Asif Mahmut, who moved to Denmark from Pakistan and has lived in Mjölnerparken for 27 years. "This is my home, and I want to fight for it".
> *(Keyton, 2020)*

In contrast to the dominant narrative, it is seen from the history of the neighbourhood from *Byggeren* to Superkilen, or to the *Almen Modstand* (Common Resistance), activists and residents have seen the neighbourhood public space in Nørrebro as an agonistic public sphere, which can lead to autonomous landscape subjectivities and insurgent citizenships that resist conforming with gendered, racialized, and classed citizenship definitions from above. They have a capacity to struggle to oppose the power structure. They have a capacity to explore the limits of their agency in the imagining and making of common spaces in need. They can build community and solidarity across categories forced upon people. They have a capacity to explore new forms of belonging with a progressive cosmopolitan consciousness in contrast to the mainstream culture. Furthermore, they unite against structural racism and housing injustices, as seen in the latest development. Such intentions and practices claim an autonomous zone interrupting the hegemonic landscapes of nationalism and neoliberalism.

In the parallel world of the architecture and design realm, Superkilen has been celebrated and touted as a showcase for social diversity, inclusivity, and democracy. For example, explaining a recent prize, the Aga Khan Award employed the 'saviour narrative' of the designers, which claims to bring social life to an abandoned space and hotbed of criminality, and to bring representation and a sense of belonging and community to those who lack these. This was repeated by the award jury, which consisted of star architects and academics from the likes of Harvard and MIT, where the Aga Khan Foundation is located (AKDN, 2016). These lines were re-romanticized and propagated by an article published in *Landscape Architecture Magazine* (Bridger, 2016). Furthermore, the same narrative is academically legitimized through publications in uncritical and celebratory fashion produced by academic jury members (Akšamija, 2020; Mostafavi, 2016). In these lines, the deep injustices and contextual socio-spatial realities are bypassed, but protests, outbursts, different expressions of dissent against injustices, and acts of self-determination are implicitly represented as a problem echoing the authorities. The sweeping out of the population—which will not be able to afford to live in the neighbourhood any more—and the dissolving of community ties and solidarity, suppressing of social and political agency, and organizational reflexes are never recognized but rather are problematized. The reanimation of colonialist ideologies through design is never revealed. Thus, injustices are implicitly normalized. Consequently, the designers' narrative has been left unquestioned, since it depends on widely shared nation-building myths that construct a protective shell around the messages delivered by the design. So, the project continues its mission without interruption.

Bloom (2020) opines that art offices like Superflex "pretend they have a critical relationship with power"; however, they exist as "internal and external propaganda function of the state" and receive a serious amount of state funding, while "groups actually doing real critical work are marginalized; they do not get funding, they are pushed to the periphery". He developed these views based on five years of observing art practices in Denmark, and on watching the Superkilen project unfold in real time in the neighbourhood in which he himself lives (ibid.). These observations urge us to question the place and agency of design practices and academia in the colonial capitalist complexes.

Conclusions

At Superkilen, the role of design is twofold. First, it fulfils its role creating a 'High Line effect' (Lang and Rothenberg, 2016). It functions to increase the land rent in the area, which had been made low through complex exclusionary processes that concentrate minorities in the area, and pathologizing the area by listing it as a ghetto while the majority of social housing is set to be privatized. These areas are forced to enter the market at a moment when the rental gap is at its highest, becoming centralized and gathering great potential to create exchange value. Nationalist discourses upsurge and legitimize the urge to bring these demarcated lands inside the borders of 'national landscapes', attacking the boundaries around the neighbourhoods with anti-ghetto laws, which will displace the people of the diaspora and exacerbate dispossession. In this bordering dynamic, the design of public space contributes to the increase in exchange value in urban areas.

Second, the design temporarily makes the subordinate groups visible, under an intentional symbolic and social regime that does not threaten the exchange value creation cycle. It signals the elimination of 'unwanted things'; it signals that the other is tamed. This temporal visibility is followed by a gradual and silent displacement and invisibility. This incremental, slow, and subtle system of bordering silences dissent and tames major outbursts. Children, youth, and adults, or neighbourhoods, political views and movements, alternative lifestyles, and landscape imaginations outside the norm are pathologized and marginalized. Solidarities cross-cutting enforced social categories and autonomous and self-determinate cultures and subjectivities in those places are rendered invisible. It fixes the question of culture under set categories and paralyzes society's reflex to build new creative progressive cultures that can challenge today's injustices. The false social justice premise of park construction (Larson, 2018) and culturalism in design works to camouflage and naturalize these cycles, confuse diaspora, and charm social democrats and liberals from the white ethnic majority, 'who try not to be racists', or 'try not to be seen as racist', while Scandinavian myths function as a protective shell to hide the truth that the design project has been a colonialist, racist border practice. Design becomes a perfect colonial, racist, and capitalist soft-war border technology in the context of Superkilen.

The design of Superkilen almost re-animates all the myths of nation-building, colonial governmentality, and biopolitics, and fulfils a pedagogical role in the normalization of internal colonization ideology. It ignores the hybridity and dynamism of one's subjectivity, deriving instead from an underlying colonialist binary hypothesis that one can only have a sense of belonging to a place, or cultural heritage according to an ethnic category tied to another land, under which he/she is classified. In the design process, the subjects are detached from their contexts, abstracted, and their social and political agency suppressed; they are then reduced to objects collected and curated within a colonial mindset. It reanimates that mindset's lines of construction of difference and alterity. This reiterates the message about where homeland is for a subject from a minority group, and how people should belong or have a sense of community in general. It ignores and silences histories, everyday practices, struggles against suppressive identity politics and racism, solidarities, and the future aspirations of those rooted in the neighbourhood, as well as those around the city and the country. It ignores how they belong to the place and to the country. It consists of confusing messages that align with national politics: Denmark is not a homeland for people who are not 'true Danes', even those with citizenship. Denmark is an anti-colonial humanistic global superpower, a saviour, and a refuge for those oppressed by other colonialists, in order to feed the denial of coloniality. The design concepts insidiously reconstruct the politics of boundaries of belonging and citizenship, which are essentialized to ethnic origin.

While borders and walls surrounding colonialist states are becoming enhanced, the loss of habitats and human displacements have been rapidly increasing due to ongoing colonialism, which unites capitalism and racism (Sassen, 2014). Harsha Walia (2012) suggests that migrants' struggle for justice, and for their right to environment and place in colonizing countries, is no different from Indigenous peoples' struggles against settler colonialism, since those migrants were once the Indigenous people only transformed into migrants under the colonial/neocolonial world order. Against the landscape ideology of settler colonialism, Tuck and Gaztambide-Fernandez (2013) put forward another polity, 'an indigenous futurity', which is a way of reformulating landscape politics of belonging and citizenship:

> It does not foreclose the inhabitation of Indigenous land by non-Indigenous people, but does foreclose settler colonialism and settler epistemologies. That

is to say that Indigenous futurity does not require the erasure of now-settlers in the ways that settler futurity requires of Indigenous peoples.

(Tuck and Gaztambide-Fernandez, 2013: 80)

Such pedagogy of landscape belonging and citizenship would defuse the colonialist nation-building myths and internal colonialism, and provide the pretext for a landscape citizenship of equality.

Finally, the analysis of Superkilen through the lenses of border theory and post/anti/decolonial theory forces us to think on ways to transform the field of urban design. Urban design is left cognitively isolated so as not to see its role in the colonial complex and must be transformed into a field which genuinely aligns with a perspective of 'indigenous futurity'. Revealing the complicity and historicity of the field can only be a start, while exploring politics, imaginations, knowledges, and practices that engage with social and environmental justice struggles could be a project for actualizing such an intention, rather than staying in the protected place in the colonialist capitalist order.

Acknowledgements

I thank Johan Pries, Ed Wall, and Tim Waterman for their helpful comments. I thank Brett Bloom for sharing his experiences and observations as an art critic, practising artist, and as a neighbourhood dweller between 2010 and 2015.

Notes

1 Only two texts could be found among the current literature that elaborates the negative consequences of the social and environmental politics of Superkilen in the larger sociopolitical context. One was written by a local activist and art critic Brett A. Bloom (2013), and the other was by two cultural geographers, E. Stanfield and Micheline van Riemsdijk (2019). Jonathan Daly from RMIT, School of Architecture and Urban Design, recently published an article in 2019 in the *Journal of Urban Design* titled "Superkilen: exploring the human–nonhuman relations of intercultural encounter". While his article takes a sceptical position regarding intercultural encounters in the park, it focuses only on the immediate behaviours and experiences of park users, ignoring the social, spatial, and political context surrounding the park and its users.

2 Some of the awards include the Aga Khan Award for Architecture (2016); "Best of the Best" Architecture & Urban Design, Red Dot Award for Product Design (2013); International Olympic Committee Award Gold Medal (2013); International Winner, Civic Trust Award (2013); Regional & Urban Design Winner, AIA Institute Honor Awards (2013); Architizer A+ Awards Honourable Mention, BDLA Prize (2013); Finalist, Mies van der Rohe Award (2013); and ARCHITECT Magazine Annual Design Review (2012).

3 Danish Architecture Centre (dac.dk/en/knowledgebase/architecture/superkilen-2/); Copenhagen Architecture Festival (cafx.dk/events/offentlig-liv-og-trivsel-pa-norrebro-byvandring-med-courban/); see more venues of celebration and dissemination of the project via BIG's website (big.dk/#projects).

4 Some of the academic publications praising the park that repeat the designers' narratives include Azra (2020); Mostafavi (2016); Brown, Dixon, and Gillham (2014); and Broto (2013).
5 Scopus search with keywords colonialism, gentrification, public space, design hit no publication; while a search with the keywords racism, gentrification, public space, and design hit only one publication: P.L Laurence's "Jane Jacob's urban ethics" (Accessed 10 September 2020).
6 Denmark ranks fourth worldwide on the list of countries with largest ecological footprint, closely following the United Arab Emirates (worldatlas.com/articles/countries-with-the-largest-ecological-footprints.html; Accessed 11 September 2020).
7 See Mastnak et al. (2014) for further elaboration of biopolitical regime of plantations in the colonies.
8 Grosfoguel (2010) and Mignolo (2010) assert that the histories of modernities, industrial revolutions, new identities and rights, liberal democracies and institutions of welfare states, and citizenship in the West actually succeed through processes of colonial exploitation of non-Western people; however, to retain the status quo and territorial authority the histories of colonial capitalist accumulations and international division of labour are always omitted from the national history making.
9 In the acknowledgements of their article, Primdahl and Kristensen (2016) thank Realdania for funding their workshop about "landscape strategy making and landscape characterization".
10 realdania.dk/projekter (Accessed 3 March 2020).
11 realdania.dk/nyheder/2017/11/bog-fra-ghetto-til-blandet-by-23112017 (Accessed 3 March 2020).
12 realdania.dk/projekter/mimersgadekvarteret (Accessed 3 March 2020).
13 See Google Earth aerial view dated 13 June 2006.
14 Rein Cano (2016) states that they intend to make the park easily found by people from other parts of the city for meeting purposes.
15 Ingels (2013) explains that with these three sharp colours, they aim to increase the signalization of the area for the people coming from other parts of the city, to help them find their way even if they do not know the environs.

Bibliography

Aga Kahn Development Network (AKDN) (2016). *Superkilen* [Online]. Available at: akdn. org/architecture/project/Superkilen (Accessed 19 May 2020).
Agustin, Óscar Garcia and Martin Bak Jørgensen (2019). '*Danes first, welfare last*'. *Jacobin* [Online]. Available at: jacobinmag.com/2019/01/denmark-social-democrats-immigration-welfare (Accessed 21 May 2020).
Akšamija, Azra (2020). *Architecture of Coexistence: Building Pluralism*. Berlin: ArchiTangle.
Anzaldúa, Gloria (1987). *Borderlands/La Frontera: The New Mestiza*, 1st ed. San Francisco: Aunt Lute Books.
Balibar, Étienne (2002). *Politics and the Other Scene*. Translated by Christine Jones, James Swenson, and Chris Turner. New York, NY and London: Verso.
Barry, Ellen and Martin Selsoe Sorensen (2018). 'In Denmark, Harsh New Laws for Immigrant "'Ghettos'"'. *New York Times* [Online]. Available at: nytimes.com/2018/07/01/world/europe/denmark-immigrant-ghettos.html (Accessed 11 March 2020).
Bendixen, Michala (2018). 'Denmark's "anti-ghetto" laws are a betrayal of our tolerant values'. *The Guardian* [Online]. Available at: theguardian.com/commentisfree/2018/jul/10/denmark-ghetto-laws-niqab-circumcision-islamophobic (Accessed 2 April 2020).

Bloom, Brett A (2013). 'Superkilen: Park med extrem borgerinddragelse!'. *Kritik*, 207, pp 47–64.

Bloom, Brett A (2020). Interviewed by Burcu Yigit Turan (10 September 2020).

Bridger, Jessica (2016). 'Life on the wedge'. *Landscape Architecture Magazine* [Online]. Available at: landscapearchitecturemagazine.org/2016/10/04/life-on-the-wedge/ (Accessed 12 January 2020).

Broto, Carles (2013). *Urban Spaces: Design and Innovation*. Barcelona: Links Books.

Brown, Lance Jay, David Dixon, and Oliver Gillham (2014). *Urban Design for an Urban Century: Shaping More Livable, Equitable, and Resilient Cities*. Hoboken, NJ: Wiley.

Certomà, Chiara (2013). 'Expanding the 'dark side of planning': Governmentality and biopolitics in urban garden planning'. *Planning Theory*, 14(1), pp 23–43.

Danbolt, Mathias (2017). 'Retro racism: Colonial ignorance and racialized affective consumption in Danish public culture'. *Nordic Journal of Migration Research*, 7(2), pp 105–113.

Danbolt, Mathias (2018). 'Art and coloniality'. *Kunstkritikk* [Online]. Available at: kunstkritikk. dk/kunst-og-kolonialitet/ (Accessed 7 August 2020).

Danish Transport, Construction and Housing Authority (2019). *Development plans* [Online]. Available at: trafikstyrelsen.dk/da/Bolig/Udsatte-boligomraader/Udviklingsplaner (Accessed 1 April 2020).

Davis, Mike (1992). 'Fortress Los Angeles: The militarization of urban space'. In: Michael Sorkin, ed. *Variations on a Theme Park: The New American City and the End of Public Space*. New York, NY: Hill and Wang, pp 154–180.

DR Nyheder (2015 [1980]). '"Byggeren'rives ned på Nørrebro - 3. maj 1980" [Video]. Available at: youtu.be/4fZznQFq8xg (Accessed 11 March 2020).

Forkert, Kirsten, Federico Oliveri, Gargi Bhattacharyya, and Janna Graham (2020). *How Media and Conflicts Make Migrants*. Manchester, UK: Manchester University Press.

General Resistance (2020). *General Resistance: Defend resident democracy!* [Online]. Available at: almenmodstand.dk/ (Accessed 1 June 2020).

Hall, Neville AT (1992). *Slave Society in the Danish West Indies: St. Thomas, St. John and St. Croix*. Baltimore, MD and London: Johns Hopkins University Press.

Hervik, Peter (2019). 'Racialization in the Nordic countries: An introduction'. In: Peter Hervik, ed. *Racialization, Racism, and Anti-Racism in the Nordic Countries*. Cham, CH: Palgrave Macmillan.

Hervik, Peter (2011). *The Annoying Difference: The Emergence of Danish Neonationalism, Neoracism, and Populism in the Post-1989 World*. London and New York, NY: Berghahn Press.

Hjort, Jonas (2013). 'Superflex: A cool urban space'. *Architecture Lab* [Online] Available at: architecturelab.net/superflex-a-cool-urban-space/ (Accessed 11 October 2020).

Hvenegård-Lassen, Kirsten (2008). 'Migrancy'. In: Poddar Prem, Rajeev S Patke, and Lars Jensen, eds. *A Historical Companion to Postcolonial Literatures - Continental Europe and its Empires*. Edinburgh: Edinburgh University Press, pp 82–84.

Ingels, Bjarke (2013). Interviewed by Citi. *Citi: 2013 FT/Citi Ingenuity Awards: Bjarke Ingels Group Finalist* [Video]. Available at: youtu.be/1FHmssV34lc (Accessed 4 January 2020).

Jensen, Lars (2008). 'Denmark and its colonies: Introduction'. In: Poddar Prem, Rajeev S Patke, and Lars Jensen, eds. *A Historical Companion to Postcolonial Literatures - Continental Europe and its Empires*. Edinburgh: Edinburgh University Press, pp 59–62.

Jensen, Lars (2015). 'Postcolonial Denmark: Beyond the rot of colonialism?' *Postcolonial Studies*, 18(4), pp 440–452.

Jensen, Tina Gudrun, Kristina Weibel, and Kathrine Vitus (2017). 'There is no racism here': Public discourses on racism, immigrants and integration in Denmark'. *Patterns of Prejudice*, 51(1), pp 51–68.

Keskinen, Suvi, Salla Tuori, Sari Irni, and Diana Mulinari (2009). *Complying With Colonialism: Gender, Race and Ethnicity in the Nordic Region*. London and New York, NY: Routledge.

Keyton, David (2020). 'Denmark residents sue over laws for dismantling "ghettos"'. *Associated Press* [Online]. Available at: apnews.com/article/4c8b59b2493a45f1567dccd417545685 (Accessed 12 June 2020).

Kipfer, Stefan (2007). 'Fanon and space: Colonization, urbanization, and liberation from the colonial to the global city'. *Environment and Planning D: Society and Space*, 25(4), pp 701–726.

Kvaale, Katja (2011). 'Something begotten in the state of Denmark? Immigrants, territorialized culture, and the Danes as an Indigenous people'. *Anthropological Theory*, 11(2), pp 223–255.

Lang, Steven and Julia Rothenberg (2016). 'Neoliberal urbanism, public space, and the greening of the growth machine: New York City's High Line park'. *Environment and Planning A: Economy and Space*, 49(8), pp 1743–1761.

Larson, Scott M (2018). 'Imagining social justice and the false promise of urban park design'. *Environment and Planning A: Economy and Space*, 50(2), pp 391–406.

Loftsdóttir, Kristín and Lars Jensen (2012). *Whiteness and Postcolonialism in the Nordic Region: Exceptionalism, Migrant Others and National Identities*. New York, NY and London: Routledge.

Madanipour, Ali (2019). 'Rethinking public space: Between rhetoric and reality'. *URBAN DESIGN International*, 24(1), pp 38–46.

Mastnak, Tomaz, Julia Elyachar, and Tom Boellstorff (2014). 'Botanical decolonization: Rethinking native plants'. *Environment and Planning D: Society and Space*, 32(2), pp 363–380.

Meader, Richard D (2009). '*Organizing Afro-Caribbean Communities: Processes of Cultural Change under Danish*'. Toledo, OH: University of Toledo. Thesis.

Mealor, Cheralyn (2008). 'Colonial exhibitions'. In: Poddar Prem, Rajeev S Patke, and Lars Jensen, eds. *A Historical Companion to Postcolonial Literatures - Continental Europe and its Empires*. Edinburgh: Edinburgh University Press, pp. 68–69.

Mikkelsen, Flemming, Knut Kjeldstadli, and Stefan Nyzell, eds. (2018). *Popular Struggle and Democracy in Scandinavia: 1700-Present*. London: Palgrave Macmillan.

Ministry of Transport, Building and Housing (2017). *Read here: The definition of a ghetto – the five criteria* [Online]. Available at: regeringen.dk/nyheder/2017/ghetto-listen-2017-to-nye-omraader-tilfoejet-fem-fjernet/ghettolisten-definition-af-en-ghetto/ (Accessed 11 March 2020).

Mitchell, Don (2017). 'People's park again: On the end and ends of public space'. *Environment and Planning A: Economy and Space*, 49(3), pp 503–518.

Mohdin, Aamna (2018). 'Denmark wants to double the punishment for crimes committed in immigrant "ghettos"'. *Quartz* [Online]. Available at: qz.com/1319659/denmark-wants-to-double-the-punishment-for-crimes-committed-in-immigrant-ghettos/ (Accessed 1 April 2020).

Mostafavi, Mohsen, ed. (2016). *Architecture and Plurality*. Zurich: Lars Muller Publishers.

Mouffe, Chantal (2013). *Agonistics: Thinking the World Politically*, 1st ed. London: Verso.

Muasya, Gabriella Isadora Nørgaard, Noella Chituka Birisawa, and Tringa Berisha (2018). 'Denmark's innocent colonial narrative'. *KULT: Racism in Denmark*, 15, pp 56–69.

Nail, Thomas (2016). *Theory of the Border*. Oxford and New York, NY: Oxford University Press.

Naum, Magdalena and Jonas M. Nordin, eds. (2013). *Scandinavian Colonialism and the Rise of Modernity: Small Time Agents in a Global Arena*. New York, NY: Springer-Verlag.

Nørgaard, Trine (2018). 'One-year-old children in ghettos must be in day care'. *DR* [Online]. Available at: dr.dk/ligetil/etaarige-boern-i-ghettoer-skal-i-dagtilbud (Accessed 2 April 2020).

Olwig, Karen Fog (1995). *Small Islands, Large Questions: Society, Culture and Resistance in the Post-Emancipation Caribbean*. London: Frank Cass.

Olwig, Kenneth R (2003). 'Natives and aliens in the national landscape'. *Landscape Research*, 28(1), pp 61–74.

Ösgård, Anton and Jonas Algers (2019). 'Denmark's shameful ghetto plan'. *Jacobin* [Online]. Available at: jacobinmag.com/2019/12/denmark-ghetto-plan-social-democrats-mette-frederiksen (Accessed 3 May 2020).

Palmberg, Mai (2009). 'The Nordic colonial mind'. In: Suvi Keskinen, Salla Tuori, Sari Irni, and Diana Mulinari, eds. *Complying with Colonialism: Gender, Race, and Ethnicity in the Nordic Region*. Farnham, UK: Ashgate, pp 35–50.

Primdahl, Jørgen and Lone S Kristensen (2016). 'Landscape strategy making and landscape characterization: Experiences from Danish experimental planning processes'. *Landscape Research*, 41(2), pp 227–238.

Ramírez, Margaret M (2020). 'City as borderland: Gentrification and the policing of Black and Latinx geographies in Oakland'. *Environment and Planning D: Society and Space*, 38(1), pp 147–166.

Regeringen (2018). *Ét Danmark uden parallelsamfund: Ingen ghettoer i 2030* [Online]. Available at: regeringen.dk/media/4937/publikation_%C3%A9t-danmark-uden-parallelsamfund.pdf (Accessed 1 April 2020).

Rein-Cano, Martin (2016). *Superkilen and Multicultural Public Spaces with Martin Rein-Cano* [Video]. Available at: youtu.be/Rvk6xeUPggE (Accessed 3 September 2020).

Roy, Ananya (2017). 'Dis/possessive collectivism: Property and personhood at city's end'. *Geoforum*, 80, pp A1–A11.

Rytter, Mikkel (2019). 'Writing against integration: Danish imaginaries of culture, race and belonging'. *Ethnos*, 84(4), pp 678–697.

Sassen, Saskia (2014). *Expulsions: Brutality and Complexity in the Global Economy*. Cambridge, MA: Harvard University Press.

Sassen, Saskia (2016). 'A massive loss of habitat'. *Sociology of Development*, 2(2), pp 204–233.

Scheinin, Martin (2004). 'What are Indigenous peoples?' In: Nazila Ghanea-Hercock and Alexandra Xanthaki, eds. *Minorities, Peoples and Self-Determination: Essays in Honour of Patrick Thornberry*. Leiden, NL: Brill Academic Publishers, pp 3–14.

Schmidt, Garbi (2011). 'Understanding and approaching Muslim visibilities: Lessons learned from a fieldwork-based study of Muslims in Copenhagen'. *Ethnic and Racial Studies*, 34(7), pp 1216–1229.

Schmidt, Garbi (2016). 'Space, politics and past–present diversities in a Copenhagen neighbourhood'. *Identities*, 23(1), pp 51–65.

Sevilla-Buitrago, Alvaro (2017). 'Gramsci and Foucault in Central Park: Environmental hegemonies, pedagogical spaces and integral state formations'. *Environment and Planning D: Society and Space*, 35(1), pp 165–183.

Sorkin, Michael (1992). *Variations on a Theme Park: The New American City and the End of Public Space*. New York, NY: Hill and Wang.

Spencer, Douglas (2016). *The Architecture of Neoliberalism: How Architecture Became an Instrument of Control and Compliance*. London: Bloomsbury.

Stanfield, E and Micheline van Riemsdijk (2019). 'Creating public space, creating "the public": Immigration politics and representation in two Copenhagen parks'. *Urban Geography*, 40(9), pp 1356–1374.

Stecher-Hansen, Marianne (2008). 'Thorkild Hansen and the critique of empire'. In: Poddar Prem, Rajeev S Patke, and Lars Jensen, eds. *A Historical Companion to Postcolonial Literatures - Continental Europe and its Empires*. Edinburgh: Edinburgh University Press, pp 74–77.

Steiner, Barbara, ed. (2013). *Superkilen*. Stockholm: Arvinius + Orfeus Publishing.

Stevnsborg, Henrik (1986). 'Byggeren 1980'. In: Jørn Vestergaard, ed. *Sociale Uroligheder. Politi Og Politik*. København: Socpol.

Superflex (2013). *Superflex Interview: A Cool Urban Space* [Video]. Available at: youtu.be/rlCo4Mg3Rdk (Accessed 21 June 2020).

Superflex (2016). *Superkilen* [Video]. Available at: vimeo.com/155427158 (Accessed 22 November 2019).

Superflex (2020). *Tools: Superkilen* [Online]. Available at: superflex.net/tools/superkilen (Accessed 19 November 2019).

Thomson, Claire (2008). 'Narratives and fictions of empire'. In: Poddar Prem, Rajeev S Patke, and Lars Jensen, eds. *A Historical Companion to Postcolonial Literatures - Continental Europe and its Empires*. Edinburgh: Edinburgh University Press, pp 88–91.

Tounsi, Fatma and Marie Northroup (2019). Interviewed by Margarida Waco. 'A resistance movement against the racist Danish "ghetto" laws'. *The Funambulist* [Podcast]. Available from: thefunambulist.net/podcast/almen-modstand-resistance-movement-racist-danish-ghetto-laws (Accessed 3 June 2020).

Tuck, Eve and Rubén A Gaztambide-Fernández (2013). 'Curriculum, replacement, and settler futurity'. *Journal of Curriculum Theorizing*, 29(1), pp 72–89.

Volquardsen, Ebbe (2014). 'Pathological escapists, passing, and the perpetual ice: Old and new trends in Danish-Greenlandic migration literature'. In: Ebbe Volquardsen and Lili-Ann Körber, eds. *The Postcolonial North Atlantic: Iceland, Greenland, and the Faroe Islands*. Berlin: Nordeuropa-Institut der Humboldt-Universität, pp 391–417.

Walia, Harsha (2012). 'Decolonizing together'. *Briarpatch* [Online]. Available at: briarpatchmagazine.com/articles/view/decolonizing-together (Accessed 31 July 2020).

Wren, Karen (2001). 'Cultural racism: Something rotten in the state of Denmark?' *Social & Cultural Geography*, 2(2), pp 141–162.

5

AVUNCULAR ARCHITECTURES

Queer futurity and life economies

Tim Waterman and Eglé Packauskaité

> [T]he figure of the spinster ... practices an avuncular form of stewardship, tending the future without contributing directly to it.
>
> *Sarah Ensor (2012: 409)*

Introduction: Spinster Ecology

Sarah Ensor's essay "Spinster Ecology: Rachel Carson, Sarah Orne Jewett, and Nonreproductive Futurity" (2012) examines the opposition between a heteronormative expectation that people's hopes for their futures are channelled through their offspring—"what kind of world are we leaving for our children?"— and turn-of-the-millennium queer theory, exemplified by Lee Edelman's *No Future: Queer Theory and the Death Drive* (2004) which declines or resists futurity, seeking a Lacanian *jouissance* in an endless present. Ensor's avuncular spinster figures a further model of "queer ecocritical practice ... [and] care that allows distance, indirection, and aloofness to persist and that transforms the vexed concept of 'enoughness' from a chastening limitation to a quietly affirmative state". This paper outlines the relevance of this stance to architectures which, though presently bound to 'development' narrowly defined as a tool for capitalist profit, could consist instead of fully elaborated development of humans, the more-than-human, and emplaced economies and ecologies. This stance presents a model of citizenship in which design practices, in particular scenario-making, are a model not only *of* citizenship, but *for* citizenship. Scenario-making is a way of imagining and prefiguring present and future relations as practiced in spaces, and when these practices are conceived as developed through economies arising from an ethic of mutual care and social ecology, questions also arise about the nature of the future practice of citizenships in those milieus. This chapter employs 'architectures' as a term that may enfold practices from the making of edifices and landscapes to the spatial work of civil engineering, which often has more impact on environments than aesthetically orientated design practices. 'Architectures' also may refer to the socio-cultural structures which provide the networks of care to support humans and the organic and inorganic ecologies of their landscapes and lifeworlds. Care, the first infrastructure, is the base of the commons and of society, and is the foundation of civility, which may be viewed as a construct of mutual practice. This infrastructure is what Patrick Geddes would call a 'life economy' (1885; and see Welter, 2002), the futurity of which—a life drive—pits itself directly against the death drive and against necropolitics (Mbembe, 2019).

Jacques Tati's Monsieur Hulot offers cues for avuncular architectures of care. His character in the film *Mon Oncle* (1958) queers the anarchic historic city and sweetly presents it as a "quietly affirmative state". Much as its lampoon of simplifying and technocratic modernist buildings, landscapes, and lifestyles is also sympathetic and gently ironic. Hulot's presence within architectural modernity is a symbol of insurgent humanity, and the *jouissance* he exemplifies is not sexualized as it is in the Lacanian sense, but is rather a Lefebvrian *jouissance*—which includes not only exuberant joyfulness, but the quietly affirmative sense of *bonheur* as well as *plaisir, volupté*, and *joie* (see Bononno, 2014: ix).

Design, citizenships, and ethics are all practices which require scenario-making imaginaries. Ethics are relational—they relate both to universal experience as they do the customary and the particular—and are embedded within customary practices and everyday life. Morals relate to situations. The 'moral imagination' describes a process in which scenarios are constructed to test whether they 'hang together' to support the maximum possible mutual care and flourishing, or, at minimum, the reduction and suppression of harm. Design is reiterative scenario-making, employing aids such as drawings, texts, and models to achieve the best spatial, use, and/or practice fit for a given situation. Citizenships are practised in places—in substantive landscapes—and require the employment of imaginaries to guide human practice. Citizenships are also composed of relations, and one of particular interest to us is a model of citizenship in which people of a certain age are seen to have a responsibility for others as 'aunties' or 'uncles', possessed of a concerned and interested distance. These aunties and uncles are accorded a level of respect which recognizes both the experience and the fallibility of the avuncular other, a usefully ironic position. This does not consist in a relation through marriage or consanguinity but as 'families' who are *gathered* in place rather than bounded or given. It bears a resemblance to notions of nationhood germane to many indigenous cultures and their associated landscapes (Driskill et al., 2011: 6–7).

Part 1: athwartness

We must return to Sarah Ensor, as the compelling exposition of her 'Spinster Ecology' also serves as our foundation. She writes:

> Attempting to discredit Rachel Carson's 1962 book *Silent Spring*, former Secretary of Agriculture Ezra Taft Benson wrote to Dwight D. Eisenhower

and asked "why a spinster with no children was so concerned about genetics". Benson's attack, its misogyny at best thinly veiled, makes legible one of the central challenges to developing a queer ecocritical practice: the status of futurity.

(2012: 409)

Rachel Carson will thus serve as the totemic 'queer auntie' for us here. Carson was a marine biologist and conservationist; *Silent Spring* (1962) details a possible dystopian future which may be avoided through the "capacity of nature to endure, and humans to change" (Garforth, 2018: 132). The book is often cited alongside the 'Blue Marble' and 'Earthrise' photographs from the Apollo missions as a catalyst for the modern environmental movement. Her earlier book, a poetic history of the oceans, *The Sea Around Us* (2011 [1951]), also brought complex ecological realities to the public. Carson's scientific work was rooted in a literary tradition of keen observation, naturalism, and environmentalism reaching back at least to the Transcendentalists. Carson never married, but her spinsterhood could only at one level be pinned to her weddedness to work. She maintained a passionate friendship and lengthy correspondence with a married woman named Dorothy Freeman. Neither woman self-identified in a way consistent with contemporary lesbianism, but a generously open definition of queerness would include Carson's unconventional affections.

Elspeth Probyn, in *Eating the Ocean* (2016), provides a valuable framework for thinking an open spectrum of queerness—in nautical terms—as 'athwartness'. She quotes Stefan Helmreich's *Alien Ocean*, in which he elaborates 'athwart theory' as

an empirical itinerary of associations and relations, a travelogue which, to draw on the nautical meaning of *athwart*, moves sidewise, tracing the contingent, drifting and bobbing, real-time, and often unexpected connections of which social action is constituted, which mixes up things and their descriptions.

(Helmreich, 2009: 23)

She compares this with Eve Kosofsky Sedgwick's etymological examination of 'queer'. "[T]he word 'queer' itself means across—it comes from the Indo-European root *-twerkw*, which also yields the German *quer* (transverse), Latin *torquere* (to twist), English *athwart*" (Sedgwick, 1994: xii). Queerness, for Probyn, is athwartness of habitus and mind, ways of thinking and acting which are troubling, eddying, recurring.

That our habitus inflects or orientates what and how we are is at one level the stuff of common sense. From the learned acquisition of manners, of how to handle knives and forks or chopsticks or with which hand to eat, to the acquired tastes we learn to like (or learn that we shouldn't like), these are practices that, as Bourdieu writes, are "something that one is".

(2016: 35–36. See also Bourdieu, 1990: 73)

Queerness, through its athwartness, its transversality,[1] its intransitiveness, shows that 'what one is' is constructed iteratively in the repetitions of daily life—and that questioning currents can flow through in unsettling ways. Probyn completes this powerful framing by quoting Loïc Wacquant: "the way society becomes deposited in persons in the form of lasting dispositions, or trained capacities and structured propensities to think, feel and act in determinant ways, which then guide them" (Wacquant, 2004: 316). These currents can be seen as the flow of collective thought or queerness: in the words of Sara Ahmed "... it is through the repetition of a shared direction that collectives are made" (Ahmed, 2006: 117).

Reproductive futurity has a direction, an extension of oneself into the future through an offspring. Looking at 'queer' as a spatial term encourages desire paths that twist and intertwine. We are not drawn towards a queer future as long as we are all orientated in a straight line. Relationships are fluid, intransitive, and often distant, as Jacques Rancière shows:

> [W]hat brings *people* together, what unites them, is nonaggregation. Let's rid ourselves of the representation of the social cement that hardened the thinking minds of the postrevolutionary age. People are united because they are people, that is to say, *distant* beings. Language doesn't unite them. On the contrary, it is the arbitrariness of language that makes them try to communicate by forcing them to translate—but also puts them in a community of intelligence.
>
> *(Rancière, 1991: 58)*

Rachel Carson's watery and intransitive work is a necessary forbear of what has come to be called hydrofeminism, itself perhaps a relative of ecofeminism and new materialism, all of which form a background for this chapter. Carson, thus is a 'queer auntie' not just for environmentalism but for many forms of relational and ecological thought that follow from her work. Carson's relation to the future is also important, as this can also be seen as one which is avuncular. "... often an aunt, the spinster stands in a kind of slanted or oblique relationship to the linear, vertical paradigms of transmission that govern familiar notions of futurity" (Ensor, 2012: 416). Ensor characterizes this slanted or oblique (athwart) relationship as one which is intransitive, in that a conclusion is not directly reached through a simple cause-and-effect, subject-to-object relationship. The queer auntie is part insider, part outsider, exerting an influence meaningfully but indirectly. The queer auntie, too, is 'out of line', as "to be 'in line' is to direct one's desires toward marriage and reproduction; to direct one's desires toward the reproduction of the family line" (Ahmed, 2006: 74). Let's return to the word 'queer' once more, as it is useful to conceive of queerness as a quality or capability, and one which may be available to all. William Turner brings 'queer' back to its 'odd' roots when he says it merely indicates "the failure to fit precisely within a category"—of which one reading might be athwartness—"and surely all persons at some point or other find themselves discomfited by the bounds of the categories that ostensibly contain their identities" (2000: 8). Fabio Cleto helps affirm that these categories, these identities, are constituted in landscape space. "Queer is, after all, a spatial term, which then gets translated into a sexual term, a term for a twisted sexuality that does not follow a 'straight line', a sexuality that is bent and crooked" (Cleto 2002: 13), and this is further corroborated by Sara Ahmed in cognitive terms: "[t]he body orientates itself in space, for instance, by differentiating between 'left' and 'right', 'up' and 'down', and 'near' and 'far', and this orientation is crucial to the sexualization of bodies" (2006: 67).

In the words of Michael Warner, "queer experience and politics might be taken as starting points rather than footnotes" (1993: vii). Ensor uses Carson's own words to illustrate a queer/athwart intransitivity: "the following springs are silent of robin song, not because we sprayed the robins *directly*, but because the poison *traveled*" (2012: 416, quoting from Carson, 1996 [1962]: 13). The "intertwined and polyvalent patterns of ecology are more a matter of process and persistence than of direct causal links, or of a linear progression (either grammatical or biological) from subject to verb to object", writes Ensor (ibid.: 418). Ecological thinking, thus, may very well require queer, intransitive, ecocritical forms of thinking and acting. Queerness as inside, outside, and athwart becomes a critical tool for thinking and a tool for observing—far more than a mere identity or its aberration.

No future

Another type of intransitivity in queer theory takes the form of refusal: a refusal to be future-oriented, a refusal to 'fight for the children', and this refusal is based in a reductive nihilist assumption that childless queers can and ought have no stake in the future. The athwartness, the intransitivity in action here is a blockage, a dead-end, which is emblematized by the millenarian exhortation to 'party like it's 1999'. Lee Edelman is the standard bearer for this party-time out-of-time, carrying the death drive into an endless present. In *No Future: Queer Theory and the Death Drive*, he delivers this *cri de coeur*:

> Fuck the social order and the Child in whose name we're collectively terrorized; fuck Annie; fuck the waif from *Les Mis*; fuck the poor, innocent kid on the Net; fuck Laws both with capital ls and with small; fuck the whole network of Symbolic relations and the future that serves as its prop.
>
> *(Edelman, 2004: 29)*

"And so", he goes on, "what is queerest about us, queerest within us, and queerest despite us is this willingness to insist intransitively—to insist that the future stop here" (ibid.: 31). Edelman seeks a "*jouissance* that at once defines and negates us" (ibid.: 5), and he uses the term *jouissance* in the Lacanian sense, which has come to be how the word is largely understood in Anglophone criticism. The death drive is evident here, and Lacanian *jouissance* seeks orgasmic fulfilment and release—the 'little death'—a sense of transcendence, perhaps, but only in the moment, an opening up into an emptying, an emptiness. *Jouissance*, says Edelman, "evokes the death drive that always insists as the void in and of the subject, beyond its fantasy of self-realization, beyond the pleasure principle" (ibid.: 25). The solipsism of this ludic void is not merely reflective of the death drive, but reflects the larger actions of necropolitical power, which for Achille Mbembe "proceeds by a sort of inversion between life and death, as if life was merely death's medium" (2019: 38). A refusal of the future plays into the hands of globalized capitalism and associated forms of imperialism and colonialism, which thrive on the orgasmic moment, on party time, on squandering, on gambling, on death. Perhaps what is most worrying about *No Future* is that the common sense of our age makes the messages of this polemic seem both believable and urgent: that people might have become so isolated and alienated within late capitalism that the very idea of mutual aid and conviviality is not just discounted, it *never even enters the picture.*

Sarah Ensor's avuncular portrait of Rachel Carson refutes the late capitalist nihilism of *No Future*. Indeed, Ensor refutes not merely the argument but the critical mode in which that argument is made: "maintaining intransitivity seems untenable given Edelman's choice to write in the polemic form, which is predicated on taking a stance *against* an existing object or position" (2012: 413). Further to this, she projects a whole realm of possibility which Edelman forecloses,

> that a queer environmentalism might become positively intransitive not because of its willingness (or structural mandate) to reject the future but instead because of how the future appears to a queer subject considering it outside the bounds of biological reproduction. What if the queer relationship to futurity is intransitive not because of how it refuses but rather because of how it facilitates a notion of the future (and of futurity) outside the realm of objects, outside the push and pull of acceptance or refusal, both outside and beyond our capacity to control? Perhaps the question is not the future, yes or no, but the future, which and whose, where and when and how.
>
> *(ibid.: 414)*

The question, for Ensor, is in the futurity of the practices of everyday life. These practices are mutual, convivial, individual, and collective, and are geared towards on the one hand the avoidance of pain and misery and on the other towards the wellness, wealth, health, and flourishing of the human species and all its planetary relations. There is so much more to the fullness of the lived moment, and this is at the core of Henri Lefebvre's criticism: "words and gestures express an *action*, and not simply some ready-made 'internal reality'. When men [sic.] speak they move forward along their line of action in a force field of possibilities" (2008 [1947]: 135). *Jouissance*, released from its narrow orgasmic sense into the fullest organic sense, is lived and practised within this force field of possibilities. The word has, as Robert Bononno tells us, a lengthy pedigree; its earliest use has been traced to the fifteenth century, where it is intended primarily as a form of usufruct. In the sixteenth century it began its association with what we may call 'pleasure', initially the pleasure of the senses generally and then, around 1589, sexual pleasure.

(2014: viii)

The term 'usufruct' is particularly telling, showing that pleasure, happiness, and well-being are found in the 'fruit of use', the products and fulfilments made manifest through emplaced practices of everyday life and everyday civility and citizenship. The concept of usufruct also interestingly situates *jouissance* in common law, where it finds itself right at home alongside the concept of a human right to the pursuit of happiness (that pursuit, of course, being not a *chase* but a *practice*). With that in mind, here is Lefebvre again: "Only an economy of enjoyment [*jouissance*] that replaces an exchange economy can end that which kills reality in the name of realism (in truth, cynicism)" (2014: 131). Edelman's cynical, orgasmic *jouissance* is precisely the engine of a capitalist exchange economy which kills not only reality, but the future. The future's impossibility is equal to neoliberal capitalism's claim that there is no alternative to economic liberalism. Capitalism, for José Esteban Muñoz:

> would have us think that it is a natural order, an inevitability, the way things would be. The "should be" of utopia, its indeterminacy and its deployment of hope, stand against capitalism's ever expanding and exhausting force field of how things "are and will be".

(Muñoz, 2009: 99)

In the next section we examine the design of the new town of Harlow in Essex, which might be easy to see as a spatial exemplar of modernist, transitive, reproductive futurity. But, rather, we wish to present Harlow Town as, at least in some ways, embodying the idea of the life drive, of Eros, the fertility of the human species and of cultural and social forms of fertility that are, rather, ecological and intransitive. This echoes instead Muñoz's reflection of a life drive:

> Queerness is also a performative because it is not simply a being but a doing for and toward the future. Queerness is essentially about the rejection of a here and now and an insistence of potentiality or concrete possibility for another world.

(2009: 1)

Part 2: Mon Oncle

From Harlow to Hulot

A growing discontent with London's Victorian laissez-faire economics and postwar disintegration of social fabric resulted in an opportunity for urban planners to use the built environment as a form of social management, which would straighten out many forms of disorder germane to the city, including queerness. Patrick Abercombie—a leading figure in reformist urban planning—laid out ambitious plans to rewrite the city's social structure by advocating the importance of belonging and community identity, considering London to be "disordered, congested, and segregated city" (Hornsey, 2010: 40). Harlow Town was conceived in this atmosphere as part of the New Towns Act of 1946, and built for the purpose of alleviating the overcrowding in London following World War II and accommodating the city's residents who had lost their homes as a result of devastation caused during the Blitz. The town saw an influx of young families in the early 1950s which gave Harlow affectionate name 'Pram Town', epitomizing the space of reproductive futurity. This badge, however, while it might have been applied to the heteronormative way in which other New Towns were conceived, described a reality commonly pictured (see Figure 5.1) but which was certainly not the whole of the ambitions of the designers.

The population of Harlow Town diversified after initially welcoming families from mainly working-class backgrounds, with the town becoming home to residents from the poorest 30% in England to middle-class citizens who could afford to buy their homes; from the baby boom of the 50s, then over the years to a simultaneously ageing population. As the demographics of the town fluctuated and changed over the years, so did its human relations. Harlow Town's planners anticipated accommodating diverse communities through mixed development in its housing schemes. Mixed development meant that the housing provided in

FIGURE 5.1 Prams in Harlow (image courtesy Harlow Museum)

Harlow Town was balanced between dwellings to rent and to sell by the local council. This ensured a diversity of occupants from various socio-economic backgrounds and family structures living in clustered communities, together. During an interview conducted on 17 November 2017, John Graham, who was one of the leading architects of the construction of Harlow Town, stated that the residential scheme provided the "freedom to be anyone you want to be in your home". An inclusive policy ensured the freedom of inhabitants to rent from the council, keeping housing available for all, and avoiding housing's monopolization by the more affluent.

Housing in Harlow Town was separated into 13 neighbourhoods, which provided frameworks for the detailed designs of architects allocated to each area. This meant that each neighbourhood was designed individually with each design being influenced by its site. Harlow's chief architect Frederick Gibberd stated that "[t]here can be no doubts that this method of housing group design has given the town variety and the people a wide choice of homes to live in" (Gibberd, 1980: 105). He believed in the importance of legible neighbourhoods: "for Gibberd and Harlow, neighbourhoods were the best option and the committees were 'Place Making', where people could be empowered to reconstruct their lives and feel the value of joy again" (Dunlea, 2016: 23). Fostering forms of neighbourhood and community kinship outside of the family were a key part of Harlow's design.

This brings us back to the notion of scenario-making in designing for citizenships. John Graham states that the town's residential scheme was inspired by "anticipating what the people would want and enjoy" (Interview, 17 November 2017). Richard Hornsey, however, reads new town development as restrictive and prescriptive:

> early new towns such as Stevenage and Harlow became mainly repositories for working-class newlyweds, who—severed from the kinship networks of the old city—often experienced social isolation and were forced to adapt to the bourgeois domestic assumptions built into the architecture of their new model homes.
>
> *(2010: 43)*

The Development Corporation adopted some basic principles for housing, such as to provide a "balanced development of houses of different standards" and a "balanced division … between houses and flats". This idea of balance "was the belief that a new town should not only contain a broad social mix but that all parts of the town should themselves be mixed" (Gibberd, 1980: 102). There was no explicit preference for nuclear families. Hornsey, however, sees Harlow's design as responding to a generalized post-war vision for new towns as a heteronormative project for reproductive futurity, paralleling and reflecting the demonization and criminalization of queer behaviour throughout society, which had increased substantially at the time (Houlbrook, 2006). The destruction of cities during the war by bombs—and after the war by demolition—and the accompanying disintegration

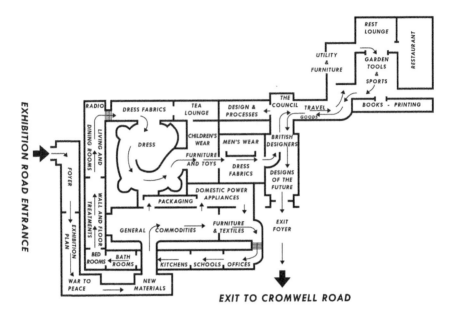

FIGURE 5.2 The prescriptive pathways of the *Britain Can Make It* exhibition prefigured the snaking pathways of the IKEA or Flying Tiger shops, as well as future exhibition design (drawing by Eglé Paçkauskaité)

of social fabric resulted in an alliance between urban design and boosterish public relations; an alliance that aimed to achieve the amelioration of poverty commonly pictured in London's slum housing. This alliance was a partnership between young architects driven by ideas of continental modernism and individuals from various fields, such as medical professionals, health reformers, volunteers from housing associations, and even eugenicists. This, to Hornsey, is the type of 'total planning' that encompasses everything from infrastructure to the fine grain of dwelling and interaction. Queer ways of inhabiting a city were out of place/out of line in a modernist scenario of a rational society.

Hornsey's critique of total planning on the urban scale extends from his emphasis on 'prescribed trajectories' observed in the layout of the 'Britain Can Make It' exhibition of 1946 (Figure 5.2), which introduced ideals of continental modernism in the UK and proposed new lifestyles for post-war Britain, to the Abercrombie plan for London.

> As the foundational supports for both a new form of metropolitan community and a self-sustaining and stable social order, Abercrombie had devised a new morality of everyday movement in which civic participation and social responsibility became identified as a matter of following the circuits and pathways to be tacitly programmed into the fabric of the postwar city.
>
> *(Hornsey, 2010: 52)*

For Hornsey, the close management of everyday spaces, evident in the exhibition layout, was also a focus of post-war urban renewal: "orchestrating the individual's daily practices and managing the conditions under which they ordinarily lived" and preventing such "disordered activities" as might be associated with queer uses of urban space (ibid.: 10). Even though Harlow, like other post-war new towns, was influenced by forms of heteronormative modernist planning for space and (consumer) citizenship such as that shown at the 'Britain Can Make It' Exhibition, its plan was adapted to a landscape vision which fitted its forms to topography while projecting possibilities for different social mixes announced by varied architectural ensembles. The rigid choreography of 'Britain Can Make It' was a product of a totalized social order of a modern way of living, which, like the map of the exhibition itself, was meant to engage the citizens in prescribed collective activities in tightly controlled spaces in "a performance of civic participation" (ibid.: 41).

> If social order was to be both established and maintained, citizens had to be reoriented away from the pernicious temptations of the unstable metropolis via the prescriptive inculcation of a more controlled—and controllable— form of metropolitan sensibility.
>
> *(ibid.: 11)*

In his landscape and urban planning, Frederick Gibberd had a gentle, sensitive, and carefully observed approach. Harlow's neighbourhoods are embedded into the existing landscape; the roads, for example, trace the topography. John Graham told us that Gibberd spent many days walking through and experiencing the terrain prior to laying out the plans for the town. His walking influenced the pedestrian and cycle routes connecting the clustered communities, providing easy access without a motorcar to any part of the town for its occupants.

The 'Britain Can Make It' Exhibition had a very specific vision for the future of urban lifestyles in post-war Britain. Harlow, on the other hand, despite being planned as a whole, was designed to allow growth and evolution over time. The town was expected to expand (especially following an influx of young families) but in the end the Conservative and neoliberal Thatcher government halted its expansion in the 1980s and a major wave of privatization of property took place, and its plan is unfortunately now hopelessly distorted by heedless development. At its core was the Market Square, its spaces designed as a market and meeting space (Figures 5.3 and 5.4).

Gibberd created scenarios for a whole society and how it would live, and the town's plan appears to incorporate a 'cellular structure', which Hornsey acknowledges reflects the teachings of Patrick Geddes:

> determining measures of "living" and "play" space were symptomatic of a wider approach to urban space that remapped it through the quotidian activities of its residents' bodies. In part, this was an inheritance from a tradition of geotechnic regional planning rooted in Patrick Geddes's work at the turn of

FIGURE 5.3   Harlow Town's Market Square called on timeless forms of association (image courtesy Harlow Museum)

the century, which understood the landscape in terms of sympathetic human use.

(Hornsey, 2010: 47)

Geddes's notion of Civic Survey is particularly relevant to Harlow Town—'diagnosis before treatment'—the town was seen as a solution to a housing crisis and disintegration of social fabric at the time; the solution was executed with utmost care not only for the citizens, but also for the landscape in which the communities could live and thrive.

> Planners already schooled in the organicist principles of Patrick Geddes began to suggest that rebuilding the built environment in adherence to these forms would create a harmonious symbiosis between mankind and nature, and ensure the optimum functional order for society's future.
>
> *(Hornsey, 2010: 70)*

Beauty and enjoyment, convivial living and good times, were some of the key features sought after by the architects who designed the buildings and neighbourhoods in the town.

> [B]eauty was expected in all features of [Harlow's] design, not only in the obvious places like parks and fine looking buildings, but when objects or 'raw materials' like street furniture, lampposts, seating and such items are placed

FIGURE 5.4 The Piazza della Erbe in front of the Medieval Palazzo della Ragione in Padua was the type of market square which influenced the design of Harlow, showing that modernism often drew its cues for space not from the hot light of invention, but from lessons of the past (IStockphoto: date of photograph unknown)

> together a series of *pleasing* compositions should emerge. Gibberd was the master planner of Harlow New Town and, as well as a man who possessed the formidable combination of being a planner, architect and a landscape architect, he had an artist's eye.
>
> *(Dunlea, 2016: 3; emphasis added) (see Figure 5.5)*

Gibberd, whom we can refer to as the 'uncle' of Harlow Town, due to the critical distance with which he walked and designed its plan rather than 'fathering' it onto a clean slate, was himself a resident in the town and stayed there until his death in 1984. Tellingly, his interest in urbanism is first in dwelling rather than buildings; an oblique relation to architecture, at least to monumental architecture.

> he has misgivings about one of his skills when speaking about his career with Roy Plomley (*Desert Island Discs,* no date) he says of himself, 'had I been less interested in people and more interested in building a monument, I'd probably have been a better architect'.
>
> *(Dunlea, 2016: 4)*

Frederick Gibberd also approached urbanism through a larger landscape-based vision, broader than the state-centric heteronormative models of citizenship and spatial order to which Hornsey refers; as a modernist architect, he did not follow the modernist

FIGURE 5.5 The Clock House, designed by John Graham, possesses an order, beauty, scale, and engagement with street life which finds roots in cues from medieval architecture through Palladianism to modernism. Henri Lefebvre describes that he observed this harmony in Padua: "In Padua, the houses are not built to present uniform facades to the gaze of passersby but to co-ordinate the succession of vaulted porticos that expand the street for pedestrians. This strictly architectural requirement results in a unity and diversity that is both pleasant and beautiful" (Lefebvre, 2014) (image courtesy Harlow Museum)

trajectories blindly or impose them in the layout of Harlow Town. His vision of total planning was that of a keen observer of people and the landscape they inhabited:

> architecture could not stand-alone and needs a tight relationship with planning to create an all-important urban scene and the architect should 'cultivate the eye of the painter and ... see the whole'.
>
> *(Gibberd, 1962: 11; as quoted in Dunlea, 2016: 23)*

Even though Frederick Gibberd worked alongside heteropatriarchal post-war urban planning, his approach worked both within and without its traditions with a sense of distance as he designed with existing landscape and socialities. Harlow Town is an example of potentialities for *athwartness* within an otherwise intransigent architectural practice.

Mon Oncle

Mon Oncle was the third feature film of Parisian director Jacques Tati, and his first in colour. It was released in 1958. It is, as David Bellos writes, not so much a stand-alone film as part of an *oeuvre* which is concerned with "many of the more striking changes that took place between 1945 and 1975: urban renewal, the growth of a leisure society, and the rise of the motor car", and which offer a material history of twentieth-century France "couched in benign and comic terms" (1999: xvi). Tati created the comic hero Monsieur Hulot (Figure 5.6), whose forwardly teetering gait, raincoat, pipe, and umbrella provide instantly recognizable imagery and body language, the power of which have often been compared (often as opposite) to Charlie Chaplin's backward-slouching, splay-footed cane-twirling. Where comic stars with roots in burlesque or mime—not just Chaplin but also Buster Keaton, Laurel and Hardy, and Max Linder—were notable for their astonishing powers of invention, Tati "was the very opposite of an inventor. He liked to say he was an observer and a realist" (ibid.: 26). His brilliance was in showing human and spatial behaviour of others in a comic light, not through harsh lampoon but through gentle exaggeration. One 14-year-old viewer of Tati's fourth film *Play Time* (1967) was quoted in the *Cahiers du cinéma* saying, "What I really liked was that, at the end of the film, when I was back in the street, the film was still going on" (Castle, 2019: 32).

Tati as Hulot is particularly interesting in the possibilities he represents for comedy and for film. His comic relationship with his surroundings is oblique. He acts as a foil for the actions of others around him as much as he is the focus of the

FIGURE 5.6 Monsieur Hulot, Jacques Tati's timeless character, on his bicycle with his nephew Gérard Arpel (played by Alain Bécourt) in the streets of the traditional town (film still: Tati, *Mon Oncle*)

comic action himself. Both Tati the creator, and Hulot the character, are queer figures: queer as in 'odd', certainly, but queer, too, to the culture of the time as Hulot is an older bachelor, as was Tati himself. Tati was a bachelor until he was 37 when he was married to a much younger woman, Micheline Winter, in 1944. An older bachelor breaks, at least until marriage and childbirth, with reproductive heteronormative expectations and futurity. That same bachelor, on the other hand, is not excluded from participation in family or society at large as an 'uncle', and that uncle can introduce interesting and useful queer or other possibilities, as shown by Eve Kosofsky Sedgwick.

> Because aunts and uncles (in either narrow or extended meanings) are adults whose intimate access to children needn't depend on their own pairing or procreation, it's very common, of course, for some of them to have the office of representing nonconforming or nonreproductive sexualities to children … But the space for non-conformity carved out by the avunculate goes beyond the important provision of role models for proto-gay kids … If having grandparents means perceiving your parents as somebody's children, then having aunts and uncles, even the most conventional of aunts and uncles, means perceiving your parents as somebody's sibs—not, that is, as alternately abject and omnipotent links in a chain of compulsion and replication that leads inevitably to you; but rather as elements in a varied, contingent, recalcitrant but re-forming seriality, as people who demonstrably could have turned out very differently—indeed as people who, in the differing, refractive relations among their own generation, can be seen already to have done so.
>
> *(1994: 63)*

The uncle occupies a space of physical and intimate proximity, yet that space is also one which allows a healthy critical distance, and one in which the powers of observation also serve to heighten the critical faculties. This is a very important message for design, as well as for the practices of citizenship. Watching the landscape and its occupants and allowing it to emerge into a flourishing space where it becomes more what its capabilities allow. For design this means prioritizing observation and evaluation over invention, not, indeed, for comic effect, but to make meaningful, to enrich, the notion of development.

Modernist development is one of the key themes of *Mon Oncle*. Monsieur Hulot lives in the timeworn traditional city where the pace of life is slow, while the family of his nine-year-old nephew Gérard, the Arpels, lives in a smart modern villa. A tumbledown wall serves as the threshold between these two worlds, and not only the look and feel of things, but the soundtrack as well, change whenever this threshold is crossed. Tati is, in his own words, sceptical of modernist architecture's means and ends, but he is not necessarily a reactionary, vehemently opposed to modernism as he is often assumed to be. In an interview for *Elle* on 3 May 1979, discussing *Play Time*, his attitude is clear:

I've never said that one should build old hospitals without sanitary equipment or that one should dress schoolchildren in black pinafores and send them to run-down schools ... In any case, the most beautiful buildings that I filmed were those in *Play Time*. I simply wanted to show that human relationships get frozen by mechanization ... Now, look what happens at Roissy [Airport] when there's a strike or a breakdown and the machinery snarls up ... Well, then people look at each other and all of a sudden they discover that they exist ... In other words, it takes some kind of unusual circumstance to get people to communicate with each other. [In *Play Time*,] there was a world of right angles with arrows telling you where to go, but you could never find anybody. Only at the end, when the nightclub is falling apart and the back wall comes down, do people really begin to have fun and the workers and waiters come out and drink with the clients.

(Castle, 2019: 28)

This same world of right angles and arrows appears in *Mon Oncle*, and the narrative is propelled by the dramatic tension between the modern city, in which rules and directions are clearly marked out, and the more aimless *joie de vivre* of the traditional city. Tati's (and Hulot's) critical distance allows him to offer a critique, to show up the humanity and the humour of those who live there, and to exercise a gentle irony in doing so. He does not damn the modern city but rather shows it as yet another circumstance in which he can show the laughter lurking everywhere in everyday life, and he pictures its inhabitants with as much affection as he does those of the traditional city. The gentleness of the romantic irony Tati employs shows, perhaps, that the irony characteristic of the postmodern art and literature of which Tati's work was a harbinger, might have taken on a happier and more trusting form than the harder and more cynical forms of irony and pastiche (which Jameson calls "devoid of laughter" and "blank parody, a statue with blind eyeballs" (1991: 17) which were to develop. Tati's films, which refuse to disbelieve in either the modern city or the traditional city, are thus immune to the dead hand of pastiche. Also, Tati's subjects wriggle themselves free from alienation in all the moments when things, as they will, fall apart. They possess a 'feigned ignorance' that makes us see the absurdity of their surroundings with a sense of amusement, rather than dissatisfaction or, perhaps, horror. In the traditional city this means the ability to make, remake, reinhabit, and jerry-build. In the modern city the breakdown of prescribed trajectories allows for the same: "the practical reconquest of a sense of place and the construction of an articulated ensemble which can be retained in memory and which the individual subject can map and remap along the moments of mobile, alternative trajectories" (Jameson, 1991: 51).

Tati's earlier films, including *Mon Oncle*, while they cannot themselves be considered postmodern, they do anticipate postmodernism through modernist forms which come to be characteristic in postmodernity, particularly the (possibly un-ironic) assemblages of incommensurate and disparate historical forms which come to form intelligible wholes. Jameson, again, sees this tendency in modernity as "uniquely corresponding to an uneven moment of social development, or to what Ernst Bloch called the 'simultaneity of the nonsimultaneous', the 'synchronicity of the non-synchronous' (*Gleichzeitigkeit des Ungleichzeitigen*)" and, as the hard, squared-off, grey forms of the modern tower blocks are glimpsed at the threshold between old and new by Hulot, "the coexistence of realities from radically different moments of history—handicrafts alongside the great cartels, peasant fields with the Krupp factories or the Ford plant in the distance" (1991: 307). We might suggest that there is a certain athwartness, queerness to these juxtapositions. Jameson also goes on to argue that postmodernism must sweep away these historic residues and in their place construct simulacra, and in this can be seen "the sense in which we can affirm, either that modernism is characterized by a situation of incomplete *modernization*, or that postmodernism is more *modern* than modernism itself". Modernization here takes on the sinister sense of a purge of culture and history, and the erasure of the Other. What is queer can only be made acceptable by reinventing it as a neutered and neutral and humourless copy of itself. This ahistorical (even anti-historical), bloodless, pastiche queer is also Edelman's futureless queer, doomed forever to party like it's 1999 and never to cross into a new millennium, denied a rightfully and joyously oblique relation with the future.

Muñoz argues that the past has agency in the present and future as well, and as a utopian presence. "[T]he field of utopian possibility is one in which multiple forms of belonging in difference adhere to a belonging in collectivity" (Muñoz, 2009: 2) he writes. "[I]t is important to call on the past, to animate it, understanding that the past has a performative nature, which is to say that rather than being static and fixed, the past does things" (ibid.: 27–28). The question of *who* animates this past is important too, and this is seen within a democratic framework that implicates not merely a majority or the prevailing common sense, but also the subaltern, minoritarian, or queer:

> Minoritarian performance—performances both theatrical and quotidian—transports us across symbolic space, inserting us in a coterminous time when we witness new formations within the present and the future ... The stage and the street ... are venues for performances that allow the spectator access to minoritarian lifeworlds that exist, importantly and dialectically, within the future and the present.
>
> *(ibid.: 56)*

The question of the inhabitation and performance of space is also taken up by Sara Ahmed, who shows, using Lefebvre, that space itself is an actor—that it has agency—and that it has what might be considered to be 'straight' meanings assigned to it through performance as well. This highlights that the minoritarian voice—that specifically debased by the majoritarian position—might well exist transversal to, athwart the straight symbolism, and this consists in a relation between action and space (Figure 5.7).

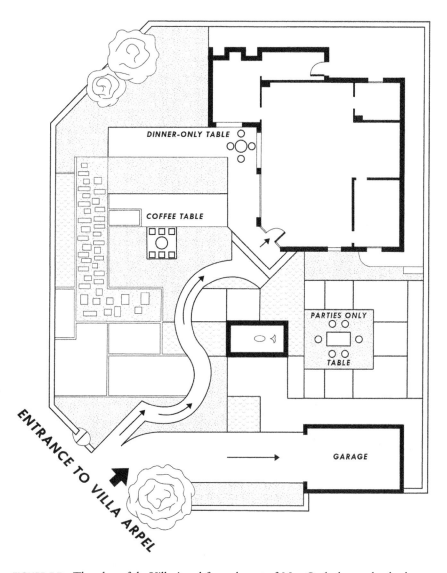

FIGURE 5.7 The plan of the Villa Arpel from the set of *Mon Oncle* shows clearly the difficulty of inhabiting this exaggerated modernist space without transgressing its prescribed routes. The Villa Arpel's regimented and prescriptive sequence of spaces are an unintended echo of the layout of the *Britain Can Make It* exhibition, illustrating as they do a typology stereotypical enough to invite parody (drawing by Eglé Pac)

It is not simply that we act in space; spatial relations between subjects and others are produced through actions, which make some things available to be reached. Or, as Lefebvre suggests: "Activity in space is restricted by that space; space 'decides' what actually may occur, but even this 'decision' has limits placed upon it" [1991: 143]. So the space of the study is shaped by a decision (that this room is for this kind of work), which itself then 'shapes' what actions 'happen' in that space. The question of action is a question then of how we inhabit space.

(Ahmed, 2006: 52)

Finally, the question here is also not merely one of the inhabitation and perform-ance of space (Figures 5.8 and 5.9) and, by extension, how these practices may be seen also as forms of citizenship, but it is also impossible to avoid mention of the fact that these spaces are very consciously the product of design. Design which projects certain possibilities and attitudes upon physical space; design which reinforces normative behaviours or alternatively which allows for alternative forms of conduct; and design in the larger sense of strategies for and imaginaries of a larger structure of society and citizenship and the spatial forms required to contain, frame, and impel these visions. The design of both the physical public realm and the sociopolitical public sphere, however, hold colliding and incommensurable visions of identity which must be able to coexist even if conflictually: or rather because democracy is, by nature, conflictual. Identity (as a constituent of citizenship) is not, for Lauren Berlant, "simply a caption for an image of an unchanging concrete

FIGURE 5.8 Jacques Tati clearly delights in showing 'queer' performance and inhabitation of [hetero]normative space. Here the spinster neighbour (in the straw hat) finds herself athwart the circulation (film still: Tati, *Mon Oncle*)

FIGURE 5.9 Father figure M. Arpel is not immune from transgression, driving his Chevrolet the wrong way in the film's closing scene (film still: Tati, *Mon Oncle*)

self", but also "a theory of the future, of history" (1997: 18). The challenge comes when "the modal form", that is, the practiced form(s) of citizenship "is called into question, when it is no longer a straight, white, reproductively inclined heterosexual but rather might be anything, any jumble of things, the logic of the national future comes into crisis" (ibid.). This problem arises, Berlant tells us, due to the infantile, intimate, and personal forms which came to signal citizenship in the Reagan years in the US. An openness to the other, fundamental to democracy, cannot arise from this inflexible model. If Jacques Tati's *Mon Oncle* is read through the lens of a more open, pre-neoliberal notion of citizenship, a conflictual public sphere and public realm is revealed, and it is here that difference is held lightly, ironically, and 'any jumble of things' is not a threat to order, but merely the kaleidoscopic variety of humanity and of urbanity (Figure 5.10). And further, Monsieur Hulot can represent a life drive invigorated by otherness and aloofness to act from humane and more-than-human nonreproductive futurity through queer ecocritical modes of practice.

From Hulot to Harlow

Frederick Gibberd's design for Harlow Town was intended as a container for such kaleidoscopic variety. Also, importantly, its genesis was not in the hot glare of invention, but rather in sensitive observation both of the site and its wood lots and

FIGURE 5.10 The border, in *Mon Oncle*, between the traditional city and the modern. Even across the tumble-down wall into the ordered new town, a horse and cart transgresses into the territory to remind the viewer of the 'simultaneity of the nonsimultaneous', and the 'synchronicity of the non-synchronous' (film still: Tati, *Mon Oncle*)

field boundaries, and also of historic and contemporary typologies, groupings, and associations—in particular of buildings, roads, and landscape (Gibberd, 1980)—which served as precedents for his townscape design. The critical, evaluative distance Gibberd displayed is far from the stereotype of the modernist designer imperiously inscribing diagram utopias onto the tabula rasa of a drained and cleared site. It is not at all a stretch to describe his stance and approach as avuncular, just like Tati's Hulot. And also, as with Hulot, the eye of the outsider, of the Other, allows for potentials that would go unseen by either the establishmentarian or the long-time resident of a landscape who is unable to see its charms or peculiarities because of a lack of distance. The avuncular designer offers a queer take on the landscape just as the queer aunt or uncle offers possibilities to the family that the heterosexual parents simply cannot. This is not to say anything about Gibberd's own sexuality or queerness, however loosely defined, but rather it posits that the avuncular architect might be a role that can be inhabited by anyone.

One can be straight and still an uncle, naturally, still a queering influence. Neither does one need to self-identify as queer in order to take a queer view or position; just as one needn't be an aunt or uncle to be avuncular. Indeed, in many cultures *anyone* of a certain age is an auntie or an uncle, and this relation can be entirely outside of the consanguineous family. It expresses instead a notion of society or nation as 'gathered' families, and the kinship of a human 'family' that together inhabits a landscape. All people have a place somewhere in this 'family'. 'Any jumble of things' becomes not a threat to order, but the basis of order, and within that jumble it is both sameness and difference that comprise what holds together to make a good life for all.[2]

Part 3: Life insurgent

One frightening vision of 'any jumble of things' for command-and-control wielding state-centric technocrats has long been the figure of the anarchist, often pictured as a swarthy, foreign, bomb-toting Other as part of a (clearly racist) tradition of image-making in state propaganda. However, it is impossible to understand the movement for New Towns, and the forms of thinking that underpin them, without recognizing their (almost completely suppressed) debt to anarchist thought—and we would like to present here too the idea that anarchism, just like queerness, provides powerful tools for thinking that one need neither be 'an anarchist' or 'a queer' to employ. Scottish planner, geographer, sociologist, and biologist Patrick Geddes is the key transitional figure in the British Isles[3] whose works and ideas can be examined to show how anarchist thinking was transported from the ideas of the early anarchist thinkers Peter Kropotkin and Elisée Reclus. While Geddes did not consider himself an anarchist, he was frank about his intellectual debt to Kropotkin and Reclus (Ryley, 2013: 158–159).

> Town and city planning as envisaged by Geddes was not governmental, but a collaborative way of developing and evolving the life of cities to promote social development and enable each individual to be a free and active citizen. It was integral to his broader emancipatory philosophy.
>
> *(ibid.: 182)*

What Kropotkin, Reclus, Geddes, and in a very oblique way Jacques Tati (who was neither an anarchist nor at all political) have in common is critical distance and observation as core to practice—which is emblematized by Geddes's Outlook Tower and Camera Obscura in Edinburgh. Anarchism may be seen here as an art of living which involves learning how societies, ecologies, cultures, and landscapes work and how they flourish, and then basing a vision of development upon enhancing their operation rather than imposing an ideal model from without. This is not a question of pragmatism versus utopianism, but rather a model of utopianism that is not totalizing in its approach; that sees utopias as plural, multidimensional, and arising from places. No two places can evolve into the same generic ideal because each landscape holds its own unique potentials.

This approach can be seen to be just as ecological, intransitive, and avuncular as the discussion of Rachel Carson's work. Patrick Geddes's 'thinking machines' (Figure 5.11) resist transitivity and seek an ecological holism of thought at the same time that this holistic approach is used as a tool to examine the particulars of specific situations. Geddes's position in these diagrams, and in his own thought,

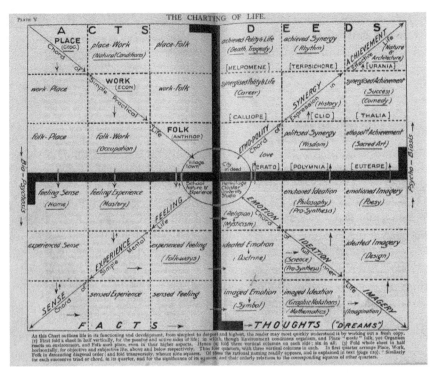

FIGURE 5.11 Patrick Geddes created diagrams he called 'thinking machines', and this is his thinking machine called 'The Notation of Life'. He strove for an all-at-onceness of thinking and observation that brought all dimensions of lived landscapes together. These are clearly ideas that do not flow in a straight line (Dumfries, Amelia (1927). 'Plate V'. In: The Interpreter Geddes: The man and his gospel)

Source: Archive and Special Collections, University of Strathclyde Library

is not in a position of command, but rather suffused throughout. His ideas are evolutionary and embroiled, arise from nature, and see humanity as nature in the electrifying terms spelled out in the motto of Elisée Reclus's 6-volume book, *L'homme et la terre*: *"l'homme est la nature prenant conscience d'elle-même"* (humanity is nature becoming conscious of itself)[4] (Reclus, 1905–1908, 2013). This motto places humanity directly as part of the 'family' of nature, but also at a conscious, critical distance from it; and also with a responsibility for it. As this book's task is to question the nature of land relations as land-*ship*, so this brings into question the idea of stewardship: that stewardship is not something done 'to' or 'over' a landscape and its associated ecologies (on its behalf, as it were), but that stewardship is undertaken as a process *with* landscapes, something Donna Haraway (2016) calls *sympoiesis*—'making with'—which we might also say, humbly, is an avuncular and queer ecocritical approach.

Geddes described such approaches as 'life insurgent', implying a powerful and inexorable life drive as a characteristic of all forms of life, including human life. And the all-at-onceness of his ideas he saw as whole 'life economies'. Economics, for Geddes, was the study, like ecology, of the whole of humanity's household: "In the domain of all the studies which directly concern man—in biology and psychology, in ethics, politics, and economics alike—it has often been pointed out how theoretic conceptions are subtly, instinctively, almost inextricably interwoven with practical considerations" (Geddes, 1885: 3). While he did seek to separate and compartmentalize the categories of economics, his work was always integrative and relational in the ways visible in 'The Notation of Life' (Figure 5.11). His habit of holding all this jumble together is a useful insurgent model for thought in the present day, when so many habits of thought and the common sense of the day are geared towards constant revolution, overthrowing, invention and innovation, restructuring, disruption, rebranding. Insurgency—life insurgent—speaks of a surging up from within pulling the past into the future, transforming the possible into the actual, always speaking of the 'not yet' and what has not arrived and what we can steward (sympoietically) into the future. This also addresses Sara Ensor's question about what it is to have 'enough'—the uncle or auntie offers superfluous possibilities above the austere minimum, about flourishing and flowering, doing and making that is also part of enoughness. For José Esteban Muñoz, "[p]erformativity and utopia both call into question what is epistemologically there and signal a highly ephemeral ontological field that can be characterized as a *doing in futurity*", and this is driven by utopian desire, that "queerness as utopian formation is a formation based on an economy of desire and desiring" (2009: 26). A key part of this life drive, this life economy, is the fullness of *jouissance*.

Avuncular architectures

We have pointed to very many provisional forms of landscape inhabitation, sociality, and citizenships as key to understanding how such forms of dwelling and association are not strict programmes or structures, as states or other authorities might wish them. Modern common sense, based as it is around the logic of the nation state and its quest for certainty, predictability, and rational order, is not easily adapted to such contingent and seemingly unstable social forms and the ways in which they hang together. It is difficult to reconcile either a utopia of rationalist order or a utopia of free market competitive individualism with more practical utopian forms based in the transformation of the everyday and the search for wellness, goodness, and *jouissance*. Sara Ensor's queer ecocritical approach requires a shift away from—or at least athwart—the dominant heteropatriarchal and nationalist orders. An avuncular distance—interested, caring, engaged, observant—allows the simultaneous existence of multiple utopias, multiple imaginaries, and their transformation. To see, as does Henri Lefebvre, cities and landscapes as *oeuvre*—as collective works over time—and, in an anarchist sense as resulting from an art of living and self-organization and management (Lefebvrian *autogestion*) means focusing not upon utopias of ultimate order or personal triumph, but on the work of mutual well-being, healthy environments for our and other species, and all the other worthwhile forms of jouissance. *Bonheur, plaisir, volupté* and *joie*; flourishing,

FIGURE 5.12 With the set for Villa Arpel from *Mon Oncle* Jacques Tati captures the essence of what Ken Olwig terms the 'BOX' in the epilogue. In fact, Tati elevates a camp, parodic, caricature of the modern architecture of the time to its Euclidean quintessence, by which the modern ironically and knowingly becomes that which was supposed to be its opposite—high kitsch. But not just kitsch in the sense of being tasteless, banal and un-genuine, but also kitsch taken to the point that it becomes kitsch's inversion, and in the process an artwork worthy of exhibition in a museum— as has been the case at Paris's Cent Quatre (film still: Tati, *Mon Oncle*)

the festal, the sensual; the full range of human togetherness. All are contained within. These forms are not exclusive to a particular style, and Jacques Tati in his timeless films shows that humans are both earnest and wry whether they inhabit traditional landscapes or modern (Figure 5.12). Looking ahead it is important to envision forms of togetherness, to see citizenships as situated, as matters for design and scenario-making, and embroiled in local, regional, planetary, and social ecologies. A life drive arises *somewhere* and in particular associations in the organic and inorganic relations in human and more-than-human lifeworlds.

Depictions of these lifeworlds arising in scholarship have more in common with premodern and indigenous lifeworlds than with those arising from Enlightenment rationality, though it is useful to see this not as a neoprimitivist inclination, but rather an evolving hybridity of thought. Rachel Carson's environmentalism, while it helped to catalyse the movement towards such forms of hybrid and relational ecological thought, was all the same part of a slightly more totalizing vision of action in its era as necessary to prevent "imminent but avoidable crisis" (Garforth, 2018: 96). Environmentalism has now become more engrained in patterns of thought, and is also now coming to be thought of as 'staying with the trouble' (Haraway, 2016) or 'arts of living on a damaged planet' (Tsing et al., 2017) at 'the end of nature' (McKibben, 2003 [1989]), or, indeed, of 'millennial visions' of 'apocalypse forever' (Garforth, 2018: 110; Skrimshire, 2010). It is now possible to imagine the future in ways that were not previously available, and these futures are not merely ominous and frightening—while there is certainly much that is—there are also new forms of hope and desire arising which are beginning to found new scenarios, new futures, new utopias. Let us return to Eve Kosofsky Sedgwick, who offers tools with which queer ecocriticism may begin to approach questions of nationalism and citizenships, and perhaps also that helps to figure new avuncular modes of citizenship, belonging, and mutual engagement:

> The work—I should more accurately say the struggle—of defamiliarizing and thereby visibilizing the nationalism that forms the overarching ideology of our age is difficult to the extent that one or another nationalism tends to become the form of last resort for *every* legitimizing political appeal— whether right or left, imperialist or anti-imperialist, religious or secular, elitist or populist, capitalist or anti-capitalist, cynical or utopian, and whether or not on behalf of an ideology that has any account whatever to offer the status of the 'nation' as such. (No analysis through Marxism, capitalism, religion, or race, for instance, offers any intelligible justification of national form, yet each of these kinds of power today takes national or nationalist shapes).
>
> *(1994: 146)*

Sedgwick points here to the possibility that national forms hang together not because of any rational or naturalistic order, but because of their own inertia and a near-universal belief in the shared imaginary. "Ways of becoming political", writes Engin Isin, "such as being citizens, strangers, outsiders, and aliens, do not exist in themselves, but only in relation to each other" (2002: 29). Design practices, both in the architectures but also in the sociopolitical and cultural structures that comprise lifeworlds, can arise from, influence, and provide new and more holistic, ecological imaginaries for future flourishing, and queer approaches athwart the normative orders are necessary for this reimagination and transformation. Design and citizenships are practices of both distance and of solidarity, and the manner in which relations to the world and to others are imagined is key to democracy. "There is no intelligence where there is aggregation, the *binding* of one mind to another", Jacques Rancière tells us, "There is intelligence where each person acts, tells what he is doing, and gives the means of verifying the reality of his action. The thing in common, placed between two minds, is the gauge of that equality" (1991: 32). The 'thing in common' echoes practices of commoning as human and landscape practices and foregrounds them as part of citizenship. Rancière shows how what we would call an avuncular mode acting upon a 'jumble of things', which all the same hangs together, informs the institution of the social and hence the political when he speaks of people united in aggregation. This is a 'hanging together' in a modernity which allows for the 'simultaneity of the nonsimultaneous', the 'synchronicity of the non-synchronous', and the queer athwartness which cuts across it transversally is what troubles the assemblage, but also part of what holds people together as distant beings.

Thus we exhort: be an auntie, be an uncle. Take on a queer ecocritical distance. See citizenships and stewardships as oblique and observational and relational. Come athwart the world and transform it for the better. And, as in Harlow and Hulot both, don't begin in the pure light of invention, but rather in the full spectrum of enlightened observation and pull all the gifts from the past into the future.

Notes

1 We are aware that the concept of transversality is elaborated in Félix Guattari's *The Three Ecologies* (2008 [1989]), but to attempt to scaffold this against the idea of athwartness here is beyond the scope of this essay, though certainly a worthy pursuit.
2 John Graham himself is an avuncular figure, who decided to reside in Harlow with his Swedish wife Britt.

> Other aspects of Swedish life which Britt might have missed in her new country were brilliantly overcome by importing part of it for months at a time in the shape of teenage 'au pair' girls. [...] Ingrid, now a grandmother, was the first in a succession of girls over four decades with many of whom we have remained in contact and they have become part of the family, because that is how they were treated. We did not have children, therefore baby-sitting was not required. The girls went everywhere with us, and that included sightseeing, shopping and theatre visits to London which, in those days, was much easier to negotiate by car. [...] With our teenage 'daughters' we had memorable theatre visits and secured autographs from 'stars' ...
>
> *(Graham, 2016: 96)*

3 Likewise in the US, Lewis Mumford's thought and immense influence was underpinned by a thorough comprehension of his anarchist forbears, and the suppression of the acknowledgement of this stream in his work means much scholarship about Mumford is incomplete.
4 Acknowledging the work of Elisée Reclus also helps to situate Rachel Carson's work in a continuous tradition of ecological, geographic, sociological, and biological thought stretching back to the nineteenth century and before, showing that her work, too, was not wholly pathbreaking as it is often thought, but rather it brought these important habits of mind to light as a crucial catalyst at a point when an American and world public had great need of such environmental awareness.

Bibliography

Ahmed, Sara (2006). *Queer Phenomenology: Orientations, Objects, Others.* Durham, NC and London: Duke University Press.

Bellos, David (1999). *Jacques Tati: His Life and Art*. London: The Harvill Press.

Berlant, Lauren (1997). *The Queen of America Goes to Washington City: Essays on Sex and Citizenship*. Durham, NC: Duke University Press.

Bononno, Robert (2014). 'Translator's note'. In: Łukasz Stanek, ed. *Toward an Architecture of Enjoyment*. Translated by Robert Bononno. Minneapolis, MN: University of Minnesota Press, pp vii–x.

Bourdieu, Pierre (1990). *The Logic of Practice*. Stanford, CA: Stanford University Press.

Carson, Rachel (1996 [1962]). *Silent Spring*. Boston: Beacon.

Carson, Rachel (2011 [1951]). *The Sea Around Us*. New York: New Road Media.

Castle, Alison, ed. (2019). *The Definitive Jacques Tati: Tati Speaks*. Köln, DE: Taschen.

Cleto, Fabio, ed. (2002). *Camp: Queer Aesthetics and the Performing Subject—A Reader*. Ann Arbor, MI: University of Michigan Press.

Driskill, Qwo-Li, et al. (2011). 'Introduction'. In: Qwo-Li Driskill, Chris Finley, Brian Joseph Gilley, and Scott Lauria Morgensen, eds. *Queer Indigenous Studies: Critical Interventions in Theory, Politics, and Literature*. Tucson, AZ: University of Arizona Press, pp 1–28.

Dunlea, Corrina (2016). *In What Ways Can the Sculpture Collection Enhance Social and Economic Development in Harlow?* London: Kingston University.

Edelman, Lee (2004). *No Future: Queer Theory and the Death Drive*. Durham, NC: Duke University Press.

Ensor, Sarah (2012). 'Spinster Ecology: Rachel Carson, Sarah Orne Jewett, and nonreproductive futurity'. *American Literature*, 84(2), pp 409–435.

Garforth, Lisa (2018). *Green Utopias: Environmental Hope Before and After Nature*. Cambridge, UK: Polity Press.

Geddes, Patrick (1885). *An Analysis of the Principles of Economics (Part I)*. London and Edinburgh: Williams and Norgate.

Gibberd, Frederick (1980). *Harlow: The Story of a New Town*. Stevenage, UK: Publications for Companies.

Graham, John (2016). *Annia Regilla - Her Cenopath - A Tale of Two Cultures*. London: Melisende.

Graham, John (2017). Interview, 17 November.

Guattari, Félix (2008 [1989]). *The Three Ecologies*. Translated by Ian Pindar and Paul Sutton. London and New York, NY: Continuum.

Haraway, Donna J (2016). *Staying with the Trouble: Making Kin in the Chthulucene*. Durham, NC and London: Duke University Press.

Helmreich, Stefan (2009). *Alien Ocean: Anthropological Voyages in Microbial Seas*. Berkeley, CA: University of California Press.

Hornsey, Richard (2010). *The Spiv and the Architect: Unruly Life in Postwar London*. Minneapolis, MN and London: University of Minnesota Press.

Houlbrook, Matt (2006). *Queer London: Perils and Pleasures in the Sexual Metropolis, 1918-1957*. Chicago: University of Chicago Press.

Isin, Engin F (2002). *Being Political: Genealogies of Citizenship*. Minneapolis, MN and London: University of Minnesota Press.

Jameson, Fredric (1991). *Postmodernism: Or, the Cultural Logic of Late Capitalism*. London: Verso.

Lefebvre, Henri (2008 [1947]). *Critique of Everyday Life Volume 1*. Translated by John Moore. London: Verso.

Lefebvre, Henri (2014). *Toward an Architecture of Enjoyment*. Translated by Robert Bononno. Minneapolis, MN: University of Minnesota Press.

Mbembe, Achille (2019). *Necropolitics*. Translated by Steven Corcoran. Durham, NC and London: Duke University Press.

McKibben, Bill (2003 [1989]). *The End of Nature: Humanity, Climate Change and the Natural World*. London: Bloomsbury.

Muñoz, José Esteban (2009). *Cruising Utopia: The Then and There of Queer Futurity*. New York, NY and London: New York University Press.

Probyn, Elspeth (2016). *Eating the Ocean*. Durham, NC and London: Duke University Press.

Rancière, Jacques (1991). *The Ignorant Schoolmaster: Five Lessons in Intellectual Emancipation*. Translated by Kristin Ross. Stanford, CA: Stanford University Press.

Reclus, Elisée (2013). *Anarchy, Geography, Modernity: Selected Writings of Elisée Reclus*. Translated by John Clark and Camille Martin. Oakland, CA: PM Press.

Reclus, Elisée (1905–1908). *L'homme et la terre*. 6 vols. Paris: Librairie Universelle.

Ryley, Peter (2013). *Making Another World Possible: Anarchism, Anti-Capitalism and Ecology in Late 19th and Early 20th Century Britain*. New York, NY and London: Bloomsbury.

Sedgwick, Eve Kosofsky (1994). 'Nationalisms and sexualities: As opposed to what?' In: *Tendencies*. London: Routledge, pp 141–153.

Skrimshire, Stefan, ed. (2010). *Future Ethics: Climate Change and Apocalyptic Imagination*. London: Bloomsbury.

Tati, Jacques (1958). *Mon Oncle* [Film]. Paris: Specta Films.

Tati, Jacques (1967). *Play Time* [Film]. Paris: Specta Films.

Tsing, Anna, et al., eds. (2017). *Arts of Living on a Damaged Planet*. Minneapolis, MN: University of Minnesota Press.

Turner, William B (2000). *A Genealogy of Queer Theory*. Philadelphia, PA: Temple University Press.

Wacquant, Loïc (2004). 'Habitus'. In: Jens Beckert and Milan Zafirovski, eds. *International Encyclopedia of Economic Sociology*. London and New York, NY: Routledge, pp 315–19.

Warner, Michael (1993). 'Introduction'. In: Michael Warner, ed. *Fear of a Queer Planet: Queer Politics and Social Theory*. Minneapolis, MN and London: University of Minnesota Press, pp vii–xxxi.

Welter, Volker M (2002). *Biopolis: Patrick Geddes and the City of Life*. Cambridge, MA and London: MIT Press.

6

SITUATING LANDSCAPE CITIZENSHIPS

Borders, margins, hybridity, and the uncanny

Joern Langhorst

This essay attempts to develop a theory of landscape citizenships based on the assumption that it is impossible to separate landscape from the people who inhabit it. It weaves together a range of critical perspectives that have been largely absent from discourses on landscape, place, identity, and belonging, and explores borders, boundaries, and edges as central locations of *landscape citizenships*. It investigates marginality and liminality as critical processes in establishing landscape citizenships and considers the uncanny as a dimension of encountering the other, the strange, and strangers. Far from looking at borders, boundaries, and edges as static constructs, it proposes overlapping, fuzzy, and nested boundaries that insist upon an openness to the other.

The goal of this essay is to set out a framework for further research and future practices. The theories and concepts discussed in the following do not claim to be complete, but offer a range of perspectives and approaches on how to read, interpret, and imagine critical aspects and dimensions of people-landscape relationships, and to establish a foundation to discuss and enact 'landscape citizenships' as a consequential, productive, and empowering concept. This concept is foundational for conceptualizing and enacting 'landscape democracy' (Egoz et al., 2018a). Landscape democracy is rooted in "a mutual understanding that landscape is an essential component to ensure everyone's wellbeing, an indisputable universal human right" (Egoz et al., 2011). Promoting and materializing landscape democracy is a critical step towards spatial justice and to enacting the 'right to landscape' (Egoz, Makhzoumi, and Pungetti, 2011), all the more pressing in current 'liquid times' (Bauman, 2000) of environmental and political uncertainties that profoundly affect spatial justice and landscape democracy.

The use of 'landscape citizenships' in the plural is intentional—this chapter argues that forms of belonging to and in a place are relational, that different people can have different forms of belonging to and in the same place, and that multiple constructions of landscape citizenships coexist in the same place at the same time. The possibility of multiple citizenships not only in one place but in one body[1] extends that principle and forms an essential foundation of this essay.

These contentions are supported by theorists and writers who are not interested in dichotomic and categorical distinctions, or in arguing for clear lines of separation and stable and rigid ordering systems—qualities traditionally associated with territorial, colonial, and capital formations of landscape and cultural identity, spatialized and materialized by stable and unambiguous borders, boundaries, and edges. Instead, this essay proposes overlapping, fuzzy, and nested boundaries that insist upon an openness to the other, and foreground marginality and liminality, referencing the following authors and texts: Homi Bhabha and his writings on hybridity, ambivalence, marginality, and the 'right to narrative' (1990, 1992, 1993, 1994, 2003) form a critical basis for this essay. He interrogates colonial and post-colonial processes of generating identity and belonging and argues that the making of cultural difference occurs in a fluid and dynamic 'thirdspace' of encounter. Encounters with 'the other', conceptualized by Bhabha (and Anthony Vidler, 1992) as the 'uncanny' are central to his thinking on the formation of individual and collective identities. In the first part of the twentieth century the Russian philosopher and literary critic Mikhail Bakhtin (1981 [1934–1941]; 1990 [1919]) explored this encounter with the other as a central dimension of the formation of language and meaning, broadening and deepening Bhabha's propositions. The writings of Gloria Anzaldúa, a scholar of Chicana cultural theory, feminist theory, and queer theory, in particular her book on growing up on the Texas-Mexican border (1987), explore the role of experiences of social and cultural marginalization in the formation of multiple simultaneous identities and forms of belonging that are inter-relational and contextual, forming richly layered hybridities.

The writings of these key thinkers are furthered by theorists who explore questions of belonging, identity, landscape, territory, borders, and rights from mostly post-colonial and Black perspectives (Gwaltney, 1993; Gilroy, 1993; Hall, 1992, 1993). In particular, following Frantz Fanon's writings on post-colonial identities and forms of belonging, Achille Mbembe's articulation of 'necropolitics' analyses how "contemporary forms of subjugation of life to the power of death" (2003: 29) reduce people to precarious conditions of life and forces them into the margins, or into exile. These perspectives have been largely absent in discourses on landscape, belonging, spatial justice, and landscape democracy. The emergent conversations on and around landscape citizenships offer not just an opportunity to include such perspectives, but clearly evidence the need to explore and acknowledge the roles of people and experiences marginalized, exiled, suppressed, or colonized in the making of landscape, landscape democracy, and society itself.

The identities of place and people are mutually constitutive, entangled, and have developed over time. Lifeworlds and lifeways[2] intersect and entangle in place and over time. Home and homeland, property and territory, garden and wilderness, culture and nature are potent constructs and material realities that are immensely consequential—livelihoods depend on them, and wars have been fought over them. The question of who belongs to a landscape—or whose landscape it is—is neither new nor easily answered.

There is a need to expand and reframe existing discourses on landscape and citizenship, as most of these, in the prevalent Western discourses, are rooted in colonial perspectives and based in understanding landscape either as physical and bounded terrain upon which the nation state is constructed, or as a country being constituted through its people and established through their actions over time (Olwig, 2002). These discourses are intertwined and rely on physical and social conditions as relatively stable. Both discourses depend on the drawing of boundaries as central and constitutive actions: here-there, us-them, cis-trans, and, maybe most importantly, self-other are some of the criteria and conditions that are used to delineate and distinguish. Implicit in these delineations is the question of who 'we' are—in the context of the places and landscapes we inhabit and call 'ours'. Landscape citizenships then are shaped by two entangled dimensions—the landscape belonging to its people, and its people belonging to the landscape. All kinds of claims and rights—but also responsibilities—are derived from these fundamental constructions of belonging to and being in place.

Cosgrove's (1998: XIV) contention that "landscape constitutes a discourse through which identifiable social groups historically have framed themselves and their relations with both the land and with other human groups, and that this discourse is closely related epistemically and technically to ways of seeing" suggests the reciprocal formation of human identity and landscape[3] as a central element in establishing a sense and fact of belonging. This framing—and formation—of (collective and individual) self and landscape then becomes spatialized, rendered visible and put into effect by the drawing and making of boundaries, borders, edges, and thresholds—categorical, conceptual, and physical.[4] While geophysical, ecological, and bioregional boundaries precede human existence in the landscape (in fact many definitions and delineations of landscape are based on these—e.g., watersheds), they are 'traced', given meaning, and assigned value and consequence by humans. This tracing is made all the more problematic by the nature of such boundaries—they are de facto zones of transition, ecotones, characterized by gradual changes over both space and time. Mathur and Da Cunha (2001), looking at the Mississippi River, clearly show that the way we draw and understand rivers does not reflect the fact that they are not bounded by banks but rather are zones of variable wetness. Other borders are preconceived by the human mind and projected onto a place, often immaterial but nonetheless consequential, and often they are made manifest by physically changing a place—by something as subtle as a sign, or incisive as a wall. International borders and property lines are some of the most significant examples. Such boundaries are mostly considered static and persistent, and as instruments to separate something (and somewhere) from something (and somewhere) else.

Hybridity and thirdspace: blurring boundaries and articulating differences

In the context of a rapidly changing world these notions need to be challenged: the philosopher Edward Casey (2011, 2017), in his investigation of La Frontera, the border between the US and Mexico, interprets boundaries not just as instruments and conditions of separation, but as something that connects, articulates, frames, and enables complex relationships over time and space. He sees the world as being made up of many different types of edges, of 'limitrophic' phenomena, that can be interacted with and inhabited in various ways.

The histories of colonialism and imperialism reveal another, darker dimension of the relationship between bounded territories and the people who inhabit them—the drawing and enforcing of boundaries onto a place and people by acts of overt or covert violence. The term 'belonging' here takes on a different connotation—one of forcible subjugation by a colonizer, and equally forced acculturation into the colonizer's cultural system. This almost by default creates multiple contested identities, and multiple contested belongings—as the territory might both represent Indigenous or colonial systems of power and signification. Chicano/a and Latino/a hybridity have been widely discussed and serve as a good example of the simultaneity of multiple identities and lifeways which may be held within one body. The writings of Gloria Anzaldúa (1987) describe such contested identities and belongings and show the difficulties and ambiguities in attempting to describe Chicano/a identities along the US–Mexican border:

> Nosotros los Chicanos straddle the borderlands. On one side of us, we are constantly exposed to the Spanish of the Mexicans, on the other side we hear the Anglos' incessant clamoring so that we forget our language ... This voluntary (yet forced) alienation makes for psychological conflict, a kind of dual identity—we don't identify with the Anglo-American cultural values and we don't totally identify with the Mexican cultural values. We are a synergy of two cultures with various degrees of Mexicanness or Angloness. I have so internalized the borderland conflict that sometimes I feel like one cancels out the other and we are zero, nothing, no one.
>
> *(Anzaldúa, 1987: 364–365)*

Anzaldúa clearly articulates here a hybridity, an existence of multiple simultaneous landscape citizenships in the same location. Landscape citizenships are not merely realms of identity and preestablished modes of belonging, as in constructs such as territory or nation states, but are realms of praxis, of the practices of everyday life, evolving within emplaced lifeworlds. Anzaldúa's hybridities fit into Edward Soja's (1996) definition of thirdspace as a 'trialectic', not either/or but both/and—a perspective that is immensely consequential for understanding the limitations and implications of the making of borders and boundaries.

The making of borders and boundaries by and large has failed to create clear, unambiguous, stable and uncontested cultural, collective, and individual identities based on the physical–political–territorial ordering of places,[5] the landscapes one dwells in. A clear and stable distinction between us-them, here-there, cis-trans, and self-other (so often encoded in statements on citizenship) appears to be as much desired (by some) as ultimately elusive.

Homi Bhabha offers another critical perspective on the colonial (and post-colonial) articulations of difference that underpins the formation of cultural identity in place from the perspective of minorities:

> It is in the emergence of the interstices—the overlap and displacement of domains of difference—that the intersubjective and collective experiences of nationness, community interest, or cultural value are negotiated. How are subjects formed 'in-between', or in excess of, the sum of the 'parts' of difference (usually intoned as race/class/gender, etc.)? How do strategies of representation or empowerment come to be formulated in the competing claims of communities where, despite shared histories of deprivation and discrimination, the exchange of values, meanings and priorities may not always be collaborative and dialogical, but may be profoundly antagonistic, conflictual and even incommensurable[6]?
>
> *(Bhabha, 1994: 2)*

He posits a 'thirdspace' between the stereotypical differences of colonial discourse, and foregrounds 'hybridity' as a positive condition that causes new forms of life and culture to come into existence (Bhabha, 1985). He argues that cultural identities cannot be ascribed to pre-given, scripted, ahistorical cultural traits that define the conventions of ethnicity or territoriality (which are often conflated in colonial discourses). 'Colonizer' and 'colonized' cannot be viewed as separate entities. Bhabha suggests instead that cultural identity, the "social articulation of difference, from the minority perspective, is a complex, on-going negotiation that seeks to authorize cultural hybridities" (1994: 2), involving an ongoing exchange of cultural performances that in turn produce a mutual and mutable recognition (or representation) of cultural difference. This negotiation creates a 'liminal' space that is a 'hybrid' site that witnesses the production—rather than just the reflection—of cultural meaning. Bhabha very much emphasized the role of boundaries when he states that "it is in this sense that the boundary becomes the place from which something begins" (ibid.: 55). His concept of liminality not only pertains to the space between cultural collectives and identities, but between historical periods, between different languages and systems of signification, describing liminal space as "in-between the designations of identity [that] becomes the process of ... inter-action, the connective tissue that constructs the difference" (ibid.: 4).

Bruno Latour's (2005) 'assemblage', Doreen Massey's (2005) 'throwntogetherness', and Chantal Mouffe's (1992) 'agonistic pluralism' all attempt to subvert the ordering of difference beyond dichotomic, dyadic, or dialectic constructions, and are implicitly concerned with the making and unmaking of boundaries and other edges, emphasizing the processual over the final and determinate. Edges, boundaries, and borders then are considered not just as delineations and delimitations of physical and non-physical things and spaces, but as the very processes and expressions of emergence, becoming, and change that connect space to human action and thought. Such delineations and delimitations, or, to use de Certeau's (1984) analogy, connotations and denotations, are not just spatial-material nor are they static[7]—edges are 'fluid', 'vibrant' (Bennett, 2009, passim), and the way and locus in which humans make sense of the world, and of themselves in that world.

On edge and on the edge: liminality, the strange(r), and the uncanny

Edges and boundaries, whether immaterial or physically manifest in place, are some of the most important instruments of the spatial production and reproduction of power and hegemonic order—and, conversely, the very sites of resistance against that order.[8] Edges are sites of negotiation, of construction and counter-construction, highly contested, seemingly separating, but in fact connecting. While the drawing of boundaries and edges within Renaissance, colonial and capitalist thought has been about clarity, visibility, legibility, and stability (Cosgrove, 1998), their actuality is rather different: edges can be ambiguous and ambivalent, porous, and enabling. Edges are the locations of what Bhabha (1994: 2) calls 'liminal' negotiation of cultural identity across differences of race, class, gender, and cultural traditions. They are 'thirdspace', where hybridity takes place, becomes manifest, is negotiated, and continuously produced and reproduced. Edges are where the other is encountered, responds, and is responded to, they are where an ongoing dialogue between what is on either side of that edge occurs, and they are where a thirdspace is created, or is emerging, subverting colonial notions of centre and periphery. More so, edges—and in particular their subtype of border—might be more influential in the formation of identity than the centre of a territory. They are the place where hybrids and in-betweens emerge, but also where entrenchment and a sharpening of identity occurs—in short, they are as much (or more) about connection than they are about separation—and more about destabilization and change than stability and order. Bhabha considers dynamism and instability as positive aspects of hybridity, as the processes and conditions that generate "new forms of life", in particular when seen from the perspective of the minority or the marginal. The view from the majority, the centre or the hegemonic may yield a different perspective: borders, boundaries, edges, and thresholds, as much as they are needed to provide order, are neither stable and predictable nor—despite appearances—clearly delineated conditions, they are complex borderlands, full of slippages and gaps.[9]

This essay proposes that the qualities of many edges and thresholds, borders, and boundaries present in landscapes are locations of what Anthony Vidler (1992) and others explore as the 'uncanny' or 'unhomely'—as places where 'the other', the 'strange' and the 'stranger', seen as unknown, dangerous, and threatening, makes its presence known.[10] The very edge that is seen as a source of safety, order, and control (as in the edge of a clearing, the wall of a home, or the fortification walls of a city) becomes an opening, establishing a connection to the very thing it aims to keep away and at bay. What is inside forms the centre, what is outside the margin, the periphery, *terra incognita*. "*Hic sunt dracones*—here be dragons" is not just a phrase on medieval maps that delineate the unknown. The unknown may appear within and across the boundaries and edges created by humans and infiltrate the familiar and homely. Homi Bhabha has reappropriated the uncanny to speak of the return of "the migrants, the minorities, the diasporic", proposing the city "the space in which emergent identifications and new social movements of the people are played out". What he calls "the perplexity of the living" might be, in these terms, interpreted through a theory of edges and the uncanny that destabilizes traditional notions of centre and periphery[11]—the spatial forms of the national and the capital, and help to comprehend how "that boundary that secures the cohesive limits of the Western nation may imperceptibly turn into a contentious internal liminality that provides a place from which to speak both of, and as, the minority, the exilic, the marginal and emergent" (Bhabha quoted in Vidler, 1992: 10–11).

A stranger in one's own land …: life in liquid modernity and liquid fear

How does one belong in and to a landscape and place in a rapidly changing, liquid, interconnected, globalized world, whose inherent uncertainty creates for many "an atmosphere of ambient fear"?

Zygmunt Bauman (1997: 50–51) describes

> The dominant sentiment is the feeling of uncertainty – about the future shape of the world, about the right way of living in it, and about the criteria by which to judge the rights and wrongs of one's way of living. Uncertainty is not exactly a newcomer in the modern world, with its past. What is new, however, is that it is no longer seen as a mere temporary nuisance which, with due effort may be either mitigated or completely overcome. The postmodern world is bracing itself for life under a condition of uncertainty which is permanent and irreducible.

His concept of 'liquid modernity' (2000) describes the failure of the modernist project of increased human emancipation, and describes our current liquid society as characterized by instability and ambiguity and by the erosion or altogether disappearance of what appeared to be stable or solid categories of identity. This liquidity affects the very concepts and practices of territoriality and belonging, and their spatialization as edges, borders, and boundaries. Rapid change is such that social patterns and institutions no longer have time to respond and provide solid and meaningful bases of human identity and collective action. As Manuel

Castells (1989: 349) suggested, "the flows of power generate the power of flows, whose material reality imposes itself as a natural phenomenon that cannot be controlled or predicted ... People live in places, power rules through flows".[12] Bauman (2003: 14) sharpens this point by stating that "power rules because it flows, because it is able (beware ever forgetting it!) to flow—to flow away". Landscape citizenships demand that power must be returned to its foundations in substantive landscapes, and that the political is practiced in places, counteracting the capacity of power to disengage from place and people and generate fear and uncertainty.

> 'Liquid fear' analyses the nature of the fear such endemic uncertainty generates, and the consequences for our ability to engage in meaningful social action to produce a viable future. Liquid fear is without obvious source, derivative and based in an internalization of 'a vision of the world' that includes insecurity and vulnerability ... Even in the absence of a genuine threat derivative fear will create a reaction similar to the experience of real danger and will attain 'a self-propelling capacity'.
>
> *(Bauman, 2006: 3)*

This pervasive but detached fear is used by the state, to ensure compliance and social passivity from its citizens, as the state is not able to deliver on its promise to protect people. The social stability on which the state's claims rely has been worn away by the uncertainties of global capitalism, and the inherent inequities of this system have generated a milieu of physical threats to personal safety "in the form of so-called terrorist violence" (ibid.).

The fear of death is arguably one of the most potent threats any system of power can deploy. Achille Mbembe (2003) expands 'necropolitics', the use of social and political power to dictate how some people may live and how some must die, into a hegemonic 'right' to impose social or civil death, the 'right' to enslave others, and other forms of political violence. Necropolitics analyses how "contemporary forms of subjugation of life to the power of death" (ibid.: 39) and different forms of necropower over the body (e.g., racialized, a state of exception, urgency, martyrdom) reduce people to precarious conditions of life. Mbembe speaks of the creation of "death worlds—new and unique forms of social existence in which vast populations are subjected to conditions of life conferring upon them the status of living dead" (ibid.: 40), profoundly disabling any ability for political action.

A prevalent contemporary response to liquid fear is a withdrawal into a particular kind of essentialism, a nationalist, latently xenophobic, ethnocentric, and isolationistic identity politics that re-interprets and re-reifies a body politic by hardening and fortifying existing and creating new boundaries and borders.[13]

These responses are ultimately a futile attempt to keep the processes of change that affect the world away from particular places and landscapes. Often, the forces and processes of globalized and global change become visible through migration and migrants—the personifications of the other, the stranger, and the uncanny/un-homely. The change that is occurring within a place or landscape is equally seen as a threat—responded to by a nostalgic re-interpretation and re-creation of region-alist architectural forms, materials, and patterns, often in conjunction with anti-modernist sentiments that mirror the xenophobic and nationalist retrenchments.[14] Bauman (1997: 55) suggests that "the age of anthropophagic [assimilation] and anthropoemic [exclusion] strategies is over". Any successful attempt to cope with the unknown, the uncertain and the confusing needs to focus not on how to get rid of the strangers and the strange, but how to live with them—"daily and per-manently" (ibid.).

Diaspora studies may provide a model of how to reconceptualize a 'belonging' to place and landscape that is shaped by constant change and incursions, appearances of and contestations by various versions of 'the Other': notions of hybridity and antiessentialism are used to demonstrate that diasporic identities are produced through difference, a difference situated between the 'here' of the host country and the 'there' of origin, between the 'us' of a dominant community and the 'them' of multiple forms of racialized identification (Bhabha, 1993). Instead of foregrounding the perspective of the nation-state and its views of 'others', (hegemony or *Leitkultur*), antiessentialist diaspora studies tend to emphasize the creative, positive, imaginative cultural geographies, and cartographies through which migrants (or the marginalized) produce themselves, rendering notions of hybridity, borderlands, and the 'thirdspace' as crucial.[15]

The question of belonging to a place or landscape as well as notions of land-scape citizenships cannot be discussed without taking into account the impacts and performances of the boundaries and edges that delineate and order land-scape and identity, place, and people, in an increasingly interconnected, liquid, conflicted, and contested world. A dynamic and hybrid understanding of borders and boundaries challenges implicitly static and stable notions of belonging—and of landscape citizenships themselves—and runs directly counter to the necropolitical as encountered in contemporary politics, or to the erosion of the political itself in the age of neoliberal governmentality. In response, the role of boundaries, borders, and edges is shifting—they may be more important in deter-mining the identity of a place or landscape than what (and who) is contained within and between them. A 'hardening' of borders and edges is a common response in times of fear and uncertainty, but conversely so are the attempts to transgress, blur, and penetrate—conflicting processes that only increase the impact and intensity of borders.

Belongings, rights, and obligations—the case of (and for) landscape democracy

Bhabha's 'contentious internal liminality' (Vidler, 1992: 10–11) then may enable the 'right to landscape' and 'right to place' as a foundation for a critical, engaged, and inclusive landscape citizenships beyond contemporary neoliberal rhetorics of neo-territoriality, globalization, sustainability, and resilience.[16] Contentious, internal liminality—and the attendant modes of belonging—as a foundation for contemporary, empowered, inclusive landscape citizenships will have to focus on how edges and boundaries can be occupied and inhabited in new, creative, positive, and imaginative ways—instead of being seen as no-man's-lands of separation, no-go zones, and places to be avoided. Any right to landscape and right to place, extending to Lefebvre's (1968) 'right to the city' includes two dimensions—the right to access, and, maybe less obvious, the right to participation.[17] Ideally both rights would be equally available to everybody—but the realities of neoliberal capitalism are different. Lefebvre's city, and by extension, I argue landscape, can be seen as the location where hegemonic power meets transgressive incursion, where displacement and marginalization 'take place',[18] where migration and hybridity occur—a battleground for the fight for the right to landscape and place, the very location where space is produced. Who then gets to participate in the production of space, and how? De Certeau (1984: XIX) discerns spatial strategies and tactics. Strategies are only available to subjects of 'will and power', so defined because of their access to a spatial or institutional location that allows them to objectify the rest of the social environment, and actualize a schematic and stratified ordering of social reality. Spatial tactics are employed by those who lack a space of their own from which to apply strategies, and continuously transgress, re-signify and disrupt the schematic ordering of reality produced through the strategic practices of the powerful. This ongoing contestation of the social order, created through multiple strategies and challenged by multiple tactics, can be read as an expression of Bhabha's contested internal liminality. A right to landscape—as a foundation for landscape citizenships—would engage the potentials of communities to empower themselves to participate in decisions on their futures and the places and landscapes they inhabit, by helping them to expose the underlying mechanisms of their marginalization, to propose alternative future scenarios, and, first and foremost, to politicize their long tradition of systemic and systematic disempowerment and disenfranchisement. Marginalized and traumatized communities and people would have to be able to participate in the discourses on their own future, to become co-authors of the landscape they inhabit instead of being cast as or self-identified as merely victims of external and hegemonial agendas.[19] This directly counteracts the operation of necropower (Mbembe, 2003), which seeks to completely disable the agency of communities and peoples by uprooting them from place ('deracination') and disempowering them so totally that landscape citizenships are utterly impossible. The ability of landscape citizenships to flourish depends upon effectively countering necropower's absolute ability to disable by terror (Linebaugh and Rediker, 2000).

Maria Kaika, in the context of empowering communities to imagine new and alternative forms of future human–environment relationships that exceed Agamben's (1998) 'bare life', suggests that we shift from 'the building of consensus' to the 'monitoring of dissensus' (Kaika, 2017: 94)—in other words, to establish diverse discourses and discover new contentious internal liminalities, "focusing on where, how, why, and by whom conflict and disagreement are generated" (ibid.: 94). This is about nothing less than landscape democracy—with both the right and obligation to participate in the making of landscape and place, society and culture, and articulating the meaning of the relationships between landscape, democracy, space and justice.[20] It is important in this context to think of landscapes not just as physical entities, but as social, political, and legal phenomena (Olwig, 2013) that are the historical expression of their shared polities and their representative and governing institutions and economies, as well as of the oppressions, forms of exploitation, exclusions, and violence that such institutions and economies license (Mels and Mitchell, 2013; Mitchell, 2007).

Democracy is often cast in the Western tradition of liberal democracy—a position that could be considered essentially colonial. There are important and emergent critiques, such as Jala Makhzoumi (2018: 31), arguing that landscape democracy is necessarily "concomitant with the call to de-link democracy from its Western association and enable bottom-up, culture and place specific discourses" in order to move towards spatial, social, and environmental justice and to meaningfully and justly enact the discourses located in Bhabha's hybridity and thirdspace.

Right to narrative and the role of language

Such meaningful and impactful participation relies on the ability to imagine and enact alternative future human–environment relationships and includes narratives and knowledge habitually excluded and suppressed as 'non-expert'. Homi Bhabha most potently conceptualized and proposed the 'right to narrative' as a key element of an individual and collective 'right to identity' (2003: 180–181)—and by extension of the right to place and landscape. These rights establish different practices of belonging and form inextricable elements of landscape citizenships—practices that can be understood by storytelling and narrative.

Language figures heavily in Bhabha's right to narrative—being able to develop and disseminate their narrative(s) in their own language is crucial for a community, group, or individual. At the same time, to be effective and assertive it needs to be a language that is understood by institutions of power. This often creates tensions and has forced marginalized communities into a choice between being inauthentic or being excluded and ignored. In the context of hybridity Bhabha has referenced the significant impact of migrants, minorities, and marginalized communities on language and identity. Bhabha's concept relates in intriguing ways to Bakhtin's (1981) theories of dialogism and heteroglossia—theories that are highly relevant to how discourses on landscape citizenships, democracy, and belonging are played out.

Dialogism recognizes the multiplicity of perspectives and voices. A dialogical work constantly engages with and is informed by other works and voices, seeking to alter or inform them. Dialogism involves the distribution of incompatible ideas and elements within different perspectives of equal value. Bakhtin critiques the view that disagreement means at least one of the positions or people must be wrong. As multiple equal points of view exist, singular meanings are replaced by a vast multitude of contesting meanings, requiring the coexistence of many incommensurable voices. The world is fundamentally irreducible to unity,[21] but not entirely immune to universally shared human experiences and values, such as pleasure, the avoidance of pain, or cardinal virtues.

> For Bakhtin, dialogism characterizes the entire social world. Authentic human life is an open-ended dialogue. The world thus merges into an open-ended, multi-voiced, dialogical whole. Its separation (as in Marxist alienation) or splitting (as in Lacanian master-signification[22]) is overcome through awareness of its dialogical character—in effect, as one big borderland. This is a world of many worlds, all equally capable of expressing themselves and conceptualizing their objects.
>
> *(Robinson, 2011: 1)*

The polyvocal perspectives of dialogism are opposed by monologism—the exclusive assertion of non-debatable meanings and a refusal of dialogue and creativity. State actions, and the hegemonial drawing of borders and boundaries are examples of monologisms. Bakhtin (1981) expands his analysis of dialogism introducing the concept of 'heteroglossia'. Heteroglossia describes the coexistence of variety within a single language. For Bakhtin this is not simply a linguistic phenomenon, but a reflection of varying ways of evaluating, conceptualizing, and experiencing the world in language, of "specific points of view on the world, forms for conceptualizing the world in words, specific world views, each characterized by its own objects, meanings and values" (ibid.: 291). He does not view language as a closed system and criticizes the creation of a unified language ('monoglossic') as a vehicle of centralized and hegemonic power that suppresses the heteroglossia of multiple everyday speech-types. Everyday speech is required to conform to official style to be effective in dealing with hegemonic power. There are interesting parallels here between de Certeau's spatial strategies and tactics, and their language equivalents. Such 'closed' and monoglossic language, frequently associated with nationalism, attempts to solidify stable relationships between words and meanings, and attempts to suppress a free and open dialogue. Bakhtin did not believe that 'monoglossical dominance' could last for long, as it would be challenged by a return of heteroglossic dialogues, much as de Certeau's spatial tactics subvert the order created by the spatial strategies of a hegemonial elite. De Certeau's *Practice of Everyday Life* (1984), as well as Bhabha's 'contentious internal liminalities' (Vidler, 1992: 10–11), and 'perplexities of living' are aspects and expressions of heteroglossic and open-ended dialogue that is central to the making of spaces and identities. Bakhtin's conceptualization of language here centres the role of language in creating identities of people and landscape, foregrounding its role in the making and unmaking, production, and reproduction of meaning and of how we see the world and ourselves in it. This is at the very heart of landscape citizenships—language mediates the different possibilities and forms for relationships between people and the landscape they inhabit. Bakhtin suggests that language-use is mediated by social ways of seeing that are always contested, in dialogue, and changing. This is echoed by Cosgrove who states that "landscape constitutes a discourse … and that this discourse is closely related epistemically and technically to ways of seeing" (1998: XIV).

Landscape itself as a heteroglossic conversation seems to provide a critical and productive perspective, inclusive and empowering, and as such should be at the front and centre of creating meaningful landscape citizenships and landscape democracy.

Stories, histories, and methodologies—a sidebar on identity, authority, and authorship

The right to narrative, to participate in a heteroglossic discourse, is indisputably necessary to form identities and to exercise the rights to landscape. As Bakhtin and Bhabha suggest, the language, the medium and process by which people make sense of themselves and the place they are living in, is critical and cannot be monoglossic. However, the very disciplinary language that is used by many geographers, architects, landscape architects, planners, historians, ethnographers, and anthropologists is closely related to a suite of methods ('ways of seeing') that emphasize a Western perspective. The canon of academic disciplines and spatial design and planning practices requires—as described by Bakhtin above—Indigenous, marginalized, non-Western people and communities to tell their stories and describe their histories and places using the hegemonic, monoglossic language—not their own—to gain admission to a dialogue about *their* culture, identity, and place. They are forced to describe and legitimize their history and existence in a language that is overtly or covertly imposed on them.[23]

'The West', as a potent hegemonial-cultural construct, is slowly starting to erode and be dismantled as non-Western histories and narratives assert themselves, often put forward by Black writers: Stuart Hall's "The West and the Rest" (1993) plays a prominent role here, deconstructing how power informs perceptions of difference between societies, and in turn how discourse forms and maintains global hegemonic power. He outlines how the 'system of representation' (ibid.: 58) it provides serves to validate the power of the Western world, and highlights how the dissemination of discourse about Western superiority and the comparative 'otherness' of the non-Western world work to maintain power hierarchies, looking in particular at colonization as a key event. Paul Gilroy's *Black Atlantic* (1993) presents a culture whose themes and techniques transcend ethnicity and nationality to produce something new and, until recently, unremarked. He outlines both a different read and a different story, subverting the (hegemonial) master–slave dialectic as the only relevant dimension in an exceedingly complex and fluid diaspora.

Stuart and Gilroy are making an argument for a different set of methodologies, to subvert and provide alternatives to the hegemonial Western-centric approaches to construct knowledge of *non-Western* cultures and subcultures, people, and places. John Langston Gwaltney similarly observes in his landmark ethnography about (his self-portrait of) Black communities:

From these narratives—these analyses of the heavens, nature and humanity—it is evident that black people are building theory on every conceivable level. An internally derived, representative impression of core black culture can serve as an anthropological link between private pain, Indigenous communal expression and the national marketplace of issues and ideas. These people not only know the troubles they've seen, but have profound insight into the meaning of those vicissitudes.

(1993: XXVI)

Sonia Saldívar-Hull expands on the need for new, non-Western methods, and methodologies, asserting that this kind of storytelling and history is instrumental for survival itself and for social transformation. She states that 'nontraditional places', books (and voices) relegated to the margins, offer stories and theories:

Because our work has been ignored by the men and women in charge of the modes of cultural production, we must be innovative in our search. Hegemony has so constructed the ideas of method and theory that often we cannot recognize anything that is different from what the dominant discourse constructs. As a consequence, we have to look in nontraditional places for our theories: in the prefaces to anthologies, in the interstices of autobiographies, in our cultural artifacts (the cuentos), and, if we are fortunate enough to have access to a good library, in the essays published in journals not widely distributed by the dominant institutions …

(2000: 46)

Saldívar-Hull's response to hegemonial modes of cultural production leads her to move away from the modes of academic and scholarly work and, instead, to turn her attention to 'the cuentos,' counter discourses that construct theory, that piece together memory and culture.[24]

Likely the most incisive and comprehensive critique of the power of Western discourse on the modes of understanding and constructing authoritative knowledge on non-Western communities and cultures is Linda Tuhiwai Smith's (2008) *Decolonizing Methodologies*. Rooted in a review on how colonization contained and positioned history, writing, land, space, and time, she frames her analysis in critical theory and examines the power and knowledge inherent in the binaries of West/other, colonizer/colonized, and oppressor/oppressed.[25] The role of Western research in this context was to collect, represent, and categorize the social, cultural, linguistic, and natural systems of Indigenous communities. Explorers and photographers exploited extant Indigenous cultures, people, and places—and told stories or histories about them, referencing them into the monoglossic language of Western-hegemonic construction of knowledge. Already established systems of order and knowing in Indigenous nations were completely dismissed by Western 'researchers', and "decades later, Indigenous peoples remember research as a process of subjugation, dehumanization, and pain" (Malsbary, 2008). The monoglossic traditions of Western research and knowledge production—which are, despite substantive critiques, still widespread and prevalent in many disciplines—amount to what Lacan (1993: 268) calls the 'master signifier'—a construct of meaning that only references itself and is removed from any discourse. Ultimately, they disempower and objectify people by telling stories about them, in a different language.

This essay has brought together discourses rooted in Latino/a, Indigenous, Atlantic African, and Afro-Caribbean studies, and queer theory. Landscape citizenships, as proposed here, may help to navigate between those and make space for all of them. Different kinds of scholarship and practices that ascertain that non-hegemonial, non-Western, minority, and marginalized perspectives can meaningfully and impactfully participate in discourses on cultural identity, place, and landscape are a necessary precondition to establish landscape citizenships.

In lieu of a conclusion: being at home in the uncanny

The thick and rich traditions of spatial and spatialized resistance, of rituals and practices that insert, assert, and make visible alternate and subaltern identities, publics, and social orders in spaces controlled by hegemonial interests[26] are promising examples of how Bhabha's contentious internal liminalities might be discovered and inhabited and form spaces of encounter with many versions of 'the Other', the 'strange'—a 'thirdspace' in Bhabha's sense. Bhabha suggests culture to be in a continual process of hybridity that produces and reproduces, or 'enables' other positions to emerge. When he talks about "the migrants, the minorities, the diasporic", proposing the city as "the space in which emergent identifications and new social movements of the people are played out" (1990: 300) he talks primarily about migrations and diasporas as from within the city and produced by a hybrid culture that constantly differentiates and diversifies, evolves, and emerges. The location and 'source' of the uncanny, unhomely, unstable, and uncomfortable is not 'from other places', but in fact from 'here'. The uncanny is always already (t)here. Ultimately, landscape and the people who inhabit it are in a constant process of becoming and becoming different in relation to each other and larger contexts.[27] The stabilizing notions of centre-periphery, or capital-border that are so fundamental to Western-colonial territorial and national identities will generate other that "displaces the histories that constitute it, and sets up new structures of authority, new political initiatives, which are inadequately understood through received wisdom" (Bhabha, 1993: 211)—both the disintegration and reassertion of national–territorial–ethnocentric constructs of self and place illustrate this.[28]

Bakhtin states that "every cultural act lives essentially on the boundaries, and derives its seriousness and significance from this fact. Separated … from these boundaries it loses ground and becomes vacuous, arrogant, degenerates and dies" (1990: 274). This contention inverts the territorial–capitalist constructions of centre and periphery. Edges, boundaries, and borders are the very location where the destabilization of essentialist order and the production and reproduction of difference occurs, where the 'other' emerges, is encountered, and cannot be avoided. In that way, they might also be the very location where a kind of 'grounding' occurs or 'takes place'.

That place is not one of consensus, homogeneity, or similarity, but one of contentious internal liminality—a place of discourse and discovery, a place that enables critical hermeneutic exchanges as a foundation for critical, contextual, relational, extant, and emergent, and a place of coexisting, contested, and contentious forms of belonging that form the basis for landscape citizenships. A place that is made by differences, but that also produces differences. Such a discursive, fluid, processual understanding of place, landscape, and identity then requires the use of the plural—the same physical location can enable, accommodate, and require multiple simultaneous alternate citizenships and belongings.

Our obsession with the production of homogeneity, of stability and an order that produces neat and exclusive categories, may be doomed to fail. Our boundaries are leaky—in the landscape and in other places. The ability to meaningfully inhabit cities, places, landscapes, to be at home in them, to belong, to be a citizen, has often invoked Lefebvre's (1996) 'right to the city' and its multiple dimensions. Far from being a static legalistic construct, this reading proposes that the right to the city (or place or landscape) involves a knitting together of subjects, materials, and traces in a multivocal, dissonant revelation of rights, characterized by 'simultaneity and encounter' (ibid.: 148). The meaningful and effective enactment of the rights to the city, to place, to landscape, and of landscape citizenships then occurs along a persisting continuity of producing and reproducing difference, of conflict, and contestation, of encountering the other, the strange and stranger, the uncanny—and inviting them in. What is necessary for that is not just a diaglossic conversation on belonging, on landscape and on being-part-of, but a series of diaglossic landscape and cultural practices. Living on edge and on the edge, inhabiting contentious liminality, a process of continuous engagement with alterity, are then required to 'make home' to belong, to be good and engaged landscape citizens, and, in Bhabha's words, to be at home in and make home for the uncanny.

Notes

1 Gloria Anzaldúa (1987) provides compelling descriptions of multiple identities inherent in a single body, and how these affect—and are affected by—the spatio-social context encountered at the US–Mexico borders.

2 Patrick Geddes (1915) suggested a matrix of 'place, work and folk' to understand the integration of people and their livelihood into the environmental givens of the particular place and region they inhabit. For the concept of lifeworld, see Husserl (1970) and Habermas and Weber. There are also intersections with Pierre Bourdieu's notion of 'habitus' and Lefebvre's and de Certeau's concept of 'everyday life'.

3 This reciprocity is complex and multidimensional. See Jacques Lacan's concept of the 'mirror stage' and Anne Spirn's (1997) framing of nature as a mirror of and for culture.

4 There is a parallel in how communities are simultaneously inclusive and exclusive constructs that act spatiotemporally, culturally, and psychosocially.

5 Territorial practice is clearly a landscape practice, and a dominant one in colonial settings and in many discourses, but it is not the only one. Bhabha notes that territory is something one is frightened off, proposing a fascinating tension between multiple and simultaneous 'senses' of landscape.

6 It might be exactly here that the notion of the margin as a productive place finds its home—also see bell hooks (1984; 1989).

7 See also de Certeau's concept of 'trajectories', forming "into unforeseeable sentences, partly unreadable paths across a space" (1984: 479–480), and its critique by Massey (2005).

8 See also Richard Sennett (2012).

9 Port cities and bordertowns are examples of this hybridity, cosmopolitanism, and openness. Such places are immensely culturally productive despite—or because of their 'underbellies'. See 'The Stranger's Path' by JB Jackson (1957).

10 Vidler (1992) writes that the uncanny was understood with "themes of anxiety and dread, provoked by a real or imagined sense of 'unhomeliness'". This seemed particularly appropriate to a moment when, as Freud noted in 1915, the entire 'homeland' of Europe, cradle and apparently secure house of Western civilization, was in the process of barbaric regression; when the territorial security that had fostered the notion of a unified culture was broken, bringing a powerful disillusionment with the European 'fatherland'.

11 For an intriguing perspective on these concepts, see Olwig (2019a) and his ideas of 'islecentrality' and 'in-continentality'.

12 Such flows of power (and the powers themselves) are frequently invisible. The flows of power might be uninhibited or operate completely independent of traditional borders and boundaries, and the attempt to reestablish and harden international borders (e.g., as with Brexit) is certainly a (futile) attempt to stem those flows.

13 In diaspora studies this is referred to as 'boundary maintenance' (Brubaker, 2005).

14 There are many examples of this, but one of the more insidious and prevalent ones is 'New Urbanism'—an approach to town and city planning and urban design that uses the mid-nineteenth-century rural small town as a model for contemporary towns and cities. In architecture, 'traditional' architecture, as a counterpoint to the internationalist–modernist style and attendant modernist–utopian agendas, has had a resurgence in many architecture programmes and practices, and associates the materials and forms of contemporary ubiquitous steel and glass high-rises and office towers with the threat of globalization and loss of national or regional identity. See also Kenneth Olwig's (2019b) discussion of reactionary modernism in his chapter "Transcendent Space: Reactionary Modernism and Diabolic Space". His framing attempts to discard a romantic/technocratic vision that underpinned ideologies such as National Socialism but makes a point to not discard many of the valuable impulses behind the framing of critical regionalism and bioregionalism. The latter are frequently infused with traces of reactionary modernism and other exclusionary myths and ideologies.

15 See Bhabha (1993), Hall (1992, 1993), and Anzaldúa (1987).

16 Sustainability and resilience are not necessarily neoliberal constructs per se, but have certainly been co-opted and deployed by neoliberalism (Kaika, 2017).

17 The right to participation includes the ability to make, to create, and influence or change one's environment.

18 See Cresswell (1996).

19 For an example of this see Langhorst (2012, 2018, 2020) and Tonnelat (2011), looking at spatial strategies, tactics, and the right to landscape in the context of the post-Hurricane Katrina recovery in New Orleans.

20 For a thorough investigation of the concepts of 'landscape democracy', see Egoz, et al. (2018b).

21 Bakhtin's dialogism forms a major difference between to dialectics and denies the possibility of transcendence of difference.

22 Which is frequently expressed in the masterplanned versions and visions of utopia that are so common in some architectural and landscape architectural practices.

23 A disturbing example is the rebuilding process in New Orleans after Hurricane Katrina. The planning process required neighbourhoods to submit proof of their viability, using the language of physical planning and excluding local, non-expert, Indigenous lived experience (Langhorst 2012, 2020).

24 See Moreno (1998, 2002).

25 See Fanon (2004 [1961]), Foucault (1995 [1975]), Memmi (1991 [1951]), Said (2003 [1978]), and Freire (1970).

26 For examples, see Langhorst (2012, 2020), Stillman and Villmoare (2010), and Regis (1999).

27

> that difference which sets the self apart from the non-self and 'us' from 'them', is no longer determined by the preordained shape of the world, nor by command from on high. It needs to be constructed, and reconstructed, and constructed once more, and reconstructed again … [t]oday's strangers are by-products, but also the means of production, in the incessant—because never conclusive—process of identity building.
>
> *(Bauman, 1997: 54)*

See also Soja and Hooper (1993).

28 Bauman refers to "two alternative but also complementary strategies" for dealing with those who do not belong in society, i.e., "assimilation–making the different similar… [and] exclusion–confining the strangers within … visible walls … expelling the strangers beyond the frontiers of the managed and manageable territory" (1993: 48).

Bibliography

Agamben, Giorgio (1998). *Homo Sacer: Sovereign Power and Bare Life.* Translated by Daniel Heller-Roazen. Stanford, CA: Stanford University Press.

Anzaldúa, Gloria (1987). *Borderlands/La Frontera: The New Mestiza.* San Francisco, CA: Aunt Lute Books.

Bauman, Zygmunt (1997). 'The making and unmaking of strangers'. In: Prina Werbner and Tariq Modood, eds. *Debating Cultural Hybridity: Multi-Cultural Identities and the Politics of Anti-Racism.* London: Zed Books, pp 46–47.

Bauman, Zygmunt (2000). *Liquid Modernity.* Cambridge, MA: Polity Press.

Bauman, Zygmunt (2003). *City of Fears, City of Hopes* (Critical Urban Studies: Occasional Papers). Available at: www.gold.ac.uk/media/documents-by-section/departments/…/city.pdf (Accessed 23 April 2018).

Bauman, Zygmunt (2006). *Liquid Fear.* Cambridge, MA: Polity Press.

Bennett, Jane (2009). *Vibrant Matter: A Political Ecology of Things.* Durham, NC: Duke University Press.

Bakhtin, Mikhail M (1981 [1934–1941]). *The Dialogic Imagination: Four Essays.* Translated by Caryl Emerson and Michael Holquist. Austin, TX: University of Texas Press.

Bakhtin, Mikhail M (1990 [1919]). *Art and Answerability: Early Philosophical Essays.* Translation by Vadim Liapunov. Austin, TX: University of Texas Press.

Bhabha, Homi K (1985). 'Signs Taken for Wonders: Questions of Ambivalence and Authority Under a Tree Outside Delhi, May 1817'. *Critical Inquiry*, 12(1), pp 144–165.

Bhabha, Homi K (1990). 'DissemiNation: Time, Narrative, and the Margins of the Modern'. In: Homi K. Bhahba, ed. *Nation and Narration.* London: Routledge, pp 291–322.

Bhabha, Homi K (1992). 'The World and the Home'. *Social Text*, 31/32, pp 141–153.

Bhabha, Homi K (1993). 'The Third Space: Interview with Homi K. Bhabha'. In: Jonathan Rutherford, ed. *Identity: Community, Culture, Difference.* London: Lawrence and Wishart, pp 207–221.

Bhabha, Homi K (1994). *The Location of Culture.* Oxford, UK: Routledge.

Bhabha, Homi K (2003). 'On Writing Rights'. In: Matthew J Gibney, ed. *Globalizing Rights.* Oxford: Oxford University Press, pp 162–182.

Brubaker, Rogers (2005). 'The "Diaspora"'. *Diaspora. Ethnic and Racial Studies*, 28(1), pp 1–19.

Casey, Edward (2011). 'Border vs. Boundary at La Frontera'. *Environment and Planning D: Society and Space*, 29, pp 384–398.

Casey, Edward (2017). *The World on Edge.* Bloomington, IN: Indiana University Press.

Castells, Manuel (1989). *The Informational City: Information, Technology, Economic Restructuring and the Urban-Regional Process.* Oxford, UK: Blackwell.

Cresswell, Tim (1996). *In Place/Out of Place: Geography, Ideology and Transgression.* Minneapolis, MN: University of Minnesota Press.

Cosgrove, Denis (1998). *Social Formation and Symbolic Landscape*, Madison, WI: University of Wisconsin Press.

de Certeau, Michel (1984). *The Practice of Everyday Life.* Los Angeles, CA: University of California Press.

Egoz, Shelley, Jala Makhzoumi, and Gloria Pungetti Gloria, eds. (2011). *The Right to Landscape: Contesting Landscape and Human Rights.* Farnham, UK: Ashgate.

Egoz, Shelley, Karsten Jørgensen, and Deni Ruggeri, eds. (2018a). 'The Case for Landscape Democracy'. In Shelley Egoz et al., eds. *Defining Landscape Democracy.* Cheltenham, UK: Edward Elgar, pp 61–70.

Egoz, Shelley, Karsten Jørgensen, and Deni Ruggeri, eds. (2018b). *Defining Landscape Democracy.* Cheltenham: Edward Elgar.

Fanon, Frantz (2004 [1961]). *The Wretched of the Earth.* Translation by Richard Philcox. New York, NY: Grove Press.

Foucault, Michel (1995 [1975]). *Discipline and Punish.* Translation by Alan Sheridan. New York, NY: Vintage Books.

Freire, Paulo (1970). *Pedagogy of the Oppressed.* Translation by Maya Bergman Ramos. London and New York, NY: Continuum.

Geddes, Patrick (1915). *Cities in Evolution: An Introduction to the Town Planning Movement and the Study of Civics.* Edinburgh and London: Williams & Norgate.

Gilroy, Paul (1993). *The Black Atlantic: Modernity and Double Consciousness.* London: Verso Books.

Gwaltney, John Langston (1993). *Drylongso. A Self Portrait of Black America.* New York, NY: New Press.

Hall, Stuart (1992). 'What Is This "Black" in Black Popular Culture?' In: Michele Wallace and Gina Dent, ed. *Black Popular Culture.* Seattle, WA: Bay Press, pp 21–36.

Hall, Stuart (1993). 'The West and the Rest: Discourse and Power'. In: Stuart Hall and Bram Gieben, eds. *Formations of Modernity.* Cambridge, UK: Polity Press, pp 275–331.

hooks, bell (1984). *Feminist Theory: From Margin to Center.* Boston: South End Press.

hooks, bell (1989). 'Choosing the Margin as a Space of Radical Openness'. *Framework: The Journal of Cinema and Media*, 36, pp.15-23.

Husserl, Edmund (1970 [1936]). *The Crisis of European Sciences and Transcendental Phenomenology: An Introduction to Phenomenological Philosophy*. Evanston, IL: Northwestern University Press.

Kaika, Maria (2017). '"Don't call me Resilient Again!" The New Urban Agenda as Immunology ... or what happens when communities refuse to be vaccinated with "smart cities" and indicators'. *Environment & Urbanization*, 29(1), pp 89–102.

Lacan, Jacques (1993). *The Seminars of Jacques Lacan, Book III, The Psychoses, 1955-1956*. New York, NY and London: W.W. Norton.

Langhorst, Joern (2012). 'Recovering Place—On the Agency of Post-Disaster Landscapes'. *Landscape Review*, 14(2), pp 48–74.

Langhorst, Joern (2018). 'Enacting Landscape Democracy: Assembling Public Open Space and Asserting the Right to the City'. In: Shelley Egoz, Karsten Jørgensen, and Deni Ruggeri, eds. *Defining Landscape Democracy: A Path to Spatial Justice*. Cheltenham, UK: Edward Elgar, pp 106–118.

Langhorst, Joern (2020). 'Decolonizing Spatial Practice: Critical Insights in the Agency of Landscape in Post-Disaster Recovery'. In: Jonathan Bean, Susannah Dickinson, and Aletheia Ida, eds. *Critical Practices in Architecture: The Unexamined*. Cambridge, UK: Cambridge Scholars Publishing, pp 45–68.

Latour, Bruno (2005). *Reassembling the Social: An Introduction to Actor-Network-Theory*. Oxford New York: Oxford University Press.

Lefebvre, Henri (1996 [1968]). 'The Right to the City'. In: *Writing on Cities*. Translation by Eleonore Kofman and Elizabeth Lebas. Cambridge, MA: Blackwell.

Linebaugh, Peter and Marcus Rediker (2000). *The Many-Headed Hydra: Sailors, Slaves, Commoners, and the Hidden History of the Revolutionary Atlantic*. Boston: Beacon Press.

Malsbary, Christine (2008). 'Review: Decolonizing Methodologies: Research and Indigenous Peoples by Tuhiwai Smith, L'. *InterActions* [Online], 4(2), n.p. Available at: escholarship. org/uc/item/65x1s5zb (Accessed 15 August 2020).

Massey, Doreen (2005). *For Space*. London: Sage Publications.

Mathur, Anuradha and Dilip Da Cunha (2001). *Mississippi Floods: Designing a Shifting Landscape*. New Haven, CT: Yale University Press.

Mbembe, Achille (2003). 'Necropolitic'. *Public Culture*, 15(1), pp 11–40.

Makhzoumi, Jala (2018). 'Landscape Architecture and the Discourse of Democracy in the Arab Middle East". In: Shelley Egoz, Karsten Jørgensen, and Deni Ruggeri, eds. *Defining Landscape Democracy: A Path to Spatial Justice*. Cheltenham, UK: Edward Elgar, pp 29–38.

Mels, Tom and Don Mitchell (2013). 'Landscape and Justice'. In: Nuala Johnson, Richard Schein, and Jamie Winders, eds. *The Wiley-Blackwell Companion to Cultural Geography*. Hoboken, NJ: Wiley-Blackwell, pp 209–224.

Memmi, Albert (1991). *The Colonizer and the Colonized*. Boston: Beacon Press.

Mitchell, Don (2007). 'Work, Struggle, Death and Geographies of Justice: The Transformation of Landscape in and beyond California's Imperial Valley'. *Landscape Research*, 32(5), pp 559–577.

Moreno, Renee Marie (1998). *Re-Membering the Body: Pain in Collective Memory and Storytelling*. Ann Arbor, MI: University of Michigan. Dissertation.

Moreno, Renee Marie (2002). '"The Politics of Location": Text as Opposition'. *College Composition and Communication*, 54(2), pp 222–242.

Mouffe, Chantal (1992). *Dimensions of Radical Democracy: Pluralism, Citizenship, Community*. London: Verso.

Olwig, Kenneth R (2002). *Landscape, Nature, and the Body Politic. From Britain's Renaissance to America's New World.* Madison, WI: The University of Wisconsin Press.

Olwig, Kenneth R (2013). 'The Law of Landscape and the Landscape of Law: The Things that Matter'. In: Peter Howard et al., eds. *The Routledge Companion to Landscape Studies.* London: Routledge, pp 253–262.

Olwig, Kenneth R (2019a). 'Are Islands Insular? A Personal View'. In: *The Meanings of Landscape.* New York, NY: Routledge, pp 88–103.

Olwig, Kenneth R (2019b). 'Transcendent Space: Reactionary Modernism, and the Diabolic Sublime'. In: *The Meanings of Landscape.* New York, NY: Routledge, pp 172–197.

Regis, Helen (1999). 'Second Lines, Minstrelsy, and the Contested Landscapes of New Orleans Afro-Creole Festivals'. *Cultural Anthropology*, 14(4), pp 472–504.

Robinson, Andrew (2011). 'Theory Bakhtin: Dialogism, Polyphony and Heteroglossia'. *Ceasefire Magazine* [Online]. Available at: ceasefiremagazine.co.uk/in-theory-bakhtin-1/ (Accessed 22 March 2020).

Said, Edward W (2003 [1978]). *Orientalism.* London: Penguin Books.

Saldívar-Hull, Sonia (2000). *Feminism on the Border: Chicana Gender Politics and Literature.* Berkeley, CA: University of California Press.

Sennett, Richard (2012). 'Reflections on the Public Realm'. In: Gary Bridge and Sophie Watson, eds. *The New Blackwell Companion to Cities.* Malden, MA: Wiley-Blackwell, pp 390–397.

Smith, Linda Tuhiwai (2008). *Decolonizing Methodologies: Research and Indigenous Peoples, 2e.* London and New York, NY: Zed Books.

Soja, Edward W (1996). *Thirdspace: Journeys to Los Angeles and Other Real-and-Imagined Places.* Oxford, UK: Basil Blackwell.

Soja, Edward and Barbara Hooper (1993). 'The Spaces That Difference Makes'. In: Michael Keith and Steve Pile, eds. *Place and the Politics of Identity.* London: Routledge, pp 183–205.

Spirn, Anne (1997). 'The Authority of Nature: Conflict in Nature and Ideology'. In: Joachim Wolschke-Bulmahn, ed. *Nature and Ideology: Natural Garden Design in the Twentieth Century.* Washington, DC: Dumbarton Oaks Research Library and Collection, pp 249–261.

Stillman, Peter G and Adelaide H Villmoore (2010). 'Democracy Despite Government: African American Parading and Democratic Theory'. *New Political Science*, 32(4), pp 485–499.

Tonnelat, Stephane (2011). 'Making Sustainability Public: The Bayou Observation Deck in the Lower Ninth Ward of New Orleans'. *Metropolitiques* [Online]. Available at: metropolitics. org/IMG/pdf/MET-Tonnelat-en.pdf (Accessed 2 July 2011).

Vidler, Anthony (1992). *The Architectural Uncanny: Essays in the Modern Unhomely.* Cambridge, MA: MIT Press.

7

BORDER CROSSING

Landscapes of *mestizaje*, citizenship, and translation

Ewa Majewska

A hybrid citizen?

The famous quotation from Virginia Woolf, "As a woman I have no country... As a woman my country is the whole world" has for a long time been the feminist declaration of antinationalism, of not belonging to a particular nation or state (2006 [1938]: 129). In *Borderlands/La Frontera: The New Mestiza*, Gloria Anzaldúa transformed Woolf's words: "As a *mestiza* I have no country... As a lesbian I have no race, my own people disclaim me. ...I am cultureless because, as a feminist, I challenge the collective cultural/religious male-derived beliefs" (1999: 102). This subject-formed-as-hybrid approaches that of Hegel's narrative in the multiplicity of contradictions—it is not merely one, in which the claim for recognition or rights clashes with the non-belonging; there are several oppositions shaping the subject in struggle on various levels, at once and in time. This article aims at reclaiming the dialectics of citizenship. It reclaims citizenship as embodied practice of lived contradictions, where the universal right becomes the lived experience of the (multiplied) borderlands, cutting through the established divisions in a transversal (Guattari, 2000), heterogeneous, yet also commonly experienced way.

Etienne Balibar (2016) rightly reminds that the notion of citizenship is a dynamic one, always already entangled in the dialectics of two clashing principles: that of inclusion—the premise of universality, and that of exceptionality—every community is formed selectively. Unfortunately, regardless of the frequent use of the word 'dialectics' in his *Citizenship*, this contradiction does not lead to any explanation of citizenship as conflicted development formed in historical process. Balibar's dialectics is stabilized in the present, and thus flattened to one dimension, while the unfolding seems arrested in the emphasized final result (for other uses of dialectics, see: Comay and Ruda, 2018). This renders it impossible to grasp its transversal evolutions, and leads to its petrification as ahistorical. A different perspective, more historically inclined, is that of Julia Kristeva, who in *Nations without Nationalism* combines the exclusivity of historical citizenship with gender, and shows how foreignness constituted a necessary, albeit sometimes unwanted, element of the national community. As she accounts the 'return to the roots' in contemporary politics, i.e., the renewed respect for ancestry and tradition, a striking connection of roots and violence is emphasized. Kristeva claims that: "the cult of origins is a hate reaction" (Kristeva, 1993). Harsh demand of maturity understood as rejection of close ties with the parents, central in Freud's psychoanalysis, meets as its counterpart the necessary appearance of an abject, of the in-betweenness of the process of subject formation, examined by Kristeva in *Powers of the Horror*. Kristeva's notion of the 'abject' should be read as that, which constitutes the border between the 'I' and the 'other'. She gives examples of bodily fluids as those peculiar entities situated between the subject and the other, thus helping establishing the borders, but she also depicts more symbolically significant markers, which are in between, as being nationless or transgendered. This exposes everyone to the culturally rejected or foreclosed in-betweenness, leading to reactions of sympathy, but also of fear (which—while its origins might be explained by Kristeva, constitutes a socially and politically dangerous phobia). The conflicted dynamics in Kristeva's understanding of cultural production is emphasized also in *Nations without Nationalism*, where she discusses Diderot's *Rameau's Nephew* and depicts how the modern subject depicted there contains its own 'strangeness', as in Hegel's depiction of culture, "encounters themselves as other" (Hegel, 1979 [1807]).

Such entanglement of identity, roots, and violence influences the composition of the landscape, where instead of clearly dividing lines between the city districts, streets, or individuals, we tend to experience borderlands in constant process of border crossing, struggle, and *mestizaje* rather than peaceful coexistence of bordering zones, which no one tries to cross. In one of the most important accounts of the post-war vision of the urban landscape, depicted by Henri Lefebvre in his *The Right to the City*, the city is discussed synthetically and with emphasis on contradictions, as *organism* (Lefebvre, 1996). While making a distinction between the old humanism, where the city is merely a *virtual* object of study, and utopia, which has to be 'studied experimentally' by means of transduction, and which seems similar to the transversality discussed later in texts of Felix Guattari as the heterogeneous practice of combining elements of contradictory systems and traditions in order for the change to be effective and far-reaching (see: Guattari, 2000 and Lefebvre, 1996). The contradictions of the city, the contradicting needs of those who inhabit the city, have to be considered in their plethora, claims Lefebvre. This makes the city a hybrid, where borders established to differentiate the class, ethnic or gendered spaces, as well as those between nature and culture, work and leisure, are constantly blurred by the lived experiences of every citizen.

The dialectics of borderlands

The very prospect of seeing all the contradictions inhabiting the borderlands is made possible by the work of Gloria Anzaldúa, expressing the lives of the *Mestizas*.

Anzaldúa's *Borderlands* opens to what had been brutally foreclosed from the expressions of cultural experience, and gives it a political articulation and significance. Her book does that based on the contradictory, hybrid experiences of the Chicanas, living on the border of US and Mexico, of Spanish/Anglo and Native traditions, of Catholicism, Protestantism, and the excluded beliefs of multiple tribes condemned for extinction by the colonial forces. If citizenship is complicated by its internal contradiction, the heterogeneity of the borderlands multiplies the differences, thus transforming the stabilized opposition between the concepts of nation/community or humanity/universality into a possibility of dialectics following several levels and transitions of these contradictions at once. In the multiplicity of conflicting her-stories, claims, experiences, embodiments, languages, dialects, and political positions, the subject inhabiting the borderlands is being born, welcomed, rejected, exploited, deified, hegemonized, abused. Anzaldúa explains:

> in fact, the borderlands are physically present wherever two or more cultures edge each other, where people of different races occupy the same territory, where under, lower, middle and upper classes touch, where the space between two individuals shrinks with intimacy.
>
> *(Anzaldúa, 1999: 19)*

The fundamental experience of the subject living at the border, being split by various labels and expectations of clarity, only adds to its essence, which is not one of 'any'. It is one of 'particular', in all the complexity of being specific, yet one of many; common, yet individualized; related, but also alone. It is almost the exact opposite of the 'whatever' (*quodlibet*) in Agamben's *The Coming Community*. In his seminal book, Agamben (1993) portrayed the coming community as one without specific characteristics, in which everyone is an 'anyone' and it does not matter any-more, whether they are French, red, or Muslim, the only important aspect is that of belonging to the community, or for being precisely what they are. Such community however could at best contemplate or enjoy its existence, but—due to the lack of distinctions and interconnections, it could not evolve or change, which is central in Kristeva's theory of subject formation or Anzaldúa's *mestizaje*—a constant process of border crossing.

Mending the broken vessel

Anzaldúa's *Mestiza* speaks many languages at once, failing at knowing any of them well. One is invited to think with her, to understand with her, to translate with her, but also to confront the impossibility of translation with her. The passages of Castilian Spanish, Chicano Spanish, Tex-Mex, and the other languages she uses alongside English signal borders and invite the reader to cross them with her—the effect of foreignness is included in the reader's experience. Just as not every trans-lation is successful, also not every word of Anzaldúa's multilingual narrative can be understood. In *The Task of the Translator*, Walter Benjamin wrote: "All translation is only a somewhat provisional way of coming to terms with the foreignness of languages" (2000 [1923]: 78). As he famously compared the practice of translation to the "mending of a broken vessel", Benjamin also reminded that the act of translation has to proceed lovingly. This emphasis on affect was also signalized by Gayatri Spivak, who in *The Politics of Translation* writes, that while not everything can be translated, and some traumas resulting of historically established forms of exploitation and injustice, such as slavery, simply cannot, the translator should proceed with love (Spivak, 1993). To Spivak this means *without assimilative claims*. To translate by no means signifies acting as if "she was just like me". Just as the mended vessel will never be the one from before the breaking, also the translated is never merely 'like' the translated. The loving gesture of the translator, as Spivak presents it, consists in recognizing difference and expressing it without assimilation. Anzaldúa's writing is a practice of translation, a constant experience of moving between languages, for the reader as well as for the writer. She declares that her practice of switching between languages constitutes her own language, 'a new language' of a Chicana who "no longer feel[s]" that she needs to 'beg entrance' (Anzaldúa, 1999: 20).

Translation should be practiced as a politics of recognition, resisting an all too easy assimilation, as a process of overcoming, even sublating the foreignness of this hybrid, monstrous language, the sense of fragmentation and lack is replaced with integrity and dignity. Norma Alarcón writes that 'la traddutora' is actually a "paradigmatic figure of Chicana feminism", which supports my interpretation of Anzaldúa's work in terms of the politics of translation (Alarcón, 1989). The painful ambivalences of *Malintzin* emphasized by Alarcón allow her to express the contradictions hidden in translation. As a translator and lover of the Spanish colonizer Cortés, *Malintzin*—the Indian woman working for the enemy—became the figure of a female traitor, although, as has been emphasized by feminist authors, her role was complex. Translation was seen as a site of utopia or otherwise inspiring social change, by many authors in different disciplines (see: Venuti, 2000). Emphasizing the importance of translation, in the 1999 edition of *Gender Trouble*, Judith Butler declares that she sees a possible form of universality as a "future-oriented labor of cultural translation" (Butler, 1999: xviii). In her later writing, Butler explicitly refers to Anzaldúa as her inspiration for understanding social change and diversity (Butler, 2004).

Translation can also appear as destruction—of traditions for example. Anzaldúa wrote of being a Mestiza as 'putting chili to the borscht' (coming from East Europe I assure you, no one would do such a thing), and other mistaken, incorrect practices, leading to the *mestizaje*, the mixing of culture, by the process of failed repetitions of the cultural codes specific to particular ethnic groups (Anzaldúa, 1999: 216). Such *mestizaje* proceeds by mistakes and failures, and thus it reminds of Jack Halberstam's inspiring argument from *The Queer Art of Failure*, where they write that the mistaken, failed repetitions of gender identities can be seen as forms of resistance to neoliberal capitalism's biopolitical regime of success and productivity (Halberstam, 2011). Halberstam's argument follows Stuart Hall and forms 'low theory', referencing the experiences of the ordinary, everyday, and common cultural practices. Halberstam's narrative shows, how in popular films and cartoons the failure to perform straight, white, powerful masculinity opens ways for weak identities to present themselves. It is an argument focusing on gender, however it can very well be applied to the multiplicity of contexts of the *Chicanas*, and thus show it as a practice of (perhaps weak) resistance to the Anglo, white, macho, straight, privileged norm.

Interestingly, I believe the *Mestiza* is not solely one who resists, she is also claiming the Chicana culture, and building it as hegemonic. In the short article *Border Arte: Nepantla, el lugar de la frontera*, Anzaldúa discusses several issues appearing in the efforts to build, establish, and institutionalize the Chicana culture, thus entering the museum as another borderland, a landscape of the cultural production of imagined community (Anzaldúa, 2009; see also: Anderson, 1983). Discussing the colonial, abusive practices of (mainly North American) museums exhibiting and selling Aztec artefacts, she embraces the critical position of resisting this appropriation, however this stance is somewhat too modest, if compared with the affirmative claims from *Borderlands/La Frontera*. Her position in *Border Arte...* is still one hesitating, whether to become mainstream, and negotiating the risks such a process might bring about, as appropriation, co-optation, etc. While these precautions sound perfectly legitimate, they foreclose the tangible possibility of *Mestiza* politics becoming not only mainstream, but also hegemonic. Such foreclosure contradicts one of the possibilities of development within the field of culture—that in which borderlands, heterogeneity and *mestizaje* actually became core positions. The sudden popularity of deconstruction, the postmodern turn, and abject art constitute examples of such hegemonic positioning, and they have to be considered as one of the historical moments of the *Mestiza* as well. From such, more holistic and historically mediated perspective, the minoritarian position of the *Mestiza* also presents itself as one of the (many possible and actual) moments in the development, or—to avoid immediately linear vision of progress—in the process of becoming of the *mestizaje*. I believe only such an expansive perspective offers an understanding the politics of the *Mestiza* in its historical, political, and cultural situatedness, thus allowing its confrontation with the issue of citizenship as well, as its different resolution.

Citizenship at the permeable borders

In closing chapter of her book, *The Rights of Others*, Seyla Benhabib claims that she does not demand that borders be entirely opened. What she wants are porous, permeable borders (Benhabib, 2015: 231). This is a 'monstrous' demand, acknowledging the ontological possibility of constant movement, thus: change, but also transit, through the borders. Such ontological openness for practiced, embodied, processual, transversal borders, rather than fixed ones, is a tough argument for legal frameworks, should they try to embrace it in their policies. Let us just imagine a legal formula of who is in the state territory if the border is permeable, and a human being is just in between two, three, or even more territories (there are places like this on world map, and as exceptional, they enlighten our understanding of the less complex zones). As we try to imagine such a situation, we are perhaps already thinking what is already present in our world. Benhabib's image of a porous border is powerful first and foremost because it unveils the already existing permeability and complexity of any border. Her idea of porous borders embraces divisions, hybridity, and distinctions, thus making her proposition one not of a utopian, but rather a heterotopic kind: passage and understanding of permeability is there, yet the existing divisions as well.

Anzaldúa tells us in detail how the border crossing is not idyllic, how it is always a struggle between conflicted normative, legal, ethnic, religious, sexual, and other orders. Her tale is thus not one of the assimilated multiculturalism, where differences have already been regulated and conflicts annihilated. The story of *mestizaje* is one of a constant struggle between various claims to recognition, systems of value, and ontologies. And it is such a hybrid subject that we need to imagine citizenship. Not one announced *via* geography, religion, or ethnicity understood as separated. We need to take into account the colonial epistemic violence (Spivak), as well as interdisciplinary resistances of identity and status struggles in order to nuance and grasp the perplexed ontologies of citizenship, its epistemology and politics in ways allowing the production of legal norms addressing the actual beings, the lives of which they are made to regulate, and the rights of which they are supposed to express. The flat recognition of my being Polish and living in Germany for example does not say much about my citizenship. The plethora of my class, gender, sexual identities, ethnic origins, and cultural position might do it a bit better, yet there are no legal frameworks that would take such a hybrid mix into consideration. For European law, as well as the legal frameworks of the two countries I come from and live in, only my nationality and current address matter. Citizenship recognizing 'the right to have rights', emphasized by Hannah Arendt, Julia Kristeva, Seyla Benhabib, and Etienne Balibar, to mention just a few authors, needs to embrace the experience and cultural construction of the *mestizaje*, otherwise it will reiterate the inadequate ontology of a stabilized, self-transparent, and coherent subject.

The *mestizaje* is not an additive practice; it should be seen as an embodied intersection, where an accident happens, involving various parts of the landscape, vehicles, and agents involved, as well as laws, customs, and policing regimes (Crenshaw, 1994). Kimberlé Crenshaw depicts intersectionality as follows:

> My objective there was to illustrate that many of the experiences Black women face are not subsumed within the traditional boundaries of race or gender discrimination as these boundaries are currently understood, and that the intersection of racism and sexism factors into Black women's lives in ways that cannot be captured wholly by looking at the women race or gender dimensions of those experiences separately. I build on those observations here by exploring the various ways in which race and gender intersect in shaping structural and political aspects of violence against women of color.
>
> *(ibid.: 1244)*

The intersectional paradigm already obliges to take into consideration the two aspects of life, while the *mestizaje* is a multilayered hybrid of a multiplicity of elements, including embodiment, *habitus*, and socialization that are at play.

Perhaps we should begin this search from a perspective on the excluded, which—while announcing an important and unfortunately still possible position, logically and historically, of those stripped of all rights, whose lives become 'bare', the *Homo Sacer*. This influential concept was discussed in the last decades by Giorgio Agamben. It depicts those whose legal non-belonging lead to experiences of state violence that would otherwise be impossible. Agamben's image of the *Homo Sacer*, who "may be killed, yet not sacrificed", who is in a position of exclusion within the political state of exception, seems particularly useful for discussions of the situation of so-called illegal immigrants in Europe, especially the ones enclosed in deportation camps, where refugees face indefinite detention and a lack of regulations, resulting in the direct application of the sovereign's powers (Agamben, 1998: 8). Agamben's thinking critically addresses the logics of 'the state of exception' imposed in Carl Schmitt's concept of absolute sovereignty, central for the fascist doctrine of the Third Reich. The exceptional sovereign (one forming the law, yet always excluded from its limitations) has as its counterpart, the 'enemy'—the group stripped of their rights, recognition, protections, property to an extent of becoming a *Homo Sacer*, and then eliminated. Jews were one of the groups to be eliminated in such way, and the fascists of early twentieth century pushed this process to the extreme. However today—Agamben claims—other groups, particularly those of undocumented migrants, occupy this position and become such 'enemies' of national states (for further development of the state of exception see *Necropolitics* (Mbembe, 2019)).

The accuracy of such description of the *sans-papiers* of today's capitalism has been contested on many grounds. Most importantly, it seems impossible to compare the most disempowered prisoners of Auschwitz, with those who cross borders, work, or are enslaved, but still act against their oppression by means of active resistance or multiple forms of persistence, survival, or escape. These are incomparable situations, and the reductive effort to situate them under the common denominator of *Homo Sacer* meets legitimate controversies. Certain critics of Agamben, such as Judith Butler and Gayatri Spivak, claim that by giving too much credit to the state and its institutions, he feeds the illusion of the state being the unique distributor of political agency (Butler and Spivak, 2007). In their book, *Who Sings the Nation State?,* they argue for acknowledging the agency of the oppressed. I agree with them, and I believe, that while the logic of the exceptional sovereignty might indeed be that of treating all disempowered as today's *Homo Sacer*, there is in the practice of resilience, resistance, and survival a multiplicity of forms of agency. These practices—while not spectacular or 'strong'—fulfil the definition of 'weak resistance' I try to build elsewhere, inspired by James C. Scott's *Weapons of the Weak* (Majewska, 2018; Scott, 1985). From the perspective of the oppressed, which might not be accessible to the theorizing subject (but this does not preclude its existence), it might however undermine theoretical claims to legitimacy. Butler and Spivak begin their analysis with the image of anti-war protesters singing the US-American national anthem in Spanish, and being criticized for that by (then) President George W. Bush. This image of singing a song as a practice of resistance, but also—consolation and creating a territory—resonates with Gilles Deleuze and Felix Guattari's image of the child singing in the dark to overcome fear in *A Thousand Plateaus* (Deleuze and Guattari, 1980 [1972]: 382). This kind of agency—one of consolation and resistance—differs from the usual image of political agency, which most often combines bravery, heroism, and strength. Here it is weak agency, weak resistance, mainly aimed at persisting, resiliently, in unwelcoming conditions.

Agamben's analysis not only neglects the political agency of the oppressed, it does not encompass the gender, sexuality, or class differences marking the victims of the oppressive state's acts, nor the micropolitics of racist individuals acting out their biases. Preoccupied with the agency of the (absolute) sovereign, it maintains the invisibility of the unheroic resistance, the micro-scale landscapes of politics orchestrated by other means than those fitting the heroic political narratives and expression. Anzaldúa's discussion of the borderlands can be read as a reminder of the limits of such perspective, it also enhances a change—from the macro-scale, down to small-scale landscapes, or even bodies. A popular Chicano/a slogan reads: "it was not me who crossed the border, the border crossed me". Thus an idea of borderland happening in the small scale of one's own body changes the perspective, allowing localizing micro-practices, both of resistance and of subjugation or oppression, invisible from the so much criticized 'Archimedean point' (see: Jaggar, 1988: 357). Anzaldúa wrote in *Borderlands*: "This is my home/this thin edge of/ barbwire" (Anzaldúa, 1999: 25).

In *Nations without Nationalism* Kristeva (1993) offers the definition of the metic based in the ancient Greek experience. She writes: "[the] metic is a resident foreigner, *Homo oeconomicus*, the status of one who agreed to contribute to the Greek economy, but lived in the suburbs and had no succession rights" (ibid.: 19). This could also be a depiction of the situation of many contemporary refugees, perhaps with some transformation of the inheritance rights, and exclusion of political rights in majority of countries. This is not a condition of complete deprivation of any political agency, however in order to grasp such agency of often exploited and severely discriminated immigrants, we need to cross the borders of routine political depictions of agency, which focus on deliberative, violent or otherwise 'active' and 'visible' expressions. The 'weapons of the weak' depicted by Scott in his study of South-East peasants movements allow imagining speechless, non-violent movements, which nevertheless articulate claims, struggle, and sometimes also achieve the realization of their demands.

Liberating borderlands

Anzaldúa shows that, and how, the process of border crossing can be emancipatory. She sees binary oppositions as existing parts of culture, which nevertheless can be overcome, yet—just as in GFW Hegel's dialectics—it is not only the immediate result (the overcoming of contradictions) but also the very process of sublating (*aufhebung*) that counts as productive for an individual's herstory. When she writes: "Who, me, confused? Ambivalent? Not so. Only your labels split me", it is clear that she expresses the fatigue of being labelled. Yet it is also the capacity of naming the cultural and social contradictions that make her and the subjectivity of the mestiza so particular (Anzaldúa, 2009: 46). Here what could be called the *loqueria* of a lesbian born Catholic and Chicana, a feminist attached to la Virgen de Guadalupe, is seen not as a disruptive set of oppositions but as a part of socialized individual development that can be integrated and work in an empowering way for a rebellious subjectivity. Some queer couples living in Poland try to stay Catholic, and they pay the price of exclusion not only from the church but also the labour market and family networks, not to mention in progressive queer circles, which are usually opposed to religion. They are facing borderlands of their own, and for some of them, Anzaldúa's texts sound like a healing narrative, not just a good theory.

Central European borderlands

My immediate association of the borderlands between the US and Mexico with the borderlands between Poland, Ukraine, and Belarus, and my effort to comparatively analyse them, might seem quite surprising, yet it was the first thought I had after reading Anzaldúa. For so many Belarusians, Ukrainians, and Russians, the experience of the Polish border, which became the border of the European Union (often critically described as 'Fortress Europe') in 2004, begins with the border guards' distinction between Poles, who are politely interpellated using 'Mr.' or 'Ms.' (*pan* or *pani*), and others, referred to disdainfully as 'you' (*ty*). This distinction comes with the accent—those coming from Eastern Europe have a milder pronunciation and sometimes also accent—and this is the border as well, one that the refugee carries with them. After the recent racist declarations from Polish conservative politicians, this border is particularly important, and people coming to Poland from Ukraine and other countries simply avoid speaking aloud in order to avoid discrimination. Although they usually speak Polish, in shops, taxis, and other places they often merely speak only the most indispensable words, and only after a friendly reaction will they speak freely. This need for a dissimulation of one's identity as foreigner clearly shows to what extend at least some refugees do feel threatened. Then another borderland is the workplace, where someone with Polish citizenship will get a contract and adequate salary, and the friendly approach of co-workers or clients, while the refugee or immigrant might not. The recent cases of death of exhaustion because of overwork are the ultimate limits of borderland. Access to medical assistance or its denial—in the time of the COVID-19 pandemic this was particularly obvious—also builds borders, as does the need for visa. In city landscapes, the borders are also particularly tangible in particular lines of public transport. In Warsaw the most spectacular border is the Poniatowski Bridge, connecting the city centre with Praga, the district with cheaper flats, where many refugees live. The ticket control becomes for some of them a version of border control, as without a valid ticket, they have to produce identity documents, which some of them do not have. In some cases this means immediate lock-down in deportation prisons, sometimes for months or years; another border landscape, where citizenship is neg(oti)ated.

The borderlands, as Anzaldúa describes them, are often spaces of economic and sexual abuse. They are the regions of *maquiladoras*, factories where the exploitation of workers reminds one of Karl Marx's descriptions of the nineteenth-century factories of early capitalism. In the 1990s Poland's eastern border became famous as a site for trafficking human beings, especially women, into prostitution, and later for the nearby deportation camps. This trafficking meant that it was in private homes, gardens, or brothels in Poland where people from the former Soviet Union, Romania, Moldova, and other countries were and are exploited and abused. Initially, when I decided to write about the mestiza from Ukraine, I thought of finding feminist writers who would express some sort of Eastern European version of *mestizaje*, which was clearly an effort typical of what Spivak once called the 'ethical' approach to translation—"the good-willing attitude 'she is just like me'" as opposed to the 'more erotic' surrender in translation (Spivak, 1993: 183). At first, it seemed obvious that the only imaginable way to accept Anzaldúa's invitation would be to provide a similar narrative—but that might actually be a misunderstanding. Is there only one version of joining the borderlands? I believe Anzaldúa would disagree. Spivak questions the ability of the colonizers to hear the subaltern's speech, and Anzaldúa provides a form of speech that in my opinion overcomes the limits of submission: it is a narrative of border crossers, of the subordinated yet resistant, of the muted yet speaking other. Spivak's critical reading of the subaltern prevents from romanticizing the borderlands as ultimately powerful, and thus painlessly political, however Anzaldúa shows how a decolonial subjectivity can actually speak. In *Feminism without Borders* (2003), Chandra Talpade Mohanty claims: "the work of *mestiza* consciousness is to break down the subject–object duality that keeps her a prisoner and to show in the flesh and through the images in her work how duality is transcended" (ibid.: 80–81). The problem of such synthetic reading of Anzaldúa's perspective, signalized already in the title of Mohanty's book, is that it might be conveniently obliterating contradictions, as well as borders, which are perhaps contradicted or even sublated in Anzaldúa's work, but never unproblematically. The Chicana lives on the border, the border crosses her, and although she overcomes the degradation, delegalization, or oppression caused by such politics of location, it is not a "feminism without borders", it is a feminism of overcoming the borders and of the process of transformation, not annihilation, of contradictions.

Borderland monsters

During the implementation of neoliberal capitalism in the 1990s, after 50 years of so-called communism, Poland was flooded with the belief that it was coming back to Europe where it "had always belonged". This paradox of coming back to where one always belonged summarizes the contradictions at the core of any country in Central Europe. There is, however, a set of theoretical tools developed by Warsaw historians such as Witold Kula and Marian Małowist, later elaborated by Immanuel Wallerstein, Tadeusz Kowalik, and Przemek Wielgosz, relating to the centre/periphery dichotomy (Kula, 2009 [1983]; Małowist, 1993). This framework situates Poland in a peripheral position while at the same time deconstructing its image as situated at the centre (Wielgosz, 2007: 198). Wallerstein's notion of 'semi-periphery', created in the 1970s and applied to the countries of East Europe, large parts of Asia and South America, as well as some countries in Africa, explores their in-betweenness. This notion, rooted in earlier comparisons of the peripheral countries offered by Małowist and Kula, resembles Anzaldúa's concept of *Mestiza* in multiple ways, as well as the notion of 'monsters', used by Antonio Gramsci to depict the 'times of interregnum', where the new is not yet here, and the past is already over, thus whatever forms of agency and culture appear at such moment, they seem 'monstrous' mostly because there is no given context to situate them in. Antonio Negri and Michael Hardt also use the reference to monsters to discuss the crisis of neoliberal capitalism and the 'lines of flight' from it, thus discussing the Caliban and how his disagreements with the enlightened Prospero (in Shakespeare's *The Tempest*) constitute the possible source of alternative modernity to come (Negri and Hardt, 2009). Silvia Federici pushed this argument further, discussing the exclusion of women as the origin of modernity, and thus negotiating not only 'Caliban', but also 'Sycorax' in *Caliban and the Witch* (Federici, 2004). Following the Polish historian, Bronisław Geremek, as well as Michel Foucault, she shows the gendered nature of the early modern landscape, suddenly populated by such institutions, as prisons, schools, and hospitals, but above all—keeping women in the 'privacy' of the household.

Monstrosity seems like the main condition of the life of semi-periphery. Just as Caliban is 'almost human', 'almost obedient', and 'almost cultivated', the semi-peripheries constantly claim access to the core countries, and constantly are lured or seduced towards yet another transformation, sacrifice, or effort in order to belong (where they supposedly always belonged). This issue of almost-not yet play resonates with the experience of *Mestiza*, who could be depicted as always already 'white and non-white', 'included and excluded', etc.—the difference being, that she does not wait for inclusion, she knows she would be hanging at the border forever. Those who ever applied for a citizenship usually encounter a similar atmosphere of almost/not yet as the semi-peripheries, and those with strong distinctive characteristics marking their difference from the majority of the inhabitants of the country they try to settle, even after obtaining the desired citizen status, will always be in the condition of *Mestiza*—both in and out at the same time, with small chance to fully enjoy their rights. Wallerstein claims that the semi-periphery signalizes a double function—of being colonized and colonizing; dependent and forcing others into dependency. He suggests that the semi-peripheries act like peripheries to the core countries and like core countries to the peripheral ones; they are also capable of taking more advantage from global economic crisis than the two other kinds of states. He also explains how the semi-peripheral countries successful in transforming the economic crisis into their gain, need to appropriate the surpluses of other semi-peripheries to their advantage.

Borderlands archived?

Anzaldúa's writing might also work as an instructive strategy for rereading the past, an effort that is, as Jacques Derrida reminds us in 'Archive Fever', always directed at the future (Derrida, 1995: 17). The archive of citizenship—our knowledges of how varied, contradictory, and sometimes also emancipatory past societies have been when it comes to citizenship—should be practiced with caution. Historical borderlands should not be 'assimilated' or 'mended' in ways resolving persisting contradictions, dissimulating inequalities, and protecting perpetrators. A critical perspective should be applied, serving not merely to monumentalize the most repressive past customs, but also to learn, discuss, and inspire the more diversified and open borderland policies and cultures. In the last years Poland became an unfortunate example of a radically conservative historical politics focused on the martyrological heroism and victimhood. In her letter to the Congress of Polish Culture in 2016, Maria Janion, a prominent scholar and one of the earliest Polish feminists, demanded rejection of the 'heroic messianism' and an embrace of the more quotidian, everyday, weak strategies of building society (Janion, 2016).

In the article *Secrecy, Archives and the Public Interest*, Howard Zinn offers a critical analysis of the stabilizing powers of the historical policies, selecting the archives according to the social and political elite's interest, and thus favouring male narrative over those of women, acknowledging the existence of written testimonies and excluding oral and other ones, choosing normative narratives over exceptional ones, etc. (Zinn, 1977). In the practice of archive, claims Zinn, hides the power of establishing and maintaining inequalities. Such practice of selection influences the processes of visibility, maintaining certain social practices, namely those contradicting, undermining, or shifting from the norm in shadow of the hegemonic models. This practice of the archives, particularly in case of the most powerful, state-run archives, translates as preserving and legitimizing historical or even imaginary information about the social. In post-war Poland such practice consisted mainly in erasing the *mestizaje* of the pre-war population. Such 'symbolic cleansing' and unification of what was constituted as the 'Polish nation' served several, sometimes contradicting, imperatives, and it was in many ways contested after 1989 and the transition to liberal democracy. One of the major changes of the 1990s and subsequent years was a series of 'discoveries' of the mixed, foreign, or simply unexpected origins, which appeared in multiple families, questioning the homogenous image of Poland's population. This discovery of heterogeneity was combined with the sudden shock of opened borders, the arrival of newcomers as well as the migration *from* Poland. All these processes diversified the Polish society, and demanded new translations of individual as well as collective identities, suddenly seen as hybrid. Since several years however that process of discovering the *mestizaje* meets strong resistance in narratives reinstalling homogeneity at the core of Polish identity. This process clashes with the liberal discourse of 'tolerance', undermined by such critics, as Zygmunt Bauman, himself a Jewish-Polish immigrant to Israel and the UK. Acknowledging the contradictions of social functioning of ethnic, sexual, class, and gendered differences, he contradicted 'tolerance' with 'acceptance' in the context of globalizing processes, and convincingly argued that the former will be insufficient to tame emerging social antagonisms resulting of cultural, economic, and political changes, leading to a sense of insecurity and collapse of the value systems (Bauman, 1998). Without the experience of *mestizaje* and the process of translation, which follows its dynamics, such acceptance as that demanded by Bauman might remain merely a wish at best, and ideological smoke screen at worse. Such experience of in-betweenness, of the impossibility of clean origins and straight story, is what, as Kristeva argued in *Strangers to ourselves* and as Anzaldúa demonstrated in her theoretical and poetic work, makes borders inhabitable. And as borders tend to cross us even more often than we cross them in times of neoliberal immediacy, exploitative capitalism, and globalized mobility, the lesson of *La Mestiza* can be valuable not just for individuals or oppressed groups searching for expression of their claims, but also to the institutions and political powers responsible for transforming the laws and policies according to historically transforming needs.

Acknowledgements

This article expands some topics of my earlier work, see: Majewska (2018). I would like to thank Tim Waterman for many fantastic suggestions for this article; Paola Bacchetta, for introducing me to the work of Anzaldúa; and countless people who make borderlands liveable.

Bibliography

Agamben, Giorgio (1993). *The Coming Community*. Minneapolis, MN: University of Minnesota Press.
Agamben, Giorgio (1998). *Homo Sacer: Sovereign Power and Bare Life*. Translated by Daniel Heller-Roazen. Stanford, CA: Stanford University Press.
Alarcon, Norma (1989). 'Traddutora, Traditora: A Paradigmatic Figure of Chicana Feminism'. *Cultural Critique*, 13(Autumn), pp 57–87.
Anderson, Benedict (1983). *Imagined Communities: Reflections on the Origin and Spread of Nationalism*. London: Verso.
Anzaldúa, Gloria (1999). *Borderlands: la frontera: the new Mestiza*. San Francisco, CA: Aunt Lute Books.
Anzaldúa, Gloria (2009). 'Border Arte. Nepantla, el Lugar de la Frontera'. In: AnaLouise Keating, ed. *The Gloria Anzaldúa Reader*. Durham, NC: Duke University Press.
Balibar, Etienne (2016). *Citizenship*. Translated by Thomas Scott-Railton. Cambridge, UK: Polity Press.
Bauman, Zygmunt (1998). *Globalization: The Human Consequences*. New York, NY: Columbia University Press.

Benhabib, Seyla (2015). *Prawa innych. Przybysze, rezydenci i obywatele.* Warsaw, PL: Wydawnictwo Krytyki Politycznej.

Benjamin, Walter (2000 [1923]). 'The Task of the Translator. An Introduction to the Translation of Baudelaire's Tableaux Parisiens'. Translated by Harry Zohn. *The Translation Studies Reader.* London: Routledge, pp 75–82.

Butler, Judith (1999). *Gender Trouble: Feminism and the Subversion of Identity.* London: Routledge.

Butler, Judith (2004). *Undoing Gender.* London: Routledge, pp 204–231.

Butler, Judith and Gayatri Chakravorty Spivak (2007). *Who Sings the Nation-State? Language, Politics, Belonging.* London: Seagull.

Comay, Rebecca and Frank Ruda (2018). *The Dash—The Other Side of Absolute Knowing.* Cambridge, MA: MIT Press.

Crenshaw, Kimberlé Williams (1994). 'Intersectionality, Identity Politics, and Violence Against Women of Color'. In: Martha Fineman and Roxanne Mykitiuk, eds. *The Public Nature of Private Violence.* New York, NY: Routledge.

Deleuze, Gilles and Felix Guattari (1980 [1972]). *Mille Plateaux.* Paris: Les Editions du Minuit.

Derrida, Jacques (1995). Archive Fever: A Freudian Impression. *Diacritics*, 25(2), pp 9–63.

Federici, Sylvia (2004). *Caliban and the Witch: Women, the Body and Primitive Accumulation.* Brooklyn, NY: Autonomedia.

Forgacs, David, ed. (2000). *Antonio Gramsci Reader. Selected Works 1916-1935.* New York, NY: NYU Press.

Guattari, Felix (2000). *The Three Ecologies.* London: The Athlone Press.

Halberstam, Judith J (2011). *The Queer Art of Failure.* Durham, NC: Duke University Press.

Hegel, Georg WF (1979 [1807]). *Phenomenology of Spirit.* Translated by AV Miller. Oxford, UK: Oxford University Press.

Jaggar, Alison M (1988). *Feminist Politics and Human Nature.* Totowa, NJ: Rowman & Alanheld.

Janion, Maria (2016). Letter to the Congress of Polish Culture. Available from: tvn24.pl/polska/list-prof-marii-janion-na-kongresie-kultury-2016-ra682632-3196012 (Accessed 19 May 2020).

Kristeva, Julia (1982). *The Powers of Horror. An Essay on Abjection.* Translated by Leon S. Roudiez. New York, NY: Columbia University Press.

Kristeva, Julia (1993). *Nations without Nationalism.* New York, NY: Columbia University Press.

Kula, Witold (2009 [1983]). *Teoria econoʹmica del sistema feudal.* Translated by Guillem Calaforra. Valencia, ES: Universitatde Valencia.

Lefebvre, Henri (1996 [1966]). 'The Right to the City'. In: Eleonor Kofman and Elisabeth Lebas, eds. *Writings on Cities.* Cambridge, MA: Blackwell Publishers.

Majewska, Ewa (2011). 'La Mestiza from Ukraine? Border Crossing with Gloria Anzaldúa'. *Signs: Journal of Women in Culture and Society*, 37(1), pp 34–41.

Majewska, Ewa (2018). 'Weak Resistance'. *Krisis: Journal for Contemporary Philosophy* [Online]. Available at: krisis.eu/weak-resistance/ (Accessed 15 August 2020).

Małowist, Marian (1993). *Europa i jej ekspansja w XIV–XVII wieku.* Warsaw, PL: Wydawnictwo Naukowe PWN.

Mbembe, Achille (2019). *Necropolitics.* Durham, NC: Duke University Press.

Mohanty, Chandra T (2003). *Feminism without Borders: Decolonizing Theory, Practicing Solidarity.* Durham, NC: Duke University Press.

Negri, Antonio and Michael Hardt (2009). *The Commonwealth.* Cambridge, MA: Harvard University Press.

Rancière, Jaques (2003). *Philosopher and His Poor.* Durham, NC: Duke University Press.

Scott, James C (1985). *Weapons of the Weak: Everyday Forms of Peasant Resistance*. New Haven, CT and London: Yale University Press.

Spivak, Gayatri (1993). *Outside/In the Teaching Machine*. London: Routledge.

Venuti, Lawrence, ed. (2000). *The Translation Studies Reader*. London: Routledge.

Wielgosz, Przemysław (2007). Od zacofania i spowrotem. Wprowadzenie do ekonomii politycznej peryferyjnego miasta przemysłowego. In: Martin Kaltwasser, Ewa Majewska, and Kuba Szreder, eds. *Futuryzm miast przemysłowych*. Krakow, PL: Korporacja ha! Art, pp 241–253.

Wallerstein, Immanuel (1976). 'Semi-Peripheral Countries and the Contemporary World Crisis'. *Theory and Society*, 3(4), pp 461-483.

Wallerstein, Immanuel (2004). *World-Systems Analysis: An Introduction*. Durham, NC and London: Duke University Press.

Woolf, Virginia (2006 [1938]) *Three Guineas*. Orlando, FL: Harcourt.

Zinn, Howard (1977). 'Secrecy, Archives and the Public Interest'. *The Midwestern Archivist*, II, pp 14–26.

8

SPATIAL INEQUALITIES AND MARGINALIZATION

Displaced Syrians in the Bekaa Valley, Lebanon

Maria Gabriella Trovato

Introduction

Over the past ten years, millions of people have been forced to flee their homes as a consequence of global issues like wars, conflicts, and climate change. These issues are contributing to the risk of large-scale and complex crises that will increase the suffering of vulnerable populations (Churruca-Muguruza, 2018). By the end of 2019, 79.5 million people around the world had been forcibly displaced; most of them stayed inside their own countries or crossed the border to neighbouring countries. Developing countries host 85% of the world's refugees and formal and informal encampment strategies are proliferating (UNHCR, 2019). Since the work of Giorgio Agamben (2002), some scholars have considered the camp as the political paradigm of our time.

The camp is often described as a limbo, void at its core and in constant need of filling with human material or 'biological substance' (Giaccaria and Minca, 2011, 2015; Agamben, 2002). For the most part, the makeshift spatial qualities of camps have been overlooked (Davies et al., 2019). This essay, a report on the Syrian refugee crisis in Lebanon, examines the informal or 'spontaneous' housing that mediates between the need for shelter and Lebanon's politics of non-intervention. It touches on the spatial inequality that confronts a displaced community searching for refuge in a neighbouring country. It also depicts the marginality that displaced people experience in informal settings, whether urban or rural, and highlights differences from formal camps in neighbouring countries like Turkey and Jordan.

Since the start of the war in Syria in 2011, over 6 million Syrians have been displaced within their own country, and 5.6 million more have sought asylum in bordering regions (UNHCR, 2020a). Lebanon hosts the highest number of displaced persons per capita in the world, including about a million and a half Syrians and a large community of Palestinian refugees (UN, Government of

Lebanon, 2019). Today refugees are found in more than 1,700 localities, often in communities that are already among the country's poorest. Proximity and an easy border crossing have led many Syrians to settle in the rural Bekaa Valley, the major agricultural plain of Lebanon. Informal gatherings and self-settlements have appeared in the agricultural fields around small and large villages and shaped a landscape of removable structures.

We conducted this research in the villages of Bar Elias and Saadnayel. Mapping exercises and extensive photography allowed us to delve into the everyday life and subjective experience of space across the rural Lebanon's disparate settlement types. While there are many insightful studies on Syrian refugees (Sanyal, 2017), refugee crisis and planning (Fawaz, 2017), and displacement as urbanism (Sukkar et al., 2020), the morphologies of informal encampment in rural areas and at the micro-spatial scale remain under-researched. To address the role that 'place' plays in migratory conditions, my concerns focus on borderscape practices and spatial organization in informal tented settlements (ITSs). The first aim is to depict the multiple ways individuals form relationships with and assign meaning to place; the second is to investigate the relationships among justice, place, and environment by looking at the use and control of space through bordering practices in everyday life.

Citizenship and right to landscape

The contemporary experience of citizenship is rhizomatic and overlapping. Though citizenship has historically been associated with the state, top-down ideas decontextualize and neglect social factors and forces (Goh, 2018). For people and communities on the move, citizenship concerns global efforts to promote democratic values and practices (Willis, 2016). It is rooted in social and economic rights grounded in the material and ecological landscape of the city and countryside (Hammett, 2017). Though access to citizenship of the host country is seen as key to the integration of refugees, several countries lack the legal framework to guide that process. A refugee remains "a human life without citizenship possessing only biological life and not political qualified life" (Agamben, 1998). Excluded from political status and "the normal identities and ordered spaces of the sovereign state", refugees are subject to a separate international humanitarian regime that manages their lives (Nyers, 2006: xiii–xv).

Syrians in Lebanon are non-citizens: 'displaced' people but not 'refugees' (Government of Lebanon and UNHCR, 2020a). By its negation of formal refugee or residency status, the Lebanese government has deprived Syrians of their 'right to have rights' and rendered them dependent on humanitarian aid and informal livelihoods (Fleifel, 2020; Stel, 2020). The 'displaced' are a vulnerable group denied access to housing, land, and property rights fundamental for rebuilding the lives of uprooted people. They appropriate spaces and deploy spatial and citizenship practices to navigate their position (Hammett, 2017). In Lebanon, the unequal allocation of citizen rights to goods and services is tied to political clientelist networks that extend from formalized state institutions to unofficial 'street politics'. As a result, displaced Syrians' access to goods and services relies heavily on informal networks. Since the 2014 adoption of the 'Policy Paper on Syrian Displacement', the Lebanese legal framework criminalizes the presence and labour of Syrians. Exclusionary practices like curfews, restricted mobility, and eviction ensure that marginalized Syrians become spatially invisible non-citizens (Painter and Philo, 1995).

This essay's notion of citizenship is tied to territorial behaviours and the ability to adapt to new and uncertain conditions. It addresses citizenship not as an abstract set of rights but as the inherent human need to live in a healthy and equitable environment. Scholars and intellectuals share the idea that there is a fundamental right to landscapes, and this right is often accompanied by a request for justice and political equality (Menatti, 2017). In humane societies, these principles should apply to forcibly displaced migrants, refugees, and vulnerable communities. However, 57% of displaced Syrian families live in overcrowded shelters that are below humanitarian standards (VASyR, 2019).

Spatial inequality and marginalization

Marginality is a condition of disadvantage for individuals and communities because of vulnerabilities arising from unequal or inequitable environmental, ethnic, cultural, social, political, and economic factors (Mehretu et al., 2000). Immigration status is one of the most insidious factors of vulnerability for marginalization (and exploitation). Marginalization is a socio-spatial process and a metaphor that draws attention to the geographical aspects of exclusion (Trudeau et al., 2011). In refugee camps, the world's most marginalized people are warehoused and 'managed' through ostensibly humanitarian structures of care and control. Refugees are reduced to meagre lives awaiting release from a space that allows survival but not political voice or community (Minca, 2015; Hyndman, 2000).

In Lebanon, the no-camp policy led to a privatized informal system to house refugees. In the absence of proper and fair governance and the presence of oppressive policies and practices, Syrians facing exclusion and marginalization depend on informal mechanisms to ensure their basic needs (Fleifel, 2020). Their daily lives are increasingly difficult given social, legal, and infrastructural discrimination against their presence and their rights to health, employment, education, safety, and mobility (Geha and Talhouk, 2018). These challenges impede their ability to take collective action related to security and everyday life (Boustani et al., 2016b).

Nationality, class, religion, and lack of recognition of refugee status combine to make Syrians marginal in Lebanon (Trudeau et al., 2011). The country's sectarian power-sharing system, known as the confessional system, politicizes the presence of Sunni Syrian refugees. Government is characterized by the proportional distribution of political power among different religious communities according to their demographic weight and geographic distribution. Different groups' representation in the government is guaranteed along confessional lines (Saliba, 2010). The major confessional powers perceive displaced Sunni Syrians as security, economic, and existential threats (Geha and Talhouk, 2018). A 2014 policy aimed at decreasing the number of Syrians in Lebanon had devastating effects on the well-being of refugees. As a result, a majority live illegally; mobility is limited, access to basic services and justice are contested (Fakhoury, 2017), and living conditions are harsh. Syrians are organized in relation to the local structures and networks of Lebanese society put on its margins.

The ability to absorb vast numbers of displaced persons rests on the flexibility and responsiveness of informal housing markets that react to spikes in demand (Fawaz, 2017). Informal land and housing markets operate as dynamic venues of property exchange that respond well to the needs and circumstances of many low-income city dwellers (Fawaz, 2009; Berner, 2001; Keivani and Werna, 2001; Fekade, 2000). Building on social capital has become a viable, innovative, and more just alternative for the marginalized to move forward (Sukkar et al., 2020).

Informal agglomerations of Syrians parallel existing Lebanese patterns in cities. Similar channels and mechanisms inform rural housing markets. As Sanyal points out:

> In rural areas, many live in informal tented settlements on private agricultural land where landlords provide land and housing for refugees to live in… Private agreements between landlords and refugees mean that the latter have less stability and protection in their everyday lives. Landlords, for example, can evict refugees as they see fit or impose costs and sanctions on them.
>
> *(2017: 102–103)*

The threats of eviction and deportation mean that Syrians living in ITSs remain insecure, marginal, and confined. The environment of marginalized populations also acts as a space for adaptive processes which seek to make proper use of the resources available. The process of spatial appropriation (Tapie, 2014) also involves the alteration of initial conditions and the creation of a habitable situation linked to the possibilities offered by the surroundings and the lifestyles and livelihoods of the groups (Mela and Toldo, 2019). Examples exist in both urban and rural informal settings.

The borderscape condition of displaced Syrian communities in Lebanon

Migration and borders are central issues in academic debates concerning the ideas of state and nation. In recent years, they have also come to dominate public debates about identity and justice in the era of globalization. In the late 1980s and early 1990s, a disciplinary shift in border studies from the concept of *border* to that of *bordering* allowed borders to be viewed as socially dynamic processes and practices of spatial differentiation (Newman, 2006; van Houtum and van Naerssen, 2002; Paasi, 1998). Border studies moved its focus from territorial dividing lines and political structures to socio-cultural and discursive processes and practices (Brambilla, 2015). Social and cultural geographers have begun to use boundary analysis to study the production and experience of exclusion in landscapes (Trudeau et al., 2011). The ways borders are produced, maintained, and rescaled (and the practices of governing place through borders) are central to understanding contemporary migration and refuge (Porter et al., 2019).

The idea of borderscapes was developed by academics and activists concerned with migration issues, border struggles, post-colonial studies, and critical border studies. In *Borderscapes: Hidden Geographies and Politics at Territory's Edge*, Rajaram and Grundy-Warr focus on creative border practices that invade and permeate everyday sites. They use the term borderscape as "an entry point, allowing for a study of practices, performances, and discourses that seek to capture, contain, and instrumentally use the border to affix a dominant spatiality, temporality, and political agency" (Rajaram and Grundy-Warr 2007: x).

Among Syrians in Lebanon, physical variations in borders allow marginalized groups to negotiate individuality and power. Over the past nine years, the continuous flow and displacement of Syrians have reorganized the land's morphology and generated new geographical patterns marked by the materialities, leakages, and enactments of borderscape practices (Brambilla, 2015). Some scholars have argued that 'undesirable' refugees re-shape and subvert border spaces to create various new forms of security and citizenship; this happens through their daily lives and through activism (Millner, 2011; Rygiel, 2011). Borderscapes foreground the complexity of border processes created, experienced, and contested by human beings. The borderscape notion defines new landscapes and new territorialities where refugee communities adapt and take possession of this border space. The term uses the suffix '-scape' to mean a place and polity created through people's practices of dwelling—their 'doing', 'undoing', and 'redoing' of landscape (Olwig, 2008). Syrians forced into migration are building relationships with the new places they are forced to live in and shaping new landscapes that embody their past memories and new emotional experiences (Egoz and De Nardi, 2017). Unwelcomed and threatened by their host communities, Syrians are marginalized and pushed to the edges of urban and rural contexts. For them, borders comprise a system to demarcate areas of belonging and to achieve safety in unstable conditions.

Syrians displaced in Lebanon

Lebanese–Syrian borders have long been porous, contested, and ill-defined (Picard, 2016; Picard, 2006). Lebanon has maintained a sizeable Syrian workforce since the 1950s in construction, public works, and agriculture (Chalcraft, 2009). Some authors have estimated that between 300,000 and 400,000 Syrians worked in Lebanon in 2011 (Longuenesse, 2015). Many of these workers were men who left their families in Syria while they did seasonal work (Kabbanji and Kabbanji, 2018). The Syrian war in 2011 precipitated a major influx of Syrians to Lebanon. Forced displacement has reconfigured the size and socio-demographic characteristics of Syrian flows and migrations.

For the first three years of the conflict, the Lebanese government ignored the issue of Syrian refugees and called for support from the international community. Lebanon did not develop a coherent strategy for dealing with the quickly rising numbers of displaced persons; fearing a reprise of the country's difficult experience with Palestinian refugee camps, the government refused to establish formal camps per UNHCR's suggestion (Sanyal, 2017: 120; Schmelter, 2016). As a result, Syrians settled in every corner of the country, and thousands of informal settlements materialized in rural areas where refugees lived in overcrowded conditions.

Syrians capitalized on historical networks of migration and employment and took the independence and opportunities offered by Lebanon's cities and rural informal encampments (Fábos and Kibreab, 2007; Grabska, 2006). Over 69% of Syrians in Lebanon live in residential structures like apartments, concierge rooms in residential buildings, and hotel rooms; 20% live in non-permanent shelters like tents and prefabricated units; and 11% live in non-residential structures like garages, workshops, farms, active construction sites, shops, warehouses, and schools. Informal settlements in urban and rural settings take place outside state control, permitting the displaced to manage uncertainty and poverty and move beyond the regulatory order (Kamalipour, 2020; Dovey, 2013).

While urban informality is often indistinguishable from the formal city, the rural informal settlements occurring in agricultural fields are reshaping the pattern of farming on the outskirts of villages (Dovey, 2013). In this condition of 'suspension', traditions, customs, and social-family relations define placemaking; they recall how life used to be in the country of origin (Hiruy, 2009). Despite cultural affinities with the Lebanese, religious and political affiliation confine Syrians to a secluded, marginal condition that denies basic rights and access to landscape resources (Egoz and De Nardi, 2017). Borders are constantly being produced, negotiated, challenged, and redrawn, and they remain essential to both host and hosted populations confronting the accelerating social change unleashed by the crisis (Trudeau et al., 2011).

Syrians in the Bekaa Valley

In March 2012, the Bekaa Valley became the main destination for displaced Syrians. The valley, an agricultural region mainly within the Bekaa Governorate, shares a long border with Syria, so it became a natural gathering point for those fleeing conflict. Also, the Bekaa Valley is home to 42% of Lebanon's total cultivated land, and the agricultural industry depends on low-wage workers. Despite the state's restrictions on employment, employers in the Bekaa region have welcomed and benefited from exploiting the Syrian workforce. Many Syrians have settled close to agricultural fields where work opportunities exist (Habib, 2018).

An informal settlement is defined as "…a settlement that was established in an unplanned and unmanaged manner, which means they are generally unrecognized" (Sanyal, 2017: 118). In Lebanon, these settlements are set up on land belonging to municipalities or private individuals, possibly with a formal agreement between landlords and residents (Sanyal, 2017). Spatially, the informal encampments range from a few improvised shelters between fields or along roadsides to sprawling camps equipped with makeshift schools and services (Davies et al., 2019).These settlements vary in size, security provisions, and amenities. Some are directly supported by charitable organizations and international NGOs, while others have more limited support. Landlords provide varying degrees of support (Sanyal, 2014).

Distribution depends on family and religious affiliations as well as the control of individuals known as 'Shaweesh'.These members of the Syrian refugee population who act as foremen/middlemen and oversee the employment of men, women, and children in the agricultural fields in Bekaa (Habib, 2018). Shaweesh act as community representatives, brokers, and informants, and their roles vary from one settlement to another. They are often appointed by landlords based on previous community relationships (Stel, 2020).The informal settlements of Syrians emerged instead of formal camps through a collusion of private interests, state acquiescence, and humanitarian aid (Sanyal, 2017).The ITS system stems from political inaction, and this non-intervention creates insecurity (Freedman, 2018). If the settlements are not formally approved by the government, they can theoretically be dissolved at any time (Schmelter, 2016).

Method and case study

This research combines direct observation and visual recording with macro-level analyses to examine relationships between the socio-spatial dynamics of inclusion/exclusion and coping mechanisms at the local and regional levels. This approach, focused on everyday practices, allowed exploration of the dynamic relationship between territory and new landscapes and patterns related to displacement. Cartographic interpretation, fieldwork, mapping exercises, and an extensive photographic campaign were the study tools. They enabled the understanding of the overall spatial organization of the thousands of informal settlements. Analysis conducted at the regional and district level used KMZ files of ITS locations, Excel files tabulating tents and services for ITSs, and infographics retrieved from the UNHCR website.This data was correlated to materials gathered from national and international NGO reports, and with maps, photos, sketches, and notes collected on site from January 2014 to January 2018.[1]

As Randolph Hester (1984) has suggested, observation is the best technique for discovering what people do and how they interact in a neighbourhood space. Our fieldwork aimed to assess the spatial character and quality of the informal settlements through the recognition of appropriation, adaptation, and construction processes. A mapping exercise that corroborated the reading and interpretation of these phenomena was useful for grasping the socio-cultural realities of communities, landscapes, and ecosystems (Ryan, 2011). Spatial components and uses were charted to determine relationships between interior and exterior and public and private spaces; they also helped us to understand everyday rhythms, activities, social constraints, desires, memories, and ways of living. Close observation allowed the identification of borderscape practices, 'evidence of occupation', open space uses and activities, and the liminal zone between the inside and outside of the tents (Hester, 1984). The method relies on past studies to understand the adaptation and transformation of the housing system in informal settlements as well as possibilities for upgrading the unbearable conditions.[2]

Informal tented settlements in the Bekaa Valley

The chapter reports on case studies located in the Bekaa Valley (see Figure 8.1). The Bekaa Governorate hosts 1,725 ITSs with a total of 18,693 tents and 113,422 individuals (UNHCR, 2020a). The majority of households live in makeshift tents made of worn fabric or canvas (69.7%), new fabric or canvas (43.7%), wood (52.4%), or nylon (27.2%) (Habib, 2018).

The spread and sprawl of these informal settlements in the Bekaa Valley produced new landscape patterns of field crops and shelter compounds. The formalization of living practices in agricultural areas is expressed by the rich thematic variations of private and settled space versus private and productive land. In the past nine years, several typologies of tent clusters have been erected based on the dimension of farming parcels, the presence of natural and built infrastructures, and distance from and relation to the villages and cities (see Figure 8.2). Linear settlements have been organized along main roads and agricultural channels and between fields, while compact settlements are located in bigger agricultural plots or on the outskirt of

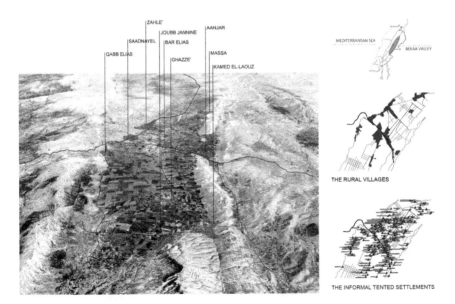

FIGURE 8.1 The Bekaa valley hosts 1,725 informal tented settlements dispersed in productive land. Located near natural and built infrastructure and sprawled on fields, the ITSs are threatening the already compromised ecological, social, and cultural stability of the region (image: Maria Gabriella Trovato)

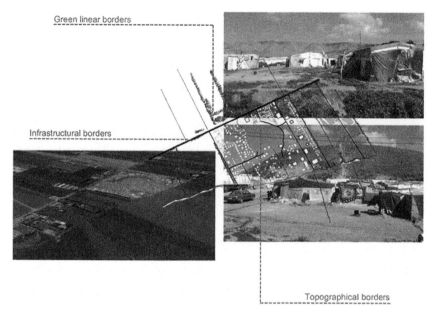

FIGURE 8.2 Compact informal tented settlements in agricultural land in the Bekaa Valley (image: Maria Gabriella Trovato)

towns. Small clusters have developed between greenhouses or scattered and hidden in farm plots; adjacent to transit nodes and infrastructures; and within walking distance to sources of employment and production.

The ITSs are redesigning borders and creating fringes of structure in an agricultural context (see Figure 8.3). Those borders, continually modified by the movement of displaced Syrians, are the results of land ownership and the manifestations of spatial and political control and confinement over the refugees (Trovato, 2018a). Salvaged materials replace natural ones in the creation of new boundaries and introduce urban artefacts into the agricultural context. Concrete slabs on fertile soil define a discontinuous limit between production and dwelling and introduce new spatial practices that change and give form to the border (see Figure 8.4). During nine years of cohabitation, collaboration has been rare. Syrians are mostly portrayed as a social and political threat to be exploited and evicted if necessary. Material and spatial boundaries embed social differences within the environment at both local and regional scales (Kosiba and Bauer, 2013). The enclosures conceal the space, size, and population of the informal settlements. They also delineate the limit between displaced persons and locals, between the informal settlement and the agricultural field.

Saadnayel and Bar Elias, Zahle district

We observed and mapped two municipalities in the Bekaa Governorate, Saadnayel and Bar Elias. Located in Zahle's district, they were selected based on accessibility

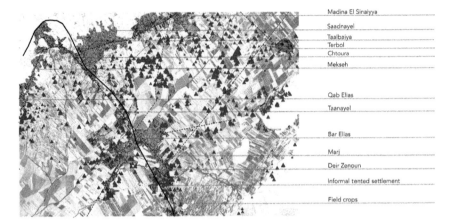

FIGURE 8.3 Informal tented settlements are scattered in the Bekaa Valley at the periphery of marginalized rural communities. Neither planned nor inserted in a strategic vision, these ITSs are redesigning village borders. The transformation of land use brings consequences for social/economic assets and the culture of these areas, profoundly changing their identity, formal composition and relationships between elements (image: Maria Gabriella Trovato)

FIGURE 8.4 The ITSs redesign the borders between rural villages and agricultural fields with temporary makeshift tents and fringes of concrete slabs on compacted soil (image: Maria Gabriella Trovato)

to the site and the availability of data collected over the past five years by the Civic Centre at the American University of Beirut in partnership with other NGOs. Saadnayel (see Figure 8.5) hosts 65 ITSs, 11 of them inactive. The number of tents per informal settlement varies from 1 to 70, with a population range from two to 425 inhabitants. In Bar Elias (see Figure 8.6), there are 58 ITSs, 7 of them no longer active. The number of tents there ranges from 3 to 146, with a population range from 20 to 870 inhabitants.

The informal settlements are mostly concentrated in the suburbs along the national road that leads to the Syrian border. Some are hidden between the scattered houses in agricultural fields, but many are visible from main roads. The organization

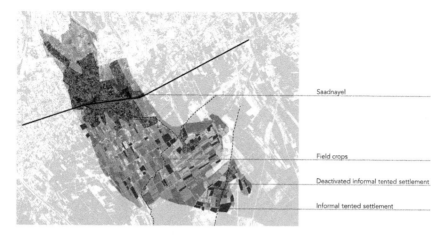

FIGURE 8.5 Saadnayel: map of the village with the active and deactivated ITSs (image: Maria Gabriella Trovato)

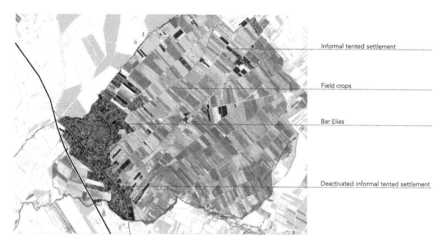

FIGURE 8.6 Bar Elias: map of the village with the active and deactivated ITSs (image: Maria Gabriella Trovato)

of tents varies from uniform rows to more organic patterns in which the addition of new space or the amalgamation of two or more shelters generates a fluid organization (Trovato, 2018b). The settlements have a relatively consistent typology of room-by-room increments based on limitations of space, access to (or lack of) long-span materials like bricks and concrete, and restrictions imposed by landowners and by the Lebanese government (Dovey, 2013). Structural improvements are often only temporary and lack authorization from the government or landlords, and the government restricts the types of materials that can be used (Save the Children, 2017). Over time, tents are adapted to seasonality, family growth, and changing needs. The typical tents are temporary timber frames covered with plastic, cardboard, and old rugs. Basic materials such as timber and plastic sheeting provide families a more

FIGURE 8.7 Inhabitants engaged in a variety of landscape practices in the interstitial spaces of their proximate areas, growing medicinal and ornamental plants, collecting recyclable materials to re-use or sell, and organizing small spaces for recreational activities (image: Maria Gabriella Trovato)

secure, private, and weather-proof living space. Material replacement generally aims to improve or repair constructions (Kamalipour, 2016).

In the Lebanese informal settlements, identity is partly provided by the reconstruction and reshaping of everyday landscapes with separation between public and private (or open and closed) spaces (see Figure 8.7). Where the ratio between tents and open spaces creates a lack of privacy, concrete slabs and bordering walls or limits are built. Protective elements extend from the interior to the threshold to prevent passers-by from getting too close to a family's space. The codes of use and appropriation of spaces adjacent to the tents are tacit, regulated by the extension of domestic activities outside the shelters, flexible, and ever-changing. Borderscape practices allow the reorganization of spatial hierarchy to recreate social and family equilibrium. This hierarchy is defined by signs denoting a nuanced passage from public to private (see Figure 8.8). Marking out space and tracing boundaries, Syrians organize protected landscapes by redefining thresholds and adjacencies; interiors and exteriors. The border is both a physical place and an expression of territoriality that reflects the basic need to live in a bounded space (Leimgruber, 1991). Boundaries affirm a sense of privacy, security, ownership, and individual rights. Borders materialize people's experiences, practices, and processes of appropriation, domination, or competition. Border areas are constantly transformed, reorganized, and marked, and social relationships are materialized in time and space as borderscapes.

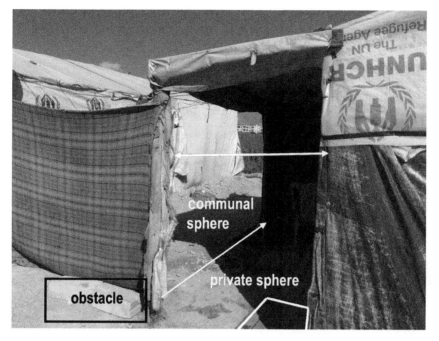

FIGURE 8.8   Spatial identity is partly provided by the reorganization of space's hierarchy to recreate a familiar landscape and produce divisions between public/private, open/closed. The codes of use and appropriation adjacent to the tent are tacit and regulate the extension of the domestic activity beyond the home (image: Maria Gabriella Trovato)

Conclusion

The complex and fluid landscape of belonging, customs, and bordering practices among displaced Syrians attests to the transformation taking place in Lebanon. Changes in land use have produced new landscapes (and a new socio-political equilibrium) based on the imposed and stigmatized legal identity of a disadvantaged minority group: Syrian refugees. The relocation of millions of individuals from the territory of their origin to thousands of informal settlements illuminates the hierarchy, difference, and alienation between citizens and non-citizens.

Practices of re-appropriation and re-adaption along borders in urban and rural landscapes offered a lens for observing the struggle of Syrians refugees to belong within these informal encampments (Rossi, 2014). Regarding borders as socio-cultural and discursive processes highlights the spatial practices those communities use to recreate places. Borders assign functions and give identity to places Syrians are living in. People shape the landscapes around them based on their experiences and actions, and those places shape our memories and emotions. That kind of connection becomes a source of essential feelings of security and identity.

The condition of people on the move calls for a different understanding of landscape: it must be addressed as the confluence of physical subsistence and psychological necessity, including the right to live in dignity and the freedom to define one's own identity (Egoz, 2018). The physical environment is central to the experience of displacement, and the idea of belonging, even under those circumstances, can be understood as a universal concept similar to human rights (Egoz et al., 2011). It transcends nation–state boundaries. As Mitchell and Breitbach argue:

> an important task in landscape research lies in recognising the power that ordinary people do have (and have had) in producing landscape. It requires learning about the way power is organized, controlled and taken, and it also requires looking at the landscape as a person engaged in the pursuit of social justice.
>
> *(2016: 216)*

Schein's work on "people who have often been written out of 'belonging', precisely through land and landscape" shines a light on the power of landscape for citizenship and community and "the right to claim belonging" (2009: 811). Despite their tenuous conditions, the ITSs in Lebanon could be the setting for the creation of new forms of refugee citizenship and solidarity, or what Sigona (2015) refers to as 'campzenship', where refugees assume a visible presence within social and geographical space (Rygiel, 2011). The term 'campzenship' captures the situated experience of membership by camp residents *and* the paradigmatic position of camps and camp-like institutions as contemporary spaces of politics (Sigona, 2015; Isin and Rygiel, 2007). Space is not the 'innocent' and 'depoliticized' context of social interactions (Soja, 1989; Lefebvre [1974], 1991). Instead, space is produced by social interaction and plays a key role in producing, and reproducing, power and identity (Sigona, 2015).

Acknowledgements

I am grateful to all the people who have contributed to this research in the past six years and who have challenged and inspired me with their thought and their practical involvement. I want to thank the Department of Landscape Design and Ecosystem Management and the Centre for Civic Engagement and Community Service at the American University of Beirut and the IFLA-Landscape Architects Without Borders working group that I have led since summer 2017 for their continuous support and critical contribution.

Notes

1 Field notes collected included most of the information on density of tents and their quantity, family gathering spaces, public/private interfaces, entrances, and access networks.
2 For studies on informal typologies, see Kamalipour (2020); Dovey and King (2011); Watson (2003). For studies on public/private interfaces in informal settlements, see Dovey and Kamalipour (2018).

Bibliography

Agamben, Giorgio (2002). *Remnants of Auschwitz: The Witness and the Archive.* London, UK: Zone Books.

Agamben, Giorgio (1998). *Homo Sacer: Sovereign Power and Bare Life.* Translated by Daniel Heller-Roazen. Stanford, CA: Stanford University Press.

Berner, Erhard (2001) 'Informal developers, patrons, and the state: Institutions and regulatory mechanisms in popular housing'. In: *Paper presented at the Naerus Conference ESF/N-AERUS Workshop* on "Coping with informality and illegality in human settlements in developing cities", Leuven, 23–26 May.

Boustani, Marwa, Estella Carpi, Hayat Gebara, and Yara Mourad (2016a). *Responding to the Syrian Crisis in Lebanon: Collaboration Between Aid Agencies and Local Governance Structures.* London: International Institute for Environment and Development.

Boustani, Marwa, Hayat Gebara, Gabriella Romanos, Anna L. Strachan, and Michael J. Warren (2016b). *Beirut, a safe Refuge? Urban Refugees Accessing Security in a Context of Plural Provision.* UNHABITAT Report. Available from: unhabitat.org/beirut-a-safe-refuge-urban-refugees-accessing-security-in-a-context-of-plural-provision/ [Accessed 15 April 2018].

Brambilla, Chiara (2015). 'Exploring the critical potential of the borderscapes concept'. *Geopolitics*, 20(1), pp 14–34.

Chalcraft, John (2009). *The Invisible Cage: Syrian Migrant Workers in Lebanon.* Chicago: The University of Chicago Press.

Churruca-Muguruza, Cristina (2018). 'The changing context of humanitarian action: Key challenges and issues'. In: Hans-Joachim Heintze and Pierre Thielbörger, eds. *International Humanitarian Action.* Basel: Springer, pp 3–18.

Coates, Robert and Jeffrey Garmany (2017). 'The ecology of citizenship: Understanding vulnerability in urban Brazil'. *International Development Planning Review*, 39, pp 37–56.

Davies, Thom, Isakjee Arshad, and Dhesi Surindar (2019). 'Informal migrant camps'. In: Katharyne Mitchell, Reece Jones, and Jennifer Fluri, eds. *Handbook on Critical Geographies of Migration.* Cheltenham, UK: Edward Elgar Publishing.

Dell'Agnese, Elena and Anne-Laure Amilhat Szary (2015). 'Borderscapes: From border landscapes to border aesthetics'. *Geopolitics*, 20(1), pp 4–13.

Dovey, Kim (2013). 'Informalising architecture: The challenge of informal settlements'. *Architectural Design*, 83, pp 82–89.

Dovey, Kim and Ross King (2011). 'Forms of informality: Morphology and visibility of informal settlements'. *Built Environment*, 37(1), pp 11–29.

Dovey, Kim and Hesam Kamalipour (2018). 'Informal/formal morphologies'. In: Kim Dovey, Elek Pafka, and Mirjana Ristic, eds. *Mapping Urbanities: Morphologies, Flows, Possibilities.* New York, NY: Routledge, pp 223–248.

Edensor, Tim (2002). *National Identity, Popular Culture and Everyday Life.* New York, NY: Berg Publishers.

Egoz, Shelley, Jala Makhzoumi, and Gloria Pungetti, eds. (2011). *The Right to Landscape.* London: Routledge.

Egoz, Shelley and Alessia De Nardi (2017). 'Defining landscape justice: The role of landscape in supporting wellbeing of migrants, a literature review'. *Landscape Research*, 42(sup1), pp S74–S89.

Egoz, Shelley (2018). 'Landscape and identity in the century of the migrant'. In: Peter Howard, Ian Thompson, and Emma Waterton, eds. *The Routledge Companion to Landscape Studies.* London: Routledge.

Fábos, Anita and Gaim Kibreab (2007). 'Urban refugees: Introduction'. *Refuge: Canada's Periodical on Refugees,* 24(1), pp 1–19.

Fakhoury, Tamirace (2017). 'Governance strategies and refugee response: Lebanon in the face of Syrian displacement'. *International Journal Middle East Studies,* 49, pp 681–700.

Fawaz, Mona (2009). 'Contracts and retaliation: Securing housing exchanges in the interstice of the formal/informal Beirut (Lebanon) housing market'. *Journal of Planning Education and Research,* 29(1), pp 90–107.

Fawaz, Mona, Nizar Saghiyeh, and Karim Nammour (2014). *Housing, Land and Property Issues in Lebanon: Implications of the Syrian Refugee Crisis.* Lebanon: UN Habitat and UNHCR. Available from: unhabitat.org/ housing-land-and-property-issues-in-lebanon-implications-of-the-syrian-refugee-crisis-august-2014/ [Accessed 23 October 2020].

Fawaz, Mona (2017). 'Planning and the refugee crisis: Informality as a framework of analysis and reflection'. *Planning Theory,* 16(1), pp 99–115.

Fekade, Wubalem (2000). Deficits of formal urban land management and informal responses under rapid urban growth, an international perspective. *Habitat International,* 24, pp 127–150.

Fleifel, Manar (2020). *The Diversity of Experiences of Syrian Displaced Persons in Lebanon: Everyday Exclusion, Informality, and Adjustment Tactics.* The Asfari Institute for Civil Society and Citizenship. Available from: activearabvoices.org/uploads/8/0/8/4/80849840/diversity_of_experiences_-_v.1.2-digital.pdf [Accessed 23 October 2020].

Freedman, Jane (2018). '"After Calais": Creating and managing (in)security for refugees in Europe'. *French Politics,* 16(4), pp 400–418.

Geha, Carmen and Joumana Talhouk (2018). *Politics and the Plight of Syrian Refugees in Lebanon: Political Brief on the Return of Syrian Refugees.* Available from: aub.edu.lb/ Documents/Politics-and-the-Plight-of-Syrian-Refugees-in-Lebanon.pdf [Accessed 23 October 2020].

Giaccaria, Paolo, and Claudio Minca (2011). 'Topographies/topologies of the camp: Auschwitz as a spatial threshold'. *Political Geography,* 30(1), pp 3–12.

Giaccaria, Paolo and Claudio Minca (2015). 'For a spatial theory of the Third Reich'. In: Paolo Giaccaria and Claudio Minca, eds. *Hitler's Geographies: The Spatialities of the Third Reich.* Chicago: Chicago University Press, pp 19–44.

Goh, Kian (2018). 'Architecture and global ethnographies'. In: Nick Axel et al., eds. *Dimensions of Citizenship: Architecture and Belonging from the Body to the Cosmos.* Los Angeles: Inventory Press.

Grabska, Katarzyna (2006). 'Marginalization in urban spaces of the global south: Urban refugees in Cairo'. *Journal of Refugee Studies,* 19(3), pp 287–307.

Habib, Rima R (2018). *Survey on Child Labour in Agriculture in the Bekaa Valley of Lebanon: The Case of Syrian Refugees.* Report for the Ministry of Labor, Lebanon. Beirut: American University of Beirut. Available from: ilo.org/wcmsp5/groups/public/---arabstates/---ro-beirut/documents/publication/wcms_711801.pdf [Accessed 30 April 2020].

Hafeda, Mohamad (2019). *Negotiating Conflict in Lebanon: A Bordering Practice in the Divided City.* London; New York, NY: I.B. Tauris.

Hammett, Daniel (2017). 'Introduction: Exploring the contested terrain of urban citizenship'. *International Development Planning Review,* 39(1), pp 1–13.

Hester, Randolph T (1984). *Planning Neighbourhood Space with People.* New York, NY: Van Nostrand Reinhold.

Hyndman, Jennifer (2000). *Managing Displacement: Refugees and the Politics of Humanitarianism.* Minneapolis, MN: University of Minnesota Press.

Hiruy, Kiros (2009). *Finding Home Away from Home: Place Attachment; Place Identity, Belonging and Resettlement among African-Australians in Hobart.* School of Geography and

Environmental Studies. Hobart, TAS: University of Tasmania. Master of Environmental Management Thesis.

Isin, Engin F. and Kim Rygiel (2007). 'Abject spaces: Frontiers, zones, camps'. In: Elizabeth Dauphinee and Cristina Masters, eds. *The Logics of Biopower and the War on Terror.* New York, NY: Palgrave Macmillan.

Kabbanji, Lama and Jad Kabbanji (2018). *Assessing the Development Displacement Nexus in Lebanon.* Working paper. International Centre for Migration Policy Development. Available from: icmpd.org/fileadmin/1_2018/Downloads_VMC2017/Assessing_the_ Development-Displacement_Nexus_in_Lebanon_final.pdf [Accessed 20 January 2020].

Kamalipour, Hesam (2016). 'Forms of informality and adaptations in informal settlements'. *International Journal of Architectural Research*, 10(3), pp 60–75.

Kamalipour, Hesam (2020). 'Improvising places: The fluidity of space in informal settlements'. *Sustainability 2020*, 12, pp 22–93.

Keivani, Ramin and Edmundo Werna (2001). 'Refocusing the housing debate in developing countries from a pluralist perspective'. *Habitat International*, 25, pp 191–208.

Kosiba, Steve and Andrew Bauer (2013). 'Mapping the political landscape: Toward a GIS analysis of environmental and social difference'. *Journal of Archaeological Method and Theory*, 20(1), pp 61–101.

Leimgruber, Walter (1991). 'Boundary, values and identity: The Swiss-Italian transborder region'. In: Dennis Rumley and Julian V. Minghi, eds. *The Geography of Border Landscapes.* London: Routledge, pp 43–62.

Lefebvre, Henry ([1974] 1991). *The Production of Space.* Oxford: Basil Blackwell.

Longuenesse, Elisabeth (2015). 'Travailleurs étrangers, réfugiés syriens et marché du travail'. *Confluences Méditerranée*, 92(1), pp 33–47.

Low, Setha M. and Irwin Altman (1992). 'Place attachment: A conceptual inquiry'. In Irwin Altman and Setha M. Low, eds. *Place Attachment.* New York, NY: Plenum Press, pp 1–12.

Mehretu, Assefa, Bruce W. Pigozzi, and Lawrence Sommers (2000). 'Concepts in social and spatial marginality'. *Geografiska Annaler*, 82B(2), pp 89–101.

Mela, Alfredo and Alessia Toldo (2019). 'Understanding social exclusion'. In: Alfredo Mela and Alessia Toldo, eds. *Socio-Spatial Inequalities in Contemporary Cities.* Cham, CH: Springer International Publishing, pp 1–6.

Menatti, Laura (2017). 'Landscape: From common good to human right'. *International Journal of the Commons*, 11(2), pp 641–683.

Mezzadra, Sandro and Brett Neilson (2013). *Border as Method, or, The Multiplication of Labor.* Durham, NC; London: Duke University Press.

Millner, Naomi (2011). 'From "refugee" to "migrant" in Calais solidarity activism: Restaging undocumented migration for a future politics of asylum'. *Political Geography*, 30, pp 320–328.

Minca, Claudio (2015). 'Geographies of the camp'. *Political Geography*, 49, pp 74–83.

Mitchell, Don and Breitbach Carrie (2018). 'Landscape'. In: John A. Agnew and James S. Duncan, eds. *The Wiley-Blackwell Companion to Human Geography.* Hoboken, NJ: John Wiley & Sons, pp 209–216.

Newman, David (2006). 'Borders and bordering: Towards an interdisciplinary dialogue'. *European Journal of Social Theory*, 9(2), pp 171–186.

Nyers, Peter (2006). *Rethinking Refugees: Beyond States of Emergency.* London: Routledge.

Olwig, Kenneth (2008). 'Performing on the landscape versus doing landscape: Preambulatory practice, sight and the sense of belonging'. In: Tim Ingold and Jo Lee Vergunst, eds. *Ways of Walking. Ethnography and Practice on Food.* Aldershot, UK: Ashgate.

Pascual-de-Sans, Àngels (2004). 'Sense of place and migration histories *idiotopy* and *idiotope*'. *Area*, 36(4), pp 348–357.

Paasi, Anssi (1998). 'Boundaries as social processes: Territoriality in the world of flows'. *Geopolitics*, 3(1), pp 69–88.

Painter, Joe and Chris Philo (1995). 'Spaces of citizenship: An introduction'. *Political Geography*, 14(2), pp 107–120.

Picard, Élizabeth (2006). 'Managing identities among expatriate businessmen across the Syrian Lebanese boundary'. In: Inga Brandell, ed. *State Frontiers: Borders and Boundaries in the Middle East*. London; New York, NY: I. B. Tauris, pp 75–100.

Picard, Élizabeth (2016). *Liban Syrie, intimes étrangers: Un siècle d'interactions sociopolitiques*. Arles, FR: Actes Sud.

Porter, Libby et al. (2019). 'Borders and refuge: Citizenship, mobility and planning in a volatile world'. *Planning Theory & Practice*, 20(1), pp 99–128.

Rajaram, Prem Kumar and Carl Grundy-Warr (2007). 'Introduction'. In: Prem Kumar Rajaram and Carl Grundy-Warr, eds. *Borderscapes: Hidden Geographies and Politics at Territory's Edge*. Minneapolis, MN: University of Minnesota Press, pp ix–xi.

Rygiel, Kim (2011). 'Bordering solidarities: Migrant activism and the politics of movement and camps at Calais'. *Citizenship Studies*, 15(1), pp 1–19.

Ryan, Robert L (2011). 'The social landscape of planning: Integrating social and perceptual research with spatial planning information'. *Landscape and Urban Planning*, 100, pp 361–363.

Rossi, Alice (2014). 'Evictions, urban displacement, and migrant re-appropriation in Turin (Northern Italy)'. *Planum. The Journal of Urbanism*, 29(II), n.p.

Saliba, Issam (2010). 'Lebanon: Constitutional law and the political rights of religious communities'. The Law Library of Congress. Available from: loc.gov/law/help/lebanon/political-rights-of-religious-communities.pdf [Accessed June 2020].

Sanyal, Romola (2017). 'A no-camp policy: Interrogating informal settlements in Lebanon'. *Geoforum*, 84, pp 117–125.

Sanyal, Romola (2014). 'Urbanizing refuge: Interrogating spaces of displacement'. *International Journal of Urban and Regional Research*, 38(2), pp 558–572.

Save the Children (2017). *Fire Risk Reduction Assessment of Vulnerable Displaced Syrian Populations and Host Community in Lebanon*. Available from: resourcecentre.savethechildren.net/node/12247/pdf/save_fireprevention_booklet_low06022017_00000002.pdf [Accessed 23 October 2020].

Schein, Richard H (2009). 'Belonging through land/scape'. *Environment and Planning A: Economy and Space*, 41(4), pp 811–826.

Schmelter, Susanne (2016). 'The question of governing Syrian refugees: An ethnography of Lebanon's humanitarian regime'. In: Natalia Ribas-Mateos, ed. *Migration, Mobilities and the Arab Spring: Spaces of Refugee Flight in the Eastern Mediterranean*. Cheltenham, UK: Edward Elgar Publishing, pp 174–190.

Sigona, Nando (2015). 'Campzenship: Reimagining the camp as a social and political space'. *Citizenship Studies*, 19(1), pp 1–15.

Soja, Edward (1989). *Postmodern Geographies: The Reassertion of Space in Critical Social Theory*. London: Verso.

Stel, Nora (2020). *Hybrid Political Order and the Politics of Uncertainty. Refugee Governance in Lebanon*. Abingdon, UK; New York, NY: Routledge.

Sukkar, Ahmad, Hani Fakhani, and Sawson Abou Zainedin (2020). *Syrian (In)formal Displacement in Lebanon*. The Asfari Institute for Civil Society and Citizenship Research Papers. Available from: activearabvoices.org/uploads/8/0/8/4/80849840/syrian__in_formal_-_en_-_v.1.3_-_digital.pdf [Accessed 23 October 2020].

Tapie, Guy (2014). *Sociologie de l'habitat contemporain: vivre l'architecture.* Marseille, FR: Parenthèses.

Trovato, Maria G. (2018a). 'A landscape perspective on the impact of Syrian refugees in Lebanon'. In Ali Asgary, ed. *Resettlement Challenges for Displaced Populations and Refugees.* Cham, CH: Springer International Publishing, pp 41–64.

Trovato, Maria G. (2018b). 'Borderscape: Forced migration and new spatial practices'. In: Laura Pedata, Enrico Porfido, and Loris Rossi, eds. *[Co]habitation Tactics: Imagining Future Spaces in Architecture, City and Landscape: Tirana, 20th-23rd September 2018: conference.* Tiranë, AL: Polis Press, pp 453–463.

Trudeau, Dan and Chris McMorran (2011). 'The geographies of marginalization'. In: Vincent Del Casino et al., eds. *A Companion to Social and Cultural Geography.* Malden, MA: Blackwell Publishing, pp 437–453.

UN and Government of Lebanon (2019). *Lebanon Crisis Response Plan 2017-2020.* Available from: unhcr.org/lb/wp-content/uploads/sites/16/2019/04/LCRP-EN-2019.pdf [Accessed 1 January 2020].

UNHCR—United Nations High Commissioner for Refugees (2020a). *Syrian Emergency.* Available from: unhcr.org/syria-emergency.html [Accessed 30 April 2020].

UNHCR—United Nations High Commissioner for Refugees (2020b). *Lebanon Map Hub.* Available from: unhcr.carto.com/builder/ [Accessed April 2020].

van Houtum, Henk and Ton van Naerssen (2002). 'Bordering, ordering and othering'. *Tijdschrift voor Economische en Sociale Geografie*, 93(2), pp 125–136.

van Houtum, Henk, Olivier Kramsch, and Wolfgang Zierhofer (2005). *B/Ordering Space.* Aldershot, UK: Ashgate.

VASyR (2019). *Vulnerability Assessment of Syrian Refugees in Lebanon.* Available from: reliefweb.int/sites/reliefweb.int/files/resources/73118.pdf [Accessed March 2020].

Watson, Donald (2003). 'How the other half builds'. In: Donald Watson, Alan Plattus, and Robert G. Shibley, eds. *Time-Saver Standards for Urban Design.* New York, NY: McGraw-Hill.

Willis, Katie (2016). 'Viewpoint: International development planning and the sustainable development goals (SDGs)'. *International Development Planning Review*, 38(2), pp 105–11.

9

DISASTER VOLUNTEERS

The constructed identity of disaster aid workers and their place in the affected community

Mary M. Nelan

Prologue

2010 marked the start of my career as a disaster volunteer. Following that January's catastrophic earthquake in Haiti, I spent a month clearing rubble, gardening, and constructing a composting toilet. I was hit with the intoxicating spirit of disaster volunteers, and I became drawn to that world and service. Though we were surrounded by crumbling grey concrete and dust, I was welcomed into a vibrant and lively community. I left Haiti a different person: being a member of that volunteer community became a part of my identity. Over the next three years, my new identity as a disaster scholar was further shaped by four additional volunteer experiences after tornadoes and hurricanes in the United States. The longer I continued my education, I more recognized that there was a dearth of knowledge on disaster volunteers. Specific questions needed answering: (a) who were they? (b) what were their experiences? and (c) how did their own identities shift during and after their time in disaster zones? To explore the phenomenon of volunteerism and to gain a better understanding of this population, I began interviewing former disaster volunteers. The following chapter is a result of that research.

Introduction

Disasters can transform the physical landscape, the population, and even the identity of a place. While a common view of the post-disaster landscape is one of chaos, panic, and violence, these perceptions are considered disaster myths as they are largely inaccurate (Tierney et al., 2006). Disaster research reveals that people help one another and that volunteers flock to the affected area to aid the survivors (Fritz and Mathewson, 1957: 195). The first helpers are often family, neighbours, and friends of those affected; they arrive before formal first responders such as police, firefighters, and paramedics (Drabek, 2013: 2).

People volunteer for different reasons: altruism, curiosity, or to fulfil their need to participate and be a part of the disaster (St. John and Fuchs, 2002: 401). Volunteer effort is not isolated to the first moments after the event; it continues for months and possibly years into the recovery effort (Helsloot and Ruitenberg, 2004). Organizations (both NGOs and government agencies) set up in the disaster area and add a new population of employees and volunteers. Though they may not settle in the area permanently, many employees and volunteers develop a sense of belonging to disaster areas, the affected places, and to the communities. They establish their own form of citizenship and identity in that geographic space.

Olwig (2005: 20) describes land as something to which people belong. Following a disaster, the land changes, causing survivors and aid workers alike to forge a new sense of belonging to this altered place. After a disaster, affected communities come together to rebuild their environment and community solidarity (Drabek, 2013). Disasters will become more and more frequent with the climate crisis, and more and more people will encounter situations altered by hazards like sea level rise. It is important to add to the growing body of work that examines how individuals are attached to altered places and how their sense of belonging changes following a disaster (Zavar and Schumann, 2019; Chamlee-Wright and Storr, 2009; Brown and Perkins, 1992; Erikson, 1976).

To increase a sense of belonging to places affected by disaster, people have produced slogans to bond the community, represent themselves as a united front to the outside world, increase media attention, and draw attention to fundraising efforts. Examples include 'Boston Strong', which followed the Boston Marathon bombing in 2013, 'Moore Strong', which mobilized support after a tornado struck Moore, Oklahoma that same year, and 'Houston Strong', which rallied people in the wake of Hurricane Harvey in 2017. This 'strong' identification is the third stage of Anthony Wallace's 'Disaster Syndrome': "the mildly euphoric identification with the damaged community, and enthusiastic participation in repair, and rehabilitation enterprises" (Wallace, 1956: 111). The 'euphoric identification' may not be limited to residents of the affected area (nor to commercial branding) but may also extend to aid workers who converge on the area following the disaster.

This chapter investigates how volunteers situate themselves during their service following disaster and how they construct their own sense of belonging and citizenship to the altered landscapes. Based on primary research that I undertook after five volunteering experiences (four in the United States, one in Haiti), this study includes data from individual semi-structured interviews with 13 other disaster volunteers. I explore how disaster aid workers belong to a disaster site and construct their identity in relation to the place and how they build citizenship within their volunteer communities and organizational communities and in relation to the affected landscape. The chapter begins with an overview of the literature currently available on the topic; the second part covers my own research into this area.

Disaster volunteers

Disaster literature discusses a variety of individuals, or 'convergers', who move into an affected area following a disaster. 'Convergers' fit within six categories: (1) the returnees, (2) the anxious, (3) the helpers, (4) the curious, (5) the exploiters, and (6) a more recently proposed group, the memorializers (Kendra and Wachtendorf, 2003; Fritz and Mathewson, 1957: 29). This paper focuses on the helpers and the curious. 'Helpers' come into the disaster area formally with relief organizations and informally on their own, without a clear organizational affiliation. The exact number of helpers is hard to calculate because of the participation of informal volunteers from elsewhere and local community members (Fritz and Mathewson, 1957). The 'curious' are highly motivated by their own desire to experience an unusual event (ibid.: 46).

Arguably, disaster volunteers exist at the intersection of 'voluntourism' and disaster (or dark) tourism. They may belong to the categories of 'curious' and 'helpers' at the same time. Evidence shows that 'curious' volunteers have low personal loss associated with the disaster, and even after the event, they will travel increasing distances to witness its impact. Both 'helpers' and 'curious' volunteers may already have an attachment to the landscape; they may travel to help to transform and rebuild it, or they may come simply to see and understand what happened to the landscape.

Tourism

John Urry's *The Tourist Gaze* highlights how tourists experience and identify with landscapes and how they exist within a visual environment. Urry (1992: 173) states that "the tourist experience involves something that is visually different and distinguished from otherwise mundane activities". Their gaze is "often visually objectified or captured through photographs, postcards, films, models, and so on" and can be "reproduced, recaptured, and redistributed over time" through these media (Urry and Larsen, 2011: 4). The gaze also creates symbols of those places, as when people kissing in Paris illustrate the romance of the city (ibid.: 61). The 'tourist gaze' has been critiqued because travellers do more than look at the land-scape: they experience it through all of their different senses (Dann and Jacobsen, 2003). Therefore, tourism has been characterized more recently as a performance (Perkins and Thorns, 2001; Edensor, 2000) or interaction with surroundings; Urry (2011) now agrees with this. The performance may include a tourist's movements through an area (either along prescribed paths or off the beaten track) and engage-ment in suggested activities at specific cultural sites. Photography is also constructed as a performance.

One theory of tourism states that the traveller interprets meanings in landscapes through discourse among tourists, locals, and intermediaries like governments, travel agents, or local tourism promotion entities (Nash, 1996). Although the physical and spatial aspects of a landscape may be fixed, each individual interacts with, experiences, understands, and relates to a landscape differently (Knudsen et al., 2008). Today's tourists use many tools, including the Internet, guidebooks, tour guides, and their own experiences, to help them decipher the identities of the places they visit and the people who live there. Each of these interpretive resources mediates the messages a tourist receives regarding the social and physical landscape.

Volunteers who populate post-disaster spaces may experience the landscape simultaneously as helpers *and* tourists (even if they don't see themselves in relation to tourism). Both roles enable their performance in the post-disaster landscape. Like all tourists, they will see and live in the landscape through the lens of their own experiences.

Research into international volunteering shows that as volunteers reside in host communities for longer periods of time, power differentials and inequalities between volunteer communities and resident communities are reduced (Devereaux, 2008). Long-term and semi-permanent volunteers live in host communities, which help to incorporate themselves into local life more easily (Devereaux, 2008; Palmer, 2002: 639). However, language can be a significant barrier in international volunteer work following disasters, and a lack of common language can impede relationships between volunteers and locals (Nelan and Grineski, 2013: 309). Following the Haitian earthquake, 51% of volunteers reported that they spent more time among other international volunteers than the host community; 78% said that they relied on fellow international volunteers to cope with the stress of the situation (ibid.). These findings suggest that international disaster volunteers find a stronger citizenship within the volunteer community than within the host community.

Volunteering after a disaster can bring up significant emotions that are hard to share with others, even fellow volunteers (Jackson and Dellinger, 2007). It can be isolating to witness such devastation. However, volunteering can also be uplifting and empowering; the volunteers have the psychic benefit of doing good, and members of affected communities see outsiders offering help to strangers. The disaster aid worker experience is complex because it exists at the intersection of multiple identities. This research explores how individual volunteers have understood their multivalent roles in the post-disaster landscape and how those roles define citizenship.

Voluntourism

Volunteer tourism, or voluntourism, promotes travel to new places with the potential for positive impacts on local populations. Research into this kind of tourism articulates two themes about the 'self of the traveller', that travel is both a way to escape from day-to-day life and a method of self-development (Wearing and Neil, 2000: 10). These experiences are advertised as a means to 'authentic' experiences and intimate encounters with local host communities (Mostafanezhad, 2014).

Research has argued that volunteer tourism can create social cohesiveness among the volunteers if they work in a group (Coghlan, 2015: 55). The individual activity of tourism becomes a group activity where individuals build a sense of belonging to a community.

Disaster tourism and thanatourism

Disaster tourism occurs when individuals travel to places where disasters and great losses to society have occurred (Miller, 2008: 118). Unlike voluntourists, disaster tourists travel not to contribute to relief efforts but to "look with interest at devastation" (ibid.). Disaster tourists may participate in organized tours that showcase devastation, like the Katrina Tours of New Orleans created by a tour bus operator following Hurricane Katrina's landfall (Pezzula, 2009).

Disaster tourism belongs to a larger category of 'dark tourism', or thanatourism, defined as:

> Travel to a location wholly, or partially, motivated by the desire for actual or symbolic encounters with death, particularly, but not exclusively, violent death, which may, to a varying degree, be activated by the person-specific features of those whose death are its focal objects.
>
> *(Seaton, 1996: 240)*

Thanatourism is "less a fascination with death per se, than feeling for the particular people who have died (personal, nationalistic, or humanitarian)" (Seaton, 1996: 243). Visiting dark sites might allow the traveller to relate to the victims, as at Nazi concentration camps, Cambodia's killing fields, and Ground Zero in New York City (Knudsen, 2011: 63).

Knudsen (2011) states that individuals engage in thanatourism to take part in history; they bear witness to the landscapes of historical catastrophes to decode what happened in the past. Knudsen (2011: 59) explains witnessing as a performance that evokes a tourist's emotions about the event and as a moral act: "while the first aspect has to do with how actual tourists perceive and feel thanatouristic sites, the second aspect deals with how the performance of witnessing expresses responsibility in relation to atrocities in the past" (ibid.). This moral aspect of thanatourism may be similar to motivations behind disaster volunteerism. Disaster volunteers combine the categories of disaster tourist and voluntourist.

Research on disaster volunteers

To explore the experiences of disaster volunteers, this study includes two research questions: first, how do disaster volunteers and relief workers construct their identities socially; and second, how does this construction of identity contribute to their understandings of their citizenship in relation to a specific event, an affected community, a particular organization, and the volunteer community at large?

Methods

I conducted 13 semi-structured phone interviews with current and former disaster volunteers, one of whom became a paid disaster relief worker. Interviewees were asked about their experiences as disaster aid workers and their lives after volunteering. I recruited participants through Facebook by posting a public flyer asking for individuals who had volunteered or been relief workers in a disaster zone in the last five years. The recruitment flyer was posted twice over the course of two weeks. It was shared by a total of 34 individuals, including members of the Team Rubicon Situation Room forum. Team Rubicon (TR) (2020), primarily composed of military veterans and based in the United States, is an international non-governmental organization that responds to provide relief after disasters.

The interview participants included ten males and three females all living in the United States; their average age was 47. The interviewees spent anywhere from four days to a year deployed as volunteers. All of the interviewees had participated in domestic volunteer efforts in the United States, and six had volunteered internationally. Fifty-three per cent had volunteered with TR. Subjects also identified work with other organizations including Citizen Emergency Response Team, AmeriCorps National Civilian Community Corps, Nechama Jewish Response to Disaster, All Hands Volunteers (now named All Hands and All Hearts), American Red Cross, International Medical Corp, J/P Haitian Relief Organization (now named CORE), Jewish Healthcare International, and BonaResponds (an organization affiliated with St. Bonaventure University). Five interviewees had volunteered with more than one organization.

My experiences

In this study, I have used my own experiences as a disaster volunteer to supplement the data I collected through telephone interviews. I was a disaster volunteer for five separate events from 2010 to 2013. My first experience was an international volunteer deployment to Haiti following the 12 January 2010 earthquake. In 2011, I deployed to Margaretsville, New York, to respond to extensive flooding following the landfall of Hurricane Irene. In March 2012, I deployed to West Liberty, Kentucky, to help the community recover from an EF-5 tornado. In November 2012 and February 2013, I travelled first to New Jersey and then to New York to respond to damage from Hurricane Sandy. In May 2013, I volunteered after a series of tornadoes hit Oklahoma City, Oklahoma and its surroundings; the most famous of these was the EF-5 that hit Moore, Oklahoma.

Findings

Examining how disaster volunteers engaged in the construction of citizenship and identity in relation to places affected by disaster revealed three distinct identities: (1) event/location identity, (2) volunteer identity, and (3) organization identity.

Event identity

Some volunteer identities were tied to the disaster itself and to the visual appre-hension of the landscapes where they were deployed. Several of the interviewees expressed their shock at the impact of the event, particularly at the first event where they volunteered; for example:

> You watch it on TV and you see the footage of Port Au Prince and you see the footage of the way that people live, it's not until the plane lands and you actually get in the car and start driving through up to where you are going to be and you actually see it, then it really sinks in.

In this case, the volunteer indicated that his expectations, based on media reports, did not prepare him for the shock of reality when he saw the devastation caused by the 2010 earthquake in Haiti.

The majority of volunteers identified their first deployment as the one with most impact. In some cases, interview subjects described this experience as a 'before and after event'; the sense of self was transformed by deployment, and subjects recognized significant changes in themselves when they returned home.

I, too, recognized this sense of change to my overall identity after my first experi-ence of volunteering. Before my deployment to Haiti, I had never travelled to a developing nation or disaster zone. The culture shock in Haiti was significant. This shock, common among travellers to developing nations, was magnified by the destruction of the earthquake. Before my volunteer work in Haiti, I identified as a young, educated woman in my mid-20s who had an interest in helping others. When I returned home, I labelled myself as a disaster volunteer.

The interviews revealed that the strength of connection volunteers had to disaster events and locations related to the duration of their time in the disaster zone. Individuals deployed to disaster zones for one week or less were less likely to feel a strong connection to the landscape itself than those who stayed for longer than a week. Those who spent more time were able to meet more local commu-nity members and had chances to travel in the area on days off. This observation correlates with previous research indicating that longer durations of time in an area decreased social barriers between volunteers and residents (Devereaux, 2008).

A volunteer's ability to connect with the local population also depended on personal safety and organizational guidelines. One interviewee stated that she socialized with the members of the local community in some places but not in others.

> You don't go out after dark; it depends on who you're with and where in the country you are. In Cap-Haïtien we go out at night, we know that we can go to certain places. We aren't going to walk in bad places, and we stay in a group, but it's ok to go out. I would not do that in downtown Port Au Prince.

The sense of safety (or danger) influences volunteers' relationships to the landscape and local community. Someone who cannot interact with the local community

FIGURE 9.1   The author working on a composting toilet in Haiti—May 2010 (photo by Mary Nelan)

outside of volunteer work may not build a strong connection there and may instead identify with the volunteer or organizational community.

As in Urry's (2002) *The Tourist Gaze*, interviewees had mixed feelings about taking photographs in disaster zones. Photographs served as tools to revisit and relive moments; such images illustrated damage from the event, relationships with local individuals, and connections within the volunteer community. In general, interviewees felt it was unethical to take pictures of survivors and homeowners without permission:

> *I think it's subjective to the time and place of the picture. In my opinion, taking a picture to reflect direness or suffering at the expense of those that are in it—consciously, I don't agree with that. But to show a smile of appreciation, of a group hug with people that you are working with to help, to which they're complicit, I think there's nothing wrong with that.*

One interviewee described flyers on display in the community saying, "Don't be a disaster photographer", the flyers aimed to stop volunteers from taking photos of the landscape and the people. In some cases, interviewees' feelings were informed by organizational guidelines that required permission from photographic subjects (and, if the photographs were to be posted on social media, from the organization itself).

> I think you have to be deliberate with what you are taking the pictures of and why you are taking them. We also had a process too … if we took photos of work being done or of people, homeowners or members of the public, we had to get permission … If it was photos of work being done or photos that were representative of [the organization] we had to seek permission before those photos were published, through our PR rep … There are situations when it is inappropriate to take photos because we're not photojournalists, we're not telling a story per se, we are there to help people …

The public posting of photographs on social media was a significant issue:

> There are definitely circumstances where you are taking photos of really significant loss for these people, and you have to be careful with how you present those photos and whether you publish them and seeking the proper permission. It's very easy to be unethical with those photos. I think inadvertently you could do a lot of harm.

Another volunteer advised, "Be careful with what you are making public because these are people's personal lives, and you need to have a filter".

Volunteer tourists use photographs to show their work and to indicate to family and friends that they can also lend a hand (Mostafanezhad, 2014). Such photos, even if they were well-intentioned, could perpetuate an othering of the survivor population; they could also appear to place the volunteers on a pedestal for 'saving' the affected community. This idea made interviewees uneasy. They did not want to appear self-promoting or self-congratulating by taking and posting pictures.

In general, interviewees said that photos of fellow volunteers were a tool for remembering their experiences and relationships with colleagues; they preserved connections to places and events. Interviewees noted that photos of other volunteers were different from tourist photos. Volunteers did not want to be associated with ordinary tourists because their purpose was help, not sightseeing.

The interviewees characterized returning home as difficult. Many stated that family, friends, and neighbours in their own communities could not understand what their work or roles had been in the disaster area. Consequently, they kept much of their experiences to themselves. One interviewee characterized the complexity of coming home:

> Among those that didn't share the experience in my social circle, there certainly is an appreciation and understanding of the experience, but their knowledge and depth of that understanding has a static limit, so it's an observable plane and not an interactive plane.

A second interviewee stated:

> *It's tough to explain to people because it's almost like trying to explain to someone what it is skydiving off a cliff and how the air feels, and how you feel/what your body goes through when you are skydiving off that cliff, and the adrenaline rush until the chute opens and you're flying. It's like trying to describe that to someone where they can't really understand it because they can't understand physiologically or visually what that feels like.*

In both cases, their disaster relief experience made an indelible mark on the volunteers. Many of the interviewees said that in those first few days after returning home, they felt closer to the disaster-stricken community than to their own communities.

> *In the first few days, you are still kind of carrying where you were with you and I think also part of that is [that] you are being asked a lot of questions by other people about what you did and what you saw and what were things like. Not only is it still fresh in your mind, but it's always being brought up, so you are still carrying that local community with you.*

A volunteer's experience and the relationship that she has with the disaster-stricken community, culture, and landscape influence her perception and personal identity after the deployment ends.

> *In the initial hours and days, it is somewhat melancholy. You miss the engagement, you miss the purpose and when you get back to your own world, it often lacks the significance that you found in those engagements.*

My own experiences mirrored this. When I returned home from a volunteer deployment, I quickly became frustrated when people complained about minor inconveniences. I felt unproductive, as if I could have more impact clearing rubble in Haiti than doing schoolwork back home. Like many volunteers returning from Haiti, I started to look at plane tickets to go back the day after I came home. I did not return, though.

While one interviewee stated that he felt more connected to the disaster-affected community than his own in those first days back, he did indicate that it was nice to return to some sense of normalcy in his home landscape:

> *It's [nice] to be back in an area that hasn't been hurt by a disaster, that's a fairly normal thriving community. It affects you seeing homes that are destroyed where everything's out on the street, closed stores and things like that. So, having that sense of normalcy is really nice.*

Data from the interviews and my own experiences show tension between the transition to home (while still processing the volunteer experience) and the effort to understand how one's identity has been transformed by volunteering.

Volunteer identity

The majority of interview participants identified with and classified themselves as 'disaster volunteers' even when they were no longer in service in disaster zones. When interviewees were asked to define the term, there were several variations. One pointed to differences in motivation:

> *I guess I think of two separate groups of people. I think about people that are going out because they legitimately think that they want to help people. Then I also think about people that are doing it because it's exciting.*

Another interviewee incorporated experience into the disaster volunteer identity and stated that a person who volunteered for even one day in a disaster zone could claim the title 'disaster volunteer'. There are different levels beyond that minimum engagement.

> *If one person, one time, commits one day to clean up their local area, they are by definition a natural disaster volunteer … my hope is to differentiate those who pursue natural disasters and engagements of the sort [and] continually ready themselves for those situations going forward.*

One-day volunteers are different from those who volunteer consistently and after multiple events. Like more time in a disaster zone, more experience also affects the disaster worker's identity. This supports previous research findings that the length of time individuals spend volunteering in a disaster zone affects how closely they connect to the host community (Devereaux, 2008). Longer periods of time in a disaster area create both a deeper connection to the volunteer identity and to the area itself.

A few interviewees did not identify with the term 'disaster volunteer' and preferred to be called volunteers or responders. Those who did identify with the term carried it with them beyond the disaster zone and did not shed in their daily lives.

Relationships built during volunteering were also maintained afterwards. Time spent together connected interview subjects to other volunteers. "A big part of it is where are people staying? I think when people are housed together, there is more of a bond that forms". Those relationships were important to interviewees when they returned home.

One interviewee felt extremely isolated from his personal network at home and said that he tried to get back into the field to volunteer as often as possible. Being a member of this community and leaving your post can produce guilt feelings. One interviewee stated that going home was hard because "you are leaving behind people who are still there doing the work". Once volunteers return home, acknowledging membership in and work shared with their volunteer communities is important.

I am always careful to frame it as a group effort … even the things that I did and the things that I am proud of and the work that I put forward there, I didn't do alone. I did it with a lot of people supporting me. Whether or not they were part of my group or not … it's not an individual accomplishment, it's a community accomplishment when it comes to rebuilding a community.

Many interviewees likened their volunteer service to military service (though all were quick to qualify it as a lesser type of service than military service). They explained that only people who were in those disaster areas and had interacted with the survivors could understand their emotions and feelings. Not all of the interviewees experienced 'post-deployment blues' (a term coined by TR), but most identified a stronger connection to the location of their service and other volunteers than to their own family, friends, and neighbours in their first few days back at home.

I too found my experience hard to put into words for others without similar experience. Following my time in Haiti, I reached out on Facebook to many volunteers I had been deployed with. After each deployment, I friended fellow volunteers to stay connected and be reminded of the identity I had found as a volunteer. Several volunteers related the same behaviours, though the type of social media may vary and has probably changed over time. One interviewee who was able to meet up with his deployment companions after returning home stated, "There are parts that you take back that nobody will understand except for the people you were there with, and I was lucky that I had people to make that transition with me".

Reunions have been important both to me and other volunteers. Even when I have volunteered for new organizations and did not know anyone in that area, I was quick to make friends and felt like I had found a community of likeminded individuals.

The vast majority of interviewees agreed that they felt a bond with other disaster volunteers (even if they had not worked with them). Several interviewees characterized the experience as 'finding my tribe'.

It depends on the volunteer I run into. If it's someone who has deployed and who has done work, either domestically or internationally, there's a closer bond there because you have a similar point of reference. Whereas if it's somebody here who [says], 'yeah I volunteered at the community centre,' they're not going to understand, they can appreciate your experience, of course, but they're not going to understand being out in the field, what that's like.

The disaster volunteer community could understand experiences that many interviewees could not share with their family, friends, and neighbours. This community existed in a post-disaster landscape. Even if all members had not been at the same event or in the same geographic region, they had some shared experience.

FIGURE 9.2 Volunteers standing on the foundations of a home in West Liberty, KY after taking it down—March 2012 (photo by Mary Nelan)

Organization identity

The majority of participants worked with TR. This bias resulted from a TR volunteer's posting of the initial recruitment flyer to the organization's online forum. That organization's tight network and strong culture may have affected this study. TR is primarily made up of military veterans, and one volunteer recognized a divide between veterans and those who had not served in the military. She saw a clear differentiation when she was deployed, and since she had not served in the military, she felt excluded from certain activities. Her organizational identity was not as strong as others from TR.

AmeriCorps also has a strong organizational identification. AmeriCorps volunteers live together before, during, and after deployments. If they are part of the National Conservation Corps or AmeriCorps St. Louis, volunteers must wear a uniform of black boots, khaki pants, and an AmeriCorps t-shirt. An interviewee from AmeriCorps self-identified as an AmeriCorps volunteer, not a disaster volunteer. Although he interacted with volunteers from other agencies, "when we are coming home at the end of the night, we are staying with other AmeriCorps volunteers". That insulation is key to organizational identity.

One Red Cross volunteer was very strongly associated with her organization. A former health care worker, she had difficulty with interview questions about her personal experiences and identity in disaster zones. She had clearly distanced herself from emotions during her deployments and used her organization and training to maintain that distance. These distancing tactics may have helped to distance her from the affected disaster area. She answered many questions with descriptions of her organization's policies and protocols rather than accounts of her own experiences. Her strong organizational identity may have limited her event/location identity and/or volunteer identity.

TR identifies differences between volunteers through their shirt colours; one interviewee explained the difference:

> *Within team Rubicon, our structure, if you are an active member and you have gone through our vetting and onboarding process, you are a grey shirt, from top to bottom, from our CEO down to a person who has only signed up and has never actually participated, you earn the title of grey shirt. A white shirt is a spontaneous volunteer. If Team Rubicon is managing or mitigating a disaster response and some of the local community comes up and wants to participate, we make them what we call a white shirt.*

The individuals I interviewed were all grey shirts. They valued the colour's meaning within and connection to TR's organization: "We are very proud of our grey shirt

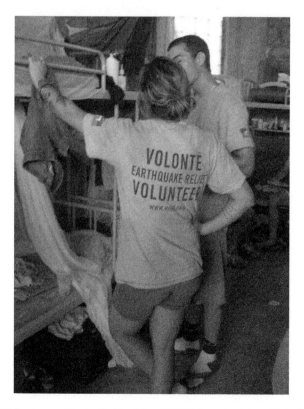

FIGURE 9.3   Volunteer t-shirt in Haiti—May 2010 (photo by Mary Nelan)

FIGURE 9.4 Volunteers removing a hay bale from a creek in upstate New York following Hurricane Irene—September 2011 (photo by Mary Nelan)

in that it's coveted and that it's earned. We take a lot of pride in that; we don't just hand it out". Even so, interviewees said that grey shirts worked hard to build a cohesive and inclusive culture on deployments irrespective of the different shirt levels.

T-shirts were strong identifiers for volunteers. Many chose to wear the organization's t-shirts (which sometimes identified them as volunteers) at work and after hours so that the local community could identify them. These t-shirts can become significant mementos for volunteers. If they contain location or event-specific information, they can contribute to all three disaster volunteer identities. On my first deployment to Haiti, I immediately bought an extra t-shirt and put it away in a locker. It was the only clean item I had after my monthlong trip, and I wore it home on the plane. I was proud of that shirt and what it signified.

Conclusion

Interviewees constructed their identities within three main categories: the event/location, the general volunteer community, and the volunteer organization they had joined. The interviewees varied in their strongest identifications and senses of belonging. Volunteers expressed strong attachments to the physical landscape, the disaster, and the local culture. When they return home, they feel a close kinship with the place where they've volunteered and to specific experiences there. Some volunteers look for plane tickets back; their goals are to feel useful and to return to a place where they felt a sense of belonging. Citizenship in the post-disaster landscape is interwoven with their organizational and volunteer communities.

However, this research also found a landscape citizenship that was not fixed in a geographic space. Interviewees expressed a bond with individuals who had had similar experiences as disaster volunteers, even if they hadn't worked together. The volunteer landscape does not have to be tied to one place, and citizenship does not have to be rooted in a single event.

Sometimes organizational identity became so strong that the interviewee's volunteer identity or connection to volunteers outside of the organization was weak. The Red Cross volunteer I interviewed was unable to answer questions as an individual but focused on organizational policy and culture. Her identity was tied up with the organization, not with any event or location or with the larger volunteer community. This differed from TR volunteers and AmeriCorps volunteers, who were more likely to go into the community and build bonds with other volunteers. This is not to say that all Red Cross volunteers have similar experiences, simply that this woman identified through the organization. Future research should investigate how organizational culture can promote or inhibit interactions and the relationship building with local community members and other volunteers. Further research could also investigate whether a lack of interaction has negative impacts on volunteers' well-being and whether the affected communities suffer negative outcomes when volunteers are isolated.

These volunteers' experiences and formations of identity and citizenship with respect to events, other volunteers, and organizations are important features of the post-disaster arena. Following disasters, the convergence of volunteers adds a new and transient community to changing landscapes. These workers can affect the recovery effort positively, but they have not been extensively studied as a population. This research seeks to build understanding so that disaster volunteers can be mobilized more effectively in future disasters and so that their needs are recognized and addressed during and after their service.

Bibliography

Brown, Barbara and Douglas Perkins (1992). 'Disruptions in place attachment'. In: Irwin Altman and Setha M Low, eds. *Place Attachment. Human Behavior and Environment Advances in Theory and Research*. Boston: Springer, pp 279–304.

Chamlee-Wright, Emily and Virgil H. Storr (2009). "There's no place like New Orleans': Sense of place and community recovery in the Ninth Ward after Hurricane Katrina'. *Journal of Urban Affairs*, 31(5), pp 615–634.

Coghlan, Alexandra (2015). 'Prosocial behavior in volunteer tourism'. *Annals of Tourism Research*, 55, pp 46–60.

Dann, Graham and Jens Kristian Steen Jacobsen (2003). 'Tourism smellscapes'. *Tourism Geographies*, 5(1), pp 3–25.

Devereaux, Peter (2008). 'International volunteering for development and sustainability: outdated paternalism or a radical response to globalisation?' *Development in Practice*, 18(3), pp 357–370.

Drabek, Thomas E (2013). *The Human Side of Disaster*, 2nd ed. Boca Raton, FL: CRC Press.

Edensor, Tim (2000). 'Staging tourism: Tourists as performers'. *Annals of Tourism Research*, 27(2), pp 322–344.

Erikson, Kai T (1976). *Everything in Its Path: Destruction of Community in the Buffalo Creek Flood*. New York, NY: Simon and Schuster.

Fritz, Charles E. and John H. Mathewson (1957). *Convergence Behaviors in Disasters; A Problem in Social Control*. Washington, DC: National Academy of Sciences – National Research Council.

Helsloot, I. and Ruitenberg, A (2004). 'Citizen response to disasters: A survey of literature and some practical implications'. *Journal of Contingencies and Crisis Management*, 12(3), pp 98–111.

Jackson, Jeffrey T. and Kirsten A. Dellinger (2007). 'Volunteer voices: Making sense of our trip to the Mississippi Gulf Coast after Katrina'. In: Danielle A. Hidalgo and Kristen Barber, eds. *Narrating the Storm: Sociological Stories of Hurricane Katrina*. Newcastle, UK: Cambridge Scholars Publishing, pp 104–127.

Kendra, James M. and Tricia Wachtendorf (2003). 'Reconsidering convergence and converger legitimacy in response to the World Trade Center disaster'. In: Lee Clark, ed. *Terrorism and Disaster: New Threats, New Ideas*. Bingley, UK: Emerald Group Publishing Limited, pp 97–122.

Knudsen, Britta Timm (2011). 'Thanatourism: Witnessing difficult pasts'. *Tourist Studies*, 11(1), pp 55–72.

Knudsen, Daniel et al. (2008). 'Landscape, tourism, and meaning: An introduction'. In: Daniel C. Knudsen et al., eds. *Landscape, Tourism, and Meaning*. New York, NY: Routledge, pp 1–7.

Miller, DeMond S (2008). 'Disaster tourism and disaster landscape attractions after Hurricane Katrina: An auto-ethnographic journey'. *International Journal of Culture, Tourism, and Hospitality Research*, 2(2), pp 115–131.

Mostafanezhad, Mary (2014). 'Volunteer tourism and the popular humanitarian gaze'. Geoforum, 54, pp 111–118.

Nash, Dennison (1996). *The Anthology of Tourism*. Tarrytown, NY: Elsevier Science.

Nelan, Mary and Sara Grineski (2013). 'Responding to Haiti's earthquake: International volunteers' health behaviors and community relationships'. *International Journal of Mass Emergencies and Disasters*, 31(2), pp 293–314.

Olwig, Kenneth R (2005). 'Representation and alienation in the political land-scape'. *Cultural Geographies*, 12(1), pp 19–40.

Palmer, Michael (2002). 'On the pros and cons of volunteering abroad'. *Development in Practice*, 12(5), pp 637–643.

Perkins, Harvey C and David C Thorns (2001). 'Gazing or performing?: Reflections on Urry's tourist gaze in the context of contemporary experience in the antipodes'. *International Sociology*, 16(2), pp 185–204.

Pezzula, Phaedra C (2009). '"This is the only tour that sells": Tourism, disaster, and national identity in New Orleans'. *Journal of Tourism and Cultural Change*, 7(2), pp 99–114.

Seaton, AV (1996). 'Guided by the dark: From thanatopsis to thanatourism'. *International Journal of Heritage Studies*, 2(4), pp 234–244.

St. John, Craigh and Jesse Fuchs (2002). 'The heartland responds to terror: Volunteering after the bombing of the Murrah Federal Building'. *Social Science Quarterly*, 83(2), pp 397–415.

Team Rubicon (2020). *Team Rubicon Disaster Response*. Available from: teamrubiconusa.org/relief/ (Accessed 15 August 2020).

Tierney, Kathleen, Christine Bevc, and Erica Kuligowski (2006). 'Metaphors matter: Disaster myths, media frames, and their consequences in Hurricane Katrina'. *The Annals of the American Academy of Political and Social Science*, 604(1), pp 57–81.

Urry, John (1992). 'The tourist gaze "revisited"'. *American Behavioral Scientist*, 36(2), pp 172–186.

Urry, John (2002). *The Tourist Gaze*, 2nd ed. Thousand Oaks, CA: Sage Publications.

Urry, John and Jonas Larsen (2011). *The Tourist Gaze 3.0*, 3rd ed. Thousand Oaks, CA: Sage Publications.

Wallace, Anthony FC (1956). *Tornado in Worcester*. Washington, DC: National Academy of Science.

Wearing, Stephen and John Neil (2000). 'Refiguring self and identity through volunteer tourism'. *Society and Leisure*, 23(2), pp 389–419.

Zavar, Elyse and Ronald Schumann (2019). 'Patterns of disaster commemoration in long-term recovery'. *Geographical Review*, 109(2), pp 157–179.

10

BEIRUT'S PUBLIC REALM AND THE DISCOURSE OF LANDSCAPE CITIZENSHIPS

Jala Makhzoumi

Introduction

Beirut's turbulent political history has made the public landscape a focus of struggle and activism. Public spaces constructed by Ottoman and French colonial regimes evolved into sites for the lived practices of citizenship through times of peace, civil war, and neoliberal privatization. Presenting at the Landscape Citizenships Symposium in 2018, I cited two examples from Beirut, the Dalieh of Raouche coastal site and the Beirut Pine Forest municipal park, where a denial of public access motivated citizens into action. I was one of a handful of active protesters in the Dalieh of Raouche, provoked by the threat of private developers to redevelop the last remaining publicly accessible, exceptionally beautiful landscape in the city (Figure 10.1). I shared with fellow protestors a sense of empowerment in enacting our rights as citizens but also frustration at the immense task of facing up to a corrupt state that was siding with powerful property developers.

A year later, thousands of protestors from all over Beirut—rich and poor, Christian, Muslim, and Druze—marched shoulder to shoulder, putting aside political affiliation and sectarian differences. The *intifada*, as the mass uprising of 17 October 2019 came to be known, was a spontaneous revolution triggered when frustration with rampant state corruption reached a boiling point. The right to publicly accessible, green, and open urban spaces, Beirut's public realm, is but one grievance in the fight against government corruption and sectarian politics that triggered the nationwide demonstrations.

As a culturally significant and emotionally charged place of everyday life, landscape enables the practice of citizenship. In this chapter, I argue that contested landscapes of the public realm create 'active citizenship' (Buijs et al., 2016) by empowering ordinary citizens to collectively protest social injustice and claim denied rights.

FIGURE 10.1   The Dalieh of Raouche (photo by Jala Makhzoumi)

In this chapter, I begin by defining the key terms, 'landscape', 'citizenship', and the 'public realm' to establish the theoretical framework for landscape citizenships. I then trace the histories of Beirut's public landscapes, applying a landscape framing to explore three contested sites: the corporate landscape of the reconstructed historic core, the Beirut Central District (BCD); coastal nature pockets threatened by privatization, of which the Dalieh of Raouche is one; and the Beirut Pine Forest, a city-scale municipal park that has been closed to the public. I conclude with the October 17th mass uprising and occupation of the BCD as the ultimate expression of active citizenship: inhabitants of the city reclaiming squares and streets from which they were excluded by the corporate takeover.

Landscape, citizenship and the public realm

'Landscape' is a word with many meanings, a way of perceiving the inhabited world that is culturally informed and place specific. As an idea, landscape implies visible, tangible setting (the *spatial* dimension), embraces invisible natural and human transformative processes (the *ecological* dimension), and incorporates intangible social values and cultural practices (the *political* dimension) (Figure 10.2). Architects, landscape architects, urban designers, and planners deal with landscape predominantly as physical space, the place of human habitation. Historians, geographers, and anthropologists explore the social, political, and legal phenomenon of landscape. Often challenging the Euclidean space sensibility of designers and planners, their research focuses on landscapes as "places [that] are governed relatively democratically and valued as the landscape of home by many ordinary citizens" (Olwig, 2018). Although planned landscapes are not uncommonly criticised as being imposed, created through "enclosure and dispossession of the commoners in the interest of the wealthy" (Olwig, 2016: 16), the spatial and the political spheres are complementary, imbedded in the idea of landscape citizenship.

The European Landscape Convention (ELC) exemplifies the expansive meaning of landscape as both physical setting and political project, just as it demonstrates landscape's potential for the cooperation of planners, researchers, scientists, and administrators. Under the overarching aim of protecting European landscape heritage and identity, the ELC expands the conventional understanding of landscape as *scenery*, emphasizing that landscape "is place and culture specific, a framework to

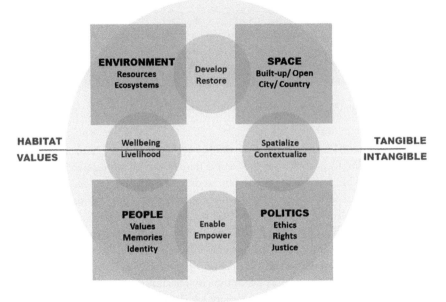

LANDSCAPE

ENVIRONMENT
Resources
Ecosystems

Develop
Restore

SPACE
Built-up/ Open
City/ Country

HABITAT

VALUES

Wellbeing
Livelihood

Spatialize
Contextualize

TANGIBLE

INTANGIBLE

PEOPLE
Values
Memories
Identity

Enable
Empower

POLITICS
Ethics
Rights
Justice

FIGURE 10.2 The layered concept of landscape embraces the tangible (environment and ecosystem, city and region) and intangible social values (shared memories, cultural practices, ethics, and rights) (diagram by Jala Makhzoumi)

address a rubric of concerns, environmental, socio-economic, cultural and political" (Makhzoumi, 2018). The ELC aims to engage people in a democratic process that reaffirms the right to enjoy and benefit from European landscapes and assume the responsibility of managing them sustainably for the benefit of future generations.

Citizenship, however, is not passive but needs to be practiced. Underlying the definition of citizenship, whether that of a country or city, and the enjoyment of the privileges offered is an acceptance of custodial responsibilities entailed. 'Active citizenship' is a term coined to denote the "ability of citizens to organize themselves", "to mobilize resources and to act" in the public interest to protect and/or claim their rights (Buijs et al., 2016: passim; Moro, 2012: passim). The ability of citizens to 'mobilize' and 'organize' is context specific because it unfolds in a tangible space—landscape—and because it changes from one place to another across cultures and over time. The practice of citizenship also involves a behavioural dimension of accepted manners and rules of conduct. These rules evolve over time, but more generally they are imposed by local/state authorities who monitor what is deemed acceptable conduct and what is not.

In its most common use, 'public realm' implies spaces that are publicly accessible, such as streets, waterfront walks, squares, and municipal parks. The universalist notion of the public realm, or public sphere, however, can be problematic because it relies on the public–private binary embedded in property regimes of a modernist view of the city. As such, the term tends to undermine the complexity of 'public', that the word has multiple meanings that yield different understandings of the nature of contemporary public space (Iveson, 1998). Iveson proposes four broad models of public space that consider more explicitly the contestation over time of public–private boundaries and the relations between different publics (ibid.: 22). Moving away from the public–private binary, a broad definition of the public realm is given by Sennett (2010: n.p.), as the "place where strangers meet", a space of social interaction, inclusive and accepting of all, not only citizens, and a place of intrigue and excitement. Despite the shortcomings, the term 'public realm' is pre-ferred to the more commonly used 'public space'. This is mainly to avoid the short-coming of public space construed simply as "the container or field of public life … inert rather than constitutive—the surface upon which the interaction of publics is played out" (Iveson, 1998: 29).

The public realm is one component of the urban landscape, albeit an important one, and both landscape and the public realm are politically charged. The diffe-rence is that *public realm* is objective, an urban typology, while *landscape* is sub-jective, embracing the totality of the urban context (of buildings and the spaces in between), of organic and inorganic more-than-human elements that modify and inflect socio-political experience and human agency.

Landscape has been used to contextualize human rights (Egoz et al., 2011), define democracy (Egoz et al., 2018) and explore human agency (Wall and Waterman, 2018). A landscape framing of the public realm links citizenship with governance. Foucault coined the term 'governmentality', combining 'government' and 'men-tality', as both a way of working and acting, but also a way of imagining relations between humans through the lens of government (Dean, 2010). Roy speaks of 'civic governmentality', arguing that "grass roots, or civic, regimes of government" have the ability to "recreate the terms of rule and citizenship", and thus expand the question of governmentality "to include not only governable spaces and disciplined subjects but also forms of self-rule in the context of the production of space in the bourgeois city" (2009: 160–161). Parallel to these is the less explored possi-bility of 'self-governmentality', a lens for viewing democratic relations as arising from people and their ability to self-manage their lives in their landscapes, which most people are already doing everywhere without government intervention or guidance. Landscape's potential to frame both scenarios, governmentality and self-governance, is explored in the discourse of landscape citizenships in the city of Beirut, Lebanon.

The emergence of public realm landscapes

Contemporary Beirut grew up on the site of the Roman Berytus at the eastern edge of the Mediterranean. By the mid-19th century, the ancient city had shrunk in size, buried under a provincial, medieval harbour town, with fields and orchards encircling the walled city. The two hills of Ashrafiyeh to the east and the Ras

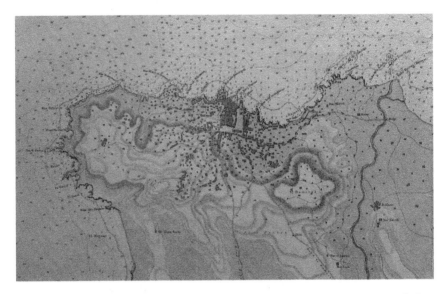

FIGURE 10.3 Beirut in 1860. The hills of Ashrafiyeh, east, and Ras Beirut, west, shelter the natural harbour of the historic city. Evidence of early urban expansion outside the medieval city walls (map by Admiralty, London, 1860)

Beirut promontory to the west sheltered the harbour town and its verdant plains (Figure 10.3). By the early 20th century, the city walls were absorbed by residential neighbourhoods that extended to both hills. The city inhabitants were religiously diverse, including Christians, Muslims, and Druze communities, European and American expatriates.

The concept of public realm was introduced during the last decades of Ottoman rule and early years of the French Mandate, part and parcel of modernizing, that is, westernizing traditional cities. In Beirut, Haussmann-style wide streets and squares were imposed on the traditional urban fabric, waterfront boulevards opened up the city to the Mediterranean Sea, and parks, squares, and amenity landscapes adorned the city. The regimental, spatially dominant landscape of the public realm undermined the traditional private–public overlap in the historic city, where commercially vibrant market streets and socially active neighbourhood alleyways were the place of interaction for business and leisure.

By the 1960s, new public spaces and the practices they suggested had been successfully integrated into the life of Beirutis. The city had a population of less than a million, low density buildings, abundant residential gardens, and views to the Mediterranean from all over the city. By this time, the modernized city centre was inclusive for all citizens, whether rich or poor, and acted as transportation hub as well as place of commerce and entertainment. The Mediterranean waterfront, the Corniche promenade, was especially successful as a new and exciting experience for the city, a place where men and women showed off their newly acquired western attire. A unified definition of the public would have been challenging because by this point multiple publics had emerged to occupy Beirut's public realm.[1]

It was at this juncture that the word 'landscape' was introduced as part of local architectural discourse. This timing explains why the two terms of *public realm* and *landscape* are used interchangeably and why the meaning of 'landscape' today continues to be associated with urban open spaces and municipal parks. This limits the perception of landscape architecture to urban beautification and undermines the versatility of a definition that has the power to frame the discourse on citizenship, in both cities and countryside (Makhzoumi, 2015).

Other capital cities, including Damascus, Baghdad, and Tehran, were similarly establishing wide streets, plazas and municipal parks to modernize their culture and identity. Lebanon, however, is singled out from neighbouring Arab countries for readily adapting to change, by virtue of its openness to the Mediterranean Sea, and in its sectarian and clientele system of governance. The system ensures that government offices are distributed proportionately to represent the religious communities, granting the Lebanese voting rights in state and parliament elections. After gaining its independence from France, a confessional political system was put in place upon declaration of the Lebanese republic in 1943. Institutionalized sectarian governance and the laissez-faire system of economic policies favoured private investment, but they were balanced by centralized state agencies that offered social, health, educational, and infrastructural services.

Tensions resulting from institutionalized sectarian politics disintegrated, leading to armed conflict and the Lebanese Civil War (1975–1990). The Taif Accord, the agreement reached in 1990 to end the civil war, failed to resolve the institutional sectarian tensions that have since, and up to the present, been at the heart Lebanon's problems. Just as worrying was the shift in state politics towards Neoliberal economies. Multinational corporate development catering to the wealthy emerged as a key driver of the economy through privatization of state institutions and appropriation of land. Neoliberal politics were engulfing not only Beirut but other Arab capital cities with the result of intensifying "issues of social equity, inclusion-exclusion, and accountability," gradually transforming them into "places of play and commodities themselves" (Daher, 2008: 47). Reduction in state social responsibilities was another facet of the political shift, as witnessed in post-civil war Lebanon with the state relinquishing its role in upholding the right of ordinary citizens, commonly siding with developers. In the words of Malley,

> Lebanon's economic and social development after independence, the 15-year civil war that almost destroyed the country, and the agreement that ultimately ended that war all provide a basis for understanding the nature of institutionalized sectarianism in the modern world and the kinds of conflicts and compromises that such sectarianism is likely to engender.
>
> *(2008: 121)*

The corporate landscape narrative of the public realm

Much of historic Beirut was destroyed during the Lebanese Civil War. Solidere,[2] a private shareholding real estate company, was entrusted by the Lebanese government to manage reconstruction and renewal of the historic core, what came to be known as the BCD. The Solidere project was the brainchild of Rafiq Hariri, and it marked the beginning of the neoliberal political-economic discourse which culminated in 1992 when he became prime minister of Lebanon. As such, Solidere combined political prowess and coercion to expropriate the historic centre in its entirety, encompassing a surface area of about 180 hectares, in addition to 60 hectares of land reclaimed from the sea by the tsunami breakwater. Solidere's master plan proposed an urban core with a mixture of office space, residential areas, commercial and retail zones, parks, and tree-lined promenades.

Criticism of the project, argues Makdisi, lies not only in that it "confuses public and private interests, but that it represents the colonization of the former by the latter" (1997: 693). This is evident in the project's disregard for the public and its dismissal of citizens whose property, inherited over generations, was expropriated by the company. With the powerful Hariri as prime minister, the state sided with Solidere justifying privatization as the means to animate the post-war economy in a time of political instability.

Solidere rightfully claimed that its proposed master plan increased the area of public realm by 25% (Gavin and Maluf, 1996). In fact, half of Municipal Beirut's public spaces today are located within the area controlled by Solidere. Although public spaces were diversified to include the Beirut Heritage Trail commemorating the city's history, and the Garden of Forgiveness as a place of healing, reconciliation, and memorializing the civil war, Solidere prioritized real estate value over built heritage, cultural values, and sentimental attachment. More problematic, Solidere's notion of open spaces in the BCD stripped *landscape* of its social and cultural significance, reducing it to a superficial design of urban scenery. Gavin and Maluf praise the high-quality landscape of Solidere and how it "incorporates all aspects of the visible appearance of the public domain and new infrastructure, including material finishes and paving, public lighting and urban street furniture", and that the "extent and quality of such finishes and planting itself will create a strong, positive image in the new city center" (ibid.: 111). Prioritizing the visual was similarly the approach of architecture, "achieved specifically and solely in *visual* terms or, to be precise, in terms of appearance and façade" (Makdisi, 1997: 686).

While the more privileged members of Beirut's society undoubtedly benefited from Solidere's corporate development, the majority, however, came to be excluded. Although ordinary citizens were not denied access, they were intimidated by the high-end, exclusive, and impeccably staged BCD, monitored by manned security and CCTV cameras. Regular signage instructed visitors of acceptable behaviours within the 'public realm' (Figure 10.4), while the price-point of commercial units in the BCD meant ordinary citizens could not afford to sit in a café or buy an ice cream. In an interview, Mohammad Ayoub, the president of NAHNOO youth-led non-governmental organization (NGO), reflects on privatization of the public realm in the city:

> everyone can't go to cafes or malls, and for those of us who are unable to leave Beirut on the weekend, what could be better than the public park to make a difference? People, though not necessarily aware of it, feel the need for public spaces.
>
> *(Lautissier and El-Mlaka, 2016)*

Food vendors, common to traditional practice of the publics elsewhere in the city, were banned from the gentrified BCD. Fishing, a traditional practice, was similarly prohibited. Invisible boundaries came to separate ordinary citizens emotionally, physically, and socially from the heart of their city, fostering a helplessness and sense that boundaries could be blithely rearranged by the wealthy (Waterman, 2018).

Alongside physical changes to the BCD, Beirut witnessed a new repressive state apparatus, which included censorship of the media, enforced by Hariri's ascension

FIGURE 10.4 The Solidere developed Zaitunay Bay Marina. The ambiguity created by listing unacceptable social practices on the site boundary, with demarcation of the waterfront as "Private Property", evidences Solidere's appropriation of the public realm (photo by Jala Makhzoumi)

to power in 1993. Up until then, Lebanon had prided itself on the freedom of press. Media censorship affected practice of the publics. For example, "a ban on street protests of any kind" was enacted after repeated clashes with trade unions seeking to organize strikes and demonstrations (Makdisi, 1997: 697). Although the ban on street protests was abandoned following the death of Hariri, securitization, both public and private, increased and compounded socio-spatial segregation, reconstructing people's geographies of the city that underlies "citizenship, entitlement and more generally the right to the city" (Fawaz et al., 2012: 191).

Beyond the BCD, Solidere's corporate model influenced state planning by shifting the focus to infrastructure-led urban development that prioritized the car (Harb, 2013). Outside the BCD, the quality of urban living was increasingly impoverished. The majority of Beirut's inhabitants found themselves hemmed into crowded middle- and low-income neighbourhoods. Vacant land, which had doubled up as breathing spaces for these communities, had shrunk from 40% of municipal Beirut in 1967 to less than 10% in 2000 (Lautissier and El-Mlaka, 2016). Drinkable water shortages and power cuts were part of everyday life, untreated sewage discharge polluted the sea, and inhabitants of the city endured the difficulty of moving through congested roads with no apparent future plan for traffic management. Beirut's last remaining inclusive public space, the waterfront, was incrementally transformed by property developers into exclusive sites, similar to Solidere's Zaitunay Bay Marina, with high-end shops and restaurants (Figure 10.5). As prime sites on the waterfront, the Ramleh al Bayda public municipal beach, the only such beach in Beirut, and the Dalieh of Raouche were targets for development. Interviewed by *The Guardian*, a fisherman from the Dalieh of Raouche speaks of "the huge bounty of fish to be caught near Solidere's Zaitunay Marina", lamenting that "private security restricts him from getting close" (Battah, 2015). Asked what he makes of projects like Zaitunay and their promise of public access, "Itani lifts his head and smiles: Those places are not made for you and me" (ibid.).

The activist landscape of the public realm

It is hard to function as a citizen when a landscape fails to support the necessities for contemporary life, or when it actively works against the practices necessary for a good life. In response to incursive development threats, civil society organisations, youth groups, and ordinary citizens reacted to the declining quality of life in Beirut (Harb, 2013), transgressions of citizen rights, and a growing sense of social injustice of which the public realm was one aspect. Well-organized and ready to mobilize, activist groups vocalized their key concerns around environmental health, but also affordable housing, public transportation, health, and privatization of the maritime public domain.

In 2015, a total breakdown in solid waste collection triggered a large-scale public protest. Discontented masses congregated in Martyrs' Square.[3] 'You Stink' protests, a

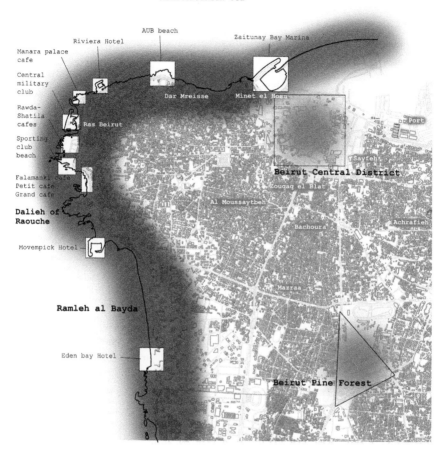

FIGURE 10.5   Incremental privatization of the Beirut waterfront (white squares). The Dalieh of Raouche and Ramleh Al Bayda (contested coastal sites), the triangular Beirut Pine Forest (bottom), and the historic city, the Beirut Central District (square frame) (map by Jala Makhzoumi)

reference to rampant state corruption and the literal stink of accumulated garbage, were rapidly and effectively dealt with by the authorities under the pretext that the demonstrators were destroying private property in the BCD.

In contrast to the totalizing landscape of the BCD, the narrative of campaigners is grounded in informal social practices in a range of public, semi-public, and private landscapes. These practices and experiences remain undesignated, with no legal categorization as 'public', argues Harb (2013), and as such, are 'politically subversive' in their potential to become appropriated and claimed through spatial practices. Two of the smaller campaigns, the Beirut Pine Forest and the Dalieh of Raouche, are herein discussed.

Protesting closure of the Beirut's Pine Forest

The Beirut Pine Forest, *Bois des Pins*, 'Horsh al-Sanawbar', or simply 'Horsh', is all that remains of the extensive 16th-century forest encircling the land side of the city. Historically, the pine forest offered a space for military training, promenading, and social gatherings during festivities. As with natural and rural landscape in urban peripheries, this forest was incrementally appropriated by the expanding urban footprint, with large tracts allocated to various state and religious institutions. What remained of the Horsh, 33 hectares, was destroyed during the civil war. Declared a municipal park and in 1990, with funding from the Paris Municipality, the Horsh was reforested and landscaped with walkways, lighting, and parking spaces along the peripheries. The park was closed to the public under the pretence of protecting the pine tree seedlings.

By 2015, the pine seedlings had matured, accentuating the prominence of the park as the only city-scale green area within the capital. However, the forest remained closed to the public. This at a time when the location of Horsh was at the heart of high density and socially segregated neighbourhoods: the communities of Sunni Tarik el-Jdideh; Christian Badaro and Mathaf; Shiite Chiyah; and the Sabra and Shatila Palestinian refugee camp. Implicitly, and unjustly, the municipality feared that these socially marginalized, low-income communities would vandalize the park. Beirut municipality's dismissive narrative and disdain towards economically challenged citizens projects the class bias that prevails in state institutions, and at a time when the Horsh is the only breathing space for these excluded communities. Ironically, exceptions to closure are made for those producing non-Lebanese identity documents and for individuals holding entry permits from the Governor of Beirut.

A local, youth led NGO, NAHNOO, spearheaded the campaign to open-up the park to the public. Organizing protests alongside planning debates with the Beirut Governor and municipality officials, NAHNOO campaigned relentlessly to raise awareness of the right of citizens to access the Horsh, a public realm landscape par excellence (Figure 10.6). Protestors organized by NAHNOO wore blonde wigs, parodying the decision by Beirut municipality to deny entry to Lebanese citizens but allow non-Lebanese, Western expatriates to enjoy the park.

In 2016 and through dialogue with the Governor of Beirut, NAHNOO succeeded in securing the opening of the park for one day a week, which continues to the present. The COVID-19 epidemic is a reminder, however, that the opening of the park is not a luxury but a public necessity—where else can citizens, especially the youth, of surrounding neighbourhoods go for exercise and fresh air?

The civic campaign for the protection of the Dalieh of Raouche

The Dalieh of Raouche is a unique natural site on the Beirut waterfront. The landscape comprises a natural fishermen's cove, marine caves, coastal terraces and rock ponds, 15-metres high coastal cliffs, and two breakaway rocks. The cliffs, a visible

FIGURE 10.6 Poster calling for a public debate on the Beirut Pine Forest, part of organized action by NAHNOO NGO that continues to protest public rights to Beirut's largest municipal park (by NAHNOO and Rena Karanouh (graphic artist))

slice into the geological formation of the city, and the precious wildlife habitats the site provides, assert the site's biodiversity significance.[4] Just as important, Dalieh is currently the only large-scale publicly accessible open space in Beirut that can hold public gatherings on holidays and feast celebrations, where people can experience nature, and where children learn how to swim in the safety of the coastal rock ponds.

The Dalieh of Raouche was popular in the decades preceding the civil war. During the war, however, the site's spatial seclusion discouraged families from visiting. By the end of the civil war, many Beirutis were not aware that the site includes more than the landmark breakaway rocks. This is partly because the water edge lies below the street level and as such is not visible, but also because there are no signs to announce the site, no formal entrance, and no paved pathways to facilitate public access. Moreover, as a publicly accessible landscape, the Dalieh was no longer popular in the decades following the civil war, as emerging richer residents of the city favoured exclusive, private sea resorts. With Beirut's Pine Forest closed and the city's waterfront incrementally appropriated, the Dalieh of Raouche and the Al Ramla Al Bayda municipal sandy beach, further south from Dalieh, were the only amenity landscapes available to low-income families, where all ages could interact regardless of gender, ethnicity, religion, and socio-economic status.

When first visiting the Dalieh of Raouche, I realized the site was a living example of the multiple dimensions of landscape—spatial, ecological, and social—and as such, a living example that landscape was not just 'scenery', the understanding that prevailed among my students.[5] I selected the Dalieh as a site for teaching many landscape design studios at the American University of Beirut, located just a 15-minute walk from the promontory. At that point, I decided that I would do all in my power to protect this unique landscape if it was ever threatened.

The Dalieh of Raouche, as indeed the entire Lebanese coast, was recognized legally as a continuous public domain by Decision No.144 of the French High Commissioner in 1925. Although designated plots were granted to Beiruti families during the Ottoman era (comprising about a 1/3 of the total area of the Dalieh of Raouche, or 11 hectares), property records dating to the 1940s indicate that the entire site was deemed *non-aedificandi*, or not designated for construction. In the lawlessness of the civil war, former Prime Minister Rafiq Hariri purchased the largest section of Beirut's maritime waterfront, as well as coastal lands along Lebanon's Mediterranean coastline. With political clout, he then proceeded to issue legal exemptions to change the planning status of coastal lands to 'buildable'. Indeed, the same procedure was followed in the land bordering Dalieh, the current location of the Movenpick Hotel complex. However, unaware that the Dalieh had been sold, the public continued to flock to the site for their regular social activities.

Private interest and public rights clashed in 2013 when the developer fenced off the site to commence construction. Alongside fencing, the developer demolished fishermen's houses and kiosks catering cooked fish to the public. Dalieh fishermen were the first to inhabit the site in the early 20th century, their trade passed down over generations. Some accepted the meagre financial compensation offered, many

FIGURE 10.7 Visitors to the Dalieh of Raouche negotiate the precarious descent to the site following the developer's fencing of the site. Concrete wave-breakers deposited by the developer in the heart of the site are visible (photo by Jala Makhzoumi)

refused. As if fencing was not enough to close off the site, concrete wave-breakers were positioned across the main plateau, occupying a large portion of the open space used for public gatherings (Figure 10.7).

The uproar that followed developed into organized resistance. I joined the protestors, many of them colleagues (architects, urban and landscape designers, planners, and graphic designers) as well as environmental activists and Lebanese NGOs. In the months that followed, we formalized our efforts under the 'Civic Campaign for the Protection of the Dalieh of Raouche' (Battah, 2015). Resistance was practiced through sustained, embodied presence on site. Defying the developer and despite Beirut Municipality positioning guards to prevent crowds from bringing down the metal fence, activists climbed through breaks in the fence to access the site. The ominous wave-breakers were covered with graffiti, and protest banners were stretched across the rocks demanding public access. The campaign organized press releases, television debates, and an international ideas competition for alternatives to privatization. The campaign prepared and published a free, bilingual booklet with facts and illustrations elaborating the archaeological, mythological, biodiversity, geological and social significance of the place, including the (il-)legal acquisition of the land, discussed below.[6] The campaign wrote an open letter to the architect, Rem Koolhaas, commissioned by the owner to develop plans for the Dalieh site, explaining the ecological significance of the Dalieh of Raouche and the social injustice entailed in privatizing the landscape. Ultimately, these actions were successful in maintaining public access and putting a hold on construction.

In the hope that international recognition would act as a warning to state authorities and cause the owner/developer to tread carefully with their plans for the site, and having exhausted all means available to us as local campaigners, we petitioned for the site to be included in the World Monument Fund's 2016 World Monuments Watch. Our petition was approved and the Dalieh of Raouche was listed, described as "a beloved coastal promenade in Beirut that is the latest victim of a development frenzy that has destroyed many of the city's open spaces" (World Monument Fund, 2016).

The precarity of the Horsh Beirut and the Dalieh of Raouche landscapes were opportunities, small as they were, for us citizens to face up to state injustice and corruption. Emboldened by our success, Dalieh campaigners expanded their mission to address transgressions beyond the capital. Working with environmentalists, NGOs, fishermen syndicates, lawyers, and ordinary citizens, the National Coastal Alliance was launched in May 2017 as a platform for citizens to pursue legal action against transgressors of the maritime public realm along the entire Lebanese coastline.[7] Two years passed before the state responded, pursuing legal action, albeit 'soft', against perpetrators of the maritime public domain, and then only to appease demand from the 17 October 2019 Lebanese civil protests against state corruption.

In summary, it is the perception and valuation of landscape as 'nature', and of landscape as a communally shared space, or commons, that underlies the discourse of activists protesting the Dalieh of Raouche and the Horsh Pine Forest. Memories of the Horsh as a communal shared space were too distant for citizens to remember. The Dalieh, however, was very much a lived-in milieu and active social space that was closed abruptly and brutally, and as Waterman states, "forbidding access or ejecting people is often a sign of the breakdown or denial of democracy in the public sphere or public realm" (2018: 143). Beyond commons, in a city that opens to the Mediterranean Sea, the Dalieh and the Al Ramla Al Bayda sandy beach are the last window affording citizens a sea view. Privatization of the maritime public domain appropriates not only the landscape but privileges access to views across the Mediterranean (Figure 10.8).

Reclaiming the public realm

The scene today, as I re-write this chapter,[8] is a far cry from that a year ago when I spoke of a handful of citizens protesting their right to the landscape of the Horsh Beirut and the Dalieh of Raouche. Today, thousands of protestors from all over Beirut, rich and poor, Christian, Muslim, and Druze, march shoulder to shoulder to occupy the heart of Beirut. Putting aside political affiliation and sectarian differences, this spontaneous revolution was triggered by frustrations with rampant state corruption that reached a boiling point. The *intifada* or *thawra*, as the city-wide

لأنهم يريدون رؤية البحر، لم نعد نرى السماء.

FIGURE 10.8 "Because they want a sea view, we can no longer see the sky"—reference to the construction of up-market residential towers at Beirut's edge on the Mediterranean Sea (by Jana Traboulsi (graphic artist))

mass uprising of 17 October 2019 came to be known, was spontaneous, unprecedented in scale—estimated at a million nationally—and leaderless. Unlike previous mass protests, this uprising was not about sectarian politics that divided the public but about adverse social and economic conditions that, for the first time, united people across the political, sectarian divide (Atallah, 2019).

Thousands of protestors congregated in Martyrs' Square (Figure 10.9) in the heart of Beirut, pouring into adjacent streets in clear view of the Saray (the council of ministers), perched on the city's ancient acropolis, and the parliament house, on the Place de l'Etoile. Protestors in Beirut were soon joined by mass protests in other large cities (Figure 10.10). Throughout, protestors were willing "to make their bodies vulnerable [as] part of the great power of resistance, and this vulnerability dictates that the resistance must be enacted in the most meaningful and visible places possible" (Waterman, 2018: 144). Spaces of the public realm, central squares, streets and highways in Beirut, as in other cities, served as the landscape of resistance. This is partly because of the sheer numbers of protestors that can only be accommodated in the streets and *saha* (central square), but also because government buildings are generally located on or in proximity of these squares. From Tripoli's Sahat Al Nour in the extreme north to Sahat Elia, in Saida, protestors were united in their demands: upturning the sectarian-based constitution.

Years of exclusion provoked 'mass trespassing' as a necessity for citizens' enactment of democratic principles (Waterman, 2018). Historically, collective sentiments, be they anti-government or in times of celebrations, were voiced in Martyrs' Square,

FIGURE 10.9 The Martyrs' Square mass uprising of 17 October 2019. The red, white and green colours of the Lebanese flag are everywhere (photo by Asian Think Tank)

and to a lesser extent in the adjacent Sahat Riyad El Solh. Protests in Beirut poured into Martyrs' Square and Riyad El Solh and spread throughout the BCD, occupying streets, camping in stylish plazas, and picnicking and fishing in exclusive marinas as an act of defiance (Figure 10.11).

The vibrant, inclusive pre-war Beirut city centre was alive in the collective memory of those that remember. For younger participants, the protests were an opportunity to lay claim to landscapes from which they were excluded. Food vendors reclaimed gentrified streets, performers sang to encourage the crowds, and graffiti artists painted the concrete barriers shielding the parliament building and the Saray. The spirit of solidarity was exceptional. Tents were erected to serve as stations representing academics, lawyers, NGOs, and civil societies to provide them with the opportunity to air their perspective of the *intifada*. Volunteer soup kitchens fed protestors and solid waste sorting was agreed with everyone pitching in to tidy the spaces at the end of each day. The protests were, for the most part, peaceful and very well organized. Protestors made use of social media, blogs, Facetime, and WhatsApp to call for rallies, announcing meeting times and locations.

For the public, relaying their frustrations and projecting their demands through the free press of TV screens across the country and worldwide was powerful and inspiring. Waterman (2018) speaks of protest as enactment, that it is 'performative and requires an audience', and that governments understand the danger of the 'power of performance', responding by seeking control of or manipulating media coverage. This was not the case in the early months of the *intifada*. However, at the time of writing this chapter, the state has resorted to heavy handed tactics of intimidation by arresting and questioning protestors, thus eroding Lebanon's long tradition of free speech and freedom of press.

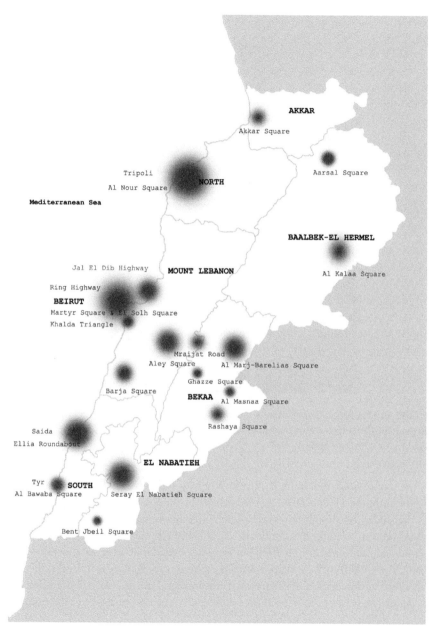

FIGURE 10.10 The intifada mass uprising of 17 October 2019 spreads to Lebanese cities. The protests were staged in public landscapes, highway and roadway junctions, and the saha(t), Arabic for the central squares of cities (by Jala Makhzoumi)

FIGURE 10.11 Embodied resistance by Beirut fishermen defying Solidere's ban on fishing at the exclusive Zaitunay Bay Marina during the intifada (photo by Howayda Al-Harithy (photographer))

A month into the intifada, Prime Minister Saad Hariri resigned. The new prime minister, Hassan Diab, formed what was supposedly a technocrat majority, as demanded by the protestors. Entrenched confessional politics, however, meant there was no avoiding the dominance of established political parties. The protests continued until early March 2020 but were suspended following state lockdown to battle the COVID-19 pandemic. With no solutions forthcoming from the authorities, further devaluation of the national currency, and the dire state of the Lebanese economy, mass protests resumed on 6 June 2020 with the battle cry: "the political virus is deadlier than the pathogen".

Conclusion

I have argued that there are as many ways to practice the publics as there are notions of citizenship. On the one hand is the staged, neatly packaged landscape of the BCD and the practices it dictates. On the other hand are historically evolved, inclusive practices in peripheral landscapes. The contrasting practices of the publics and of citizenships reflect the tensions between the meanings of "landscape as place, polity, and community; and the modern meaning of landscape as a scenic space" (Olwig, 2019: 1). Within the span of a century, Beirut's public realm landscape came to exemplify both meanings. The vibrant and socially meaningful pre-civil war historic city centre was stripped of its cultural significance, transformed by market-driven reconstruction into an exclusive scenic setting, and the social injustice of the transformation was provocative. Occupation of streets and squares in the BCD by protestors in the 17 October uprising became the ultimate expression, a collective defiance of state corruption and neoliberal hegemony.

Although limited in scale and aspiration, social practices that evolved over decades in the Horsh and in landscapes of the Beirut waterfront were equally significant, albeit limited, expressions of citizenships, and in many ways a prelude to the mass uprising of the intifada. Denying access to public realm landscapes enticed citizens to defend their right not only to access these shared communal landscapes but, just as importantly, to protect what they perceived as urban nature. Here, landscape was synonymous with nature. Construction and privatization threatened not only nature but also inherited cultural values associated with the landscape of Dalieh of Rauouche (Davis and Burke, 2011). 'Landscape as nature' empowered citizen action to protect their city's waterfront and fight to reclaim the maritime public domain for the entire Lebanese coast.

'Micro-spatial' urban practices abound in Beirut (Fawaz et al., 2015). Cities all over the world are witnessing similar action from citizens, be it guerrilla and community gardening, flash mobbing, 'empty spaces' movements, or subcultural practices like graffiti/street art (Iveson, 2013). These forms of active citizenship have in common the struggle to claim the rights to the city of the socially and economically marginalized. The public realm, semi-public spaces, and other open spaces are the arena for their action. For such practices to fulfil their potential and contribute to the establishment of democratic rights to the city, Iveson argues that "new democratic forms of authority in the city must be asserted through the formation and action of new political subjects" (ibid.: 954). Different approaches to the design and planning of urban landscapes are also necessary. Olwig explains that historically,

> landscape was a prototypical democracy defined as people and their place, as governed and shaped by customary law, and as formed by representative institutions that were concerned with things *that matter*, and hence not as defined by landscape planners and architects as things as *matter*.
>
> *(2018: XVI)*

Urban planners and designers were the tools for the Solidere project and other neoliberal wholesale development plans for the city.

Could there have been another path for reconstruction of the Beirut city centre? Another path would have undoubtedly taken much longer to plan and implement, just as it would have involved extended dialogue between owners and tenants and local and state authorities. Ironically, speedy construction and avoiding endless

stakeholder negotiations were Solidere's winning sales pitch. The latter is characteristic of the political and financial clout of market-driven urban development
that "manipulates every possible medium to propagate the message that there is no
alternative to that which it proposes" (Yigit-Turan, 2018: 211).

As more and more people live in cities, designers and planners shaping the
environments we inhabit shoulder the growing ethical responsibly of balancing the
demands of politically powerful clients with the needs and aspirations of ordinary
citizens and communities that are socially marginalized. And because landscape
embraces the "substantive legal, political and material practices through which polities shape urban and rural places" (Olwig, 2019: 4), it is more likely to uncover
disparities and address social inequalities. A landscape framing of Beirut's shrinking
public realm helped unfold socio-spatial transformations and conflicting practice
of the publics over time. Landscape also served as a platform for active citizenship,
practiced as everyday civility as in the Dalieh of Raouche (and other sites threatened
by privatization) and through occupation of the BCD, performed and enacted as
an expression of mass resistance. In both cases, landscapes of the public realm fired
social imaginaries and inspired collective dreaming of the "endless potential for
transformation" (Athique, 2008: 25).

Notes

1 Shared community spaces and landscapes have a long history in pre-colonial rural
Lebanon. Village *musha*—collectively owned/managed common lands include woodlands,
scrublands and grazing lands—is a living example of a public realm landscape. Ironically,
while creating the public realm in cities, the French Mandate failed to include *musha* as a
recognized land category outside of cities. As a result, public realm landscapes in villages
are today as threatened by privatization as the public realm in cities (Fayyad, 2018).
2 The name stands for Société Libanaise pour le Développement et la Reconstruction du
Centre-ville de Beyrouth, French for "The Lebanese Company for the Development and
Reconstruction of Beirut Central District" (Solidere, 2017).
3 English 'Martyrs' Square', also historically known as 'Place des Cannons', Arabic 'Sahat al
Shohada', is Beirut's politically charged space par excellence, comparable to Cairo's 'Tahrir
Square' and Istanbul's 'Taksim Square'.
4 "Coastal Vegetation Survey and Conservation for Lebanon", the Darwin Project 1999–
2002, selected Dalieh of Raouche as one of four case study sites measuring biodiversity of
coastal Lebanon. The project was a collaboration between the Royal Botanic Gardens at
Kew and the American University of Beirut, where I taught landscape architecture.
5 Arabic translation of 'landscape' is limited to scenery, which denies the layered meanings of
the English word, limits public understanding and perception of the landscape profession
(Makhzoumi, 2002).
6 As part of the campaign efforts, we worked on a bilingual booklet elaborating the natural,
social and cultural significance of the site and explaining that privatizing the maritime
public domain is in breach of Lebanese law (The Civil Campaign to Protect the Dalieh of
Raouche).
7 Transgressions published in the Legal Agenda (legal-agenda.com/) were: 27.4% of development with state permit exceed the designated footprint; the issued 73 article allowing
transgressions on the maritime public lands; 52% of development in the maritime public
domain did not have a permit.

8 Editors' note: This material was first presented at the *Landscape Citizenships* symposium in November of 2018. The final draft was submitted in the summer of 2020, before the massive explosion in Beirut.

Bibliography

Atallah, Sami (2019). 'Protesters are drawing their own red lines'. *Lebanese Centre for Policy Studies*. Available at: lcps-lebanon.org/featuredArticle.php?id=249 (Accessed 12 September 2019).

Athique, Adrian (2008). 'Media audiences, ethnographic practice and the notion of a cultural field'. *European Journal of Cultural Studies*, 11, pp 25–41.

Battah, Habib (2015). 'A city without a shore: Rem Koolhaas, Dalieh and the paving of Beirut's coast', *The Guardian*, 17 March. Available at: theguardian.com/cities/2015/mar/17/rem-koolhaas-dalieh-beirut-shore-coast (Accessed 10 July 2020).

Buijs, Arjen, et al. (2016). 'Active citizenship for urban green infrastructure: fostering the diversity and dynamics of citizen contributions through mosaic governance'. *Current Opinion in Environmental Sustainability*, 22, pp 1–6.

Butler, Andrew (2018). 'Landscape assessment as conflict and consensus'. In Shelley Egoz, Karsten Jorgensen, and Deni Ruggeri, eds. *Defining Landscape Democracy: Perspectives on Spatial Justice*. London: Edward Elgar Publishing, pp 85–95.

The Civil Campaign to Protect the Dalieh of Raouche (n.d.). [Online]. Available at: dalieh.org (Accessed 10 July 2020).

Daher, Rami (2008). 'Amman: Disguised genealogy and recent urban restructuring and Neoliberal threats'. In Yasser Elsheshtawy, ed. *The Evolving Arab City. Tradition, Modernity and Urban Development*. London: Routledge, pp 37–68.

Davis, Diana and Edmund Burke, eds. (2011). *Environmental Imaginaries of the Middle East and North Africa*. Athens, OH: Ohio University Press.

Dean, Mitchell (2010). *Governmentality: Power and Rule in Modern Society*. London: Sage.

Egoz, Shelley, Karsten Jorgensen, and Deni Ruggeri, eds. (2018). *Defining Landscape Democracy: Perspectives on Spatial Justice*. London: Edward Elgar Publishing.

Egoz, Shelley, Jala Makhzoumi, and Gloria Pungetti, eds. (2011). *The Right to Landscape: Contesting Landscape and Human Rights*. London: Ashgate.

Fawaz, Mona, Mona Harb, and Ahmad Gharbieh (2012). 'Living Beirut's security zones: An investigation of the modalities and practice of urban security'. *City & Society*, 24(2), pp 173–195.

Fawaz, Mona, Ahmad Gharbieh, Nadine Bekdache, and Abir Saksouk-Sasso, eds. (2015). *Practicing the Public* [Arabic]. Beirut: Beirut Urban Lab, American University of Beirut.

Fayyad, Reem (2018). *Revisiting musha lands through an ecological landscape approach: the case of Tibneen*. Masters of Urban Design Thesis, American University of Beirut, Beirut.

Gavin, Angus and Ramez Maluf (1996). *Beirut Reborn. The Restoration and Development of the Central District*. London: Academy Editions.

Harb, Mona (2013). 'Public spaces and spatial practices: claims from Beirut'. *Jadaliyya* [Online]. Available at: jadaliyya.com/Details/29684 (Accessed 2020).

Iveson, Kurt (1998). 'Putting the public back into public spaces'. *Urban Policy and Research*, 16(1), pp 21–33.

Iveson, Kurt (2013). 'Cities within the city: Do-It-Yourself Urbanism and the right to the city'. *International Journal of Urban and Regional Research*, 37(3), pp 941–956.

Lautissier, Hugo, and El Mlaka, Jeremy. (2016). 'From city to coast, taking back Beirut's public spaces'. Middle East Eye June 28, 2016. www.middleeasteye.net/features/city-coast-taking-back-beiruts-public-spaces (21/08/2020).

Malley, Maudth (2018). 'The Lebanese Civil War and the Taif Accord: conflict and compromise engendered by institutionalized sectarianism'. *The History Teacher*, 52(1), pp 121–159.

Makdisi, Saree (1997) 'Laying claim to Beirut: urban narratives and spatial identity in the age of Solidere'. *Critical Inquiry*, 23(3), pp 660–705.

Makhzoumi, Jala (2002). 'Landscape in the Middle East: an inquiry'. *Landscape Research*, 27(3), pp 213–228.

Makhzoumi, Jala (2015). 'Borrowed or rooted? The discourse of "landscape" in the Arab Middle East'. In D. Bruns, O. Kuhne, A. Schonwald and S. Theile eds. *Landscape Culture-Culturing Landscapes: The Differentiated Construction of Landscapes*. Wiesbaden: Springer Verlarg, pp 111–126.

Makhzoumi, Jala (2018). 'Landscape architecture and the discourse on democracy in the Middle East'. In Shelley Egoz, Karsten Jorgensen, and Deni Ruggeri, eds. *Defining Landscape Democracy: Perspectives on Spatial Justice*. London: Edward Elgar Publishing, pp 29–38.

Moro, Giovanni (2012). *Citizens in Europe. Civic Activism and the Community Democratic Experiment*. New York: Springer.

NAHNOO (2017). *Horsh Beirut*. Available at: nahnoo.org/our-causes/horsh-beirut/ (Accessed 10 July 2020).

Olwig, Kenneth (2016). 'Virtual enclosure, ecosystem services, landscape's character and 'rewilding' of the commons: the 'Lake District' case'. *Landscape Research*, 41(2), pp 253–265.

Olwig, Kenneth (2018). 'The landscape path to spatial justice: questioning, rather than fixing, the definition of landscape democracy'. In Shelley Egoz, Karsten Jorgensen, and Deni Ruggeri, eds. *Defining Landscape Democracy: Perspectives on Spatial Justice*. London: Edward Elgar Publishing, pp XV–XXI.

Olwig, Kenneth (2019). *The Meanings of Landscape. Essays on Place, Space, Environment and Justice*. London: Routledge.

Roy, Ananya (2009) 'Civic governmentality: the politics of inclusion in Beirut and Mumbai'. *Antipode*, 41(1), pp 159–179.

Sennett, Richard (2010). 'The public realm'. In Gary Bridge and Sophie Watson, eds. *The Blackwell City Reader*, 2nd ed. Hoboken, NJ: Wiley-Blackwell, pp 261–272.

Solidere (2017). *About*. Available at: solidere.com/corporate/about (Accessed 10 July 2020)

Wall, Ed and Tim Waterman, eds (2018). *Landscape and Agency: Critical essays*. Routledge, London.

Waterman, Tim (2018). 'Democracy and trespass: political dimensions of landscape access'. In Shelley Egoz, Karsten Jorgensen, and Deni Ruggeri, eds. *Defining Landscape Democracy: Perspectives on Spatial Justice*. London: Edward Elgar Publishing, pp 143–152.

World Monuments Fund (2016). *The 2016 World Monuments Watch Includes 50 At-Risk Cultural Heritage Sites in 36 Countries*. Available at: wmf.org/press-release/2016-world-monuments-watch-includes-50-risk-cultural-heritage-sites-36-countries (Accessed 10 July 2020).

Yigit-Turan, Burcu. (2018) 'Learning from Occupy Gezi Park: redefining landscape democracy in an age of "planetary urbanism"'. In Shelley Egoz, Karsten Jorgensen, and Deni Ruggeri, eds. *Defining Landscape Democracy: Perspectives on Spatial Justice*. London: Edward Elgar Publishing, pp 210–221.

11

WORKING WITH UNCERTAINTIES

Living with masterplanning at Elephant and Castle

Ed Wall

I have sat on various committees, in all their different guises at Southwark…
and no-one can ever tell really what is going to go on. There were tenants'
committees at that place, meeting every six weeks for about five years and
they got no-where because in the end it will be decided by the developer
[Lend Lease] and Southwark [Council].

Steven, market operator (Interview, 2012)

Introduction

Uncertainties in the progress of replanning South London's Elephant and Castle
have dominated the lives of residents, businesses, and landowners for over a century. Masterplanned redevelopments have been proposed over decades, materially transforming the neighbourhood, and facilitating new forms of appropriation
through transferring and consolidating land ownerships. Since the 1880s attempts
to regenerate Elephant and Castle have provided contexts for practices of appro-
priating buildings and land resulting in the displacement of businesses, families,
and individuals. During the latest process of regeneration, the area has also been
the focus of struggles that highlight how lives and work have been impacted.
Mapping has provided an important means of representing other potential futures,
as well as the results of populations displaced. In this chapter I explore uncertain
lives and landscapes. I focus on unbalanced relations of power within practices of
masterplanning, and I look for ways of working with uncertainty through processes
of mapping. I discuss notions of incompleteness in both masterplanning and mapping,
two overlapping methods that employ plan view representations, and I question
how uncertainties that result from their partiality can be exploited to deny or form
knowledge. Through focusing on the Elephant and Castle area of South London, a
neighbourhood that has been repeatedly mapped and masterplanned—in addition

to having been demolished through regeneration twice in the last 50 years—I aim to reveal obstacles to belonging experienced by people whose lives and work are bound up with uncertainties of urban change.

At the core of this chapter are tensions between two systems of working with plan views: first, collective processes of mapping, in how I find can strengthen relations between people and places, and second, practices of masterplanning that I recognize can repeatedly unsettle. I am interested that mapping can define new forms of citizenship, as claimed by David Matless (1999), and I am concerned how such senses of belonging are unsettled as landscapes are repeatedly contested through processes of masterplanning. But despite such conflicts, practices of mapping and masterplanning are closely related. They share many traditions: maps and masterplans begin with geographical surveys; they adopt top-down aerial views; they tend to operate at vast scales; and they employ advanced technologies to survey, measure, draw, and edit. Both techniques are bound up in different ways with relations of state power, commercial interests, and military or police control. Mapping and masterplanning are also more closely associated with finished and objective representations—maps and plans—than the more open-ended, politically motivated, and partial processes from which they are constructed.

We can also recognize mapping and masterplanning as being in some way incomplete. Both are processes that mediate and transform relations between people and places—they reveal decisions made, as maps and masterplans are differentially produced and used. Through this incompleteness masterplanning and mapping are bound up with uncertainties. The processes of mapping, Denis Cosgrove describes, involves "… sets of choices, omissions, uncertainties and intentions" (1999: 7). The narratives that are produced and communicated through mapping and masterplanning reveal a construction and reconstruction of knowledge. But this is knowledge that is never comprehensive, that is always partial, and that is produced to serve a purpose. Practices of mapping and masterplanning selectively bring together, edit, or omit knowledge, influencing ways that people can relate to their surroundings. In the contexts of prolonged periods of large-scale urban regeneration, such uncertainties can be exploited to either benefit or highlighted to challenge unbalanced relations of power.

I develop arguments from fieldwork at Elephant and Castle Market that I under-took between 2011 and 2013 in addition to findings from mapping workshops that began in 2013. During the fieldwork I employed a combination of methods, including direct observation, semi-structured interviews, and document surveys, to investigate the making and remaking of urban public spaces at Elephant and Castle. In the mapping workshops I explored techniques of working with intentionally incomplete maps—initially with participants walking, working, and shopping in the Elephant and Castle Market and Shopping Centre. While the fieldwork proved important for gathering data that pertained to the impact of local government planning practices, national government agendas, and the interests of global devel-opment corporations, the workshops opened up conceptual questions about how market traders, managers, residents, and visitors relate to urban landscapes that are facing imminent disruption through regeneration. In bringing together these two areas of research I question: first, how uncertainties in planned large-scale urban change can be amplified and exploited by politicians, planners, and developers and in the process undermine relations of belonging for residents and workers; and second, how mapping can provide opportunities for people to identify with landscapes and establish forms of citizenship.

There are three parts to this chapter, focused on uncertainty, incompleteness, and belonging. I explore uncertainties in practices of mapping and processes of masterplanning as Austin Zeiderman describes in *Uncertainty in Urban Life*: "as an obstacle and an opportunity" (Zeiderman et al., 2015: 283). I am interested in how such difficulties are navigated and to whom opportunities are afforded as people who have lived and worked in Elephant and Castle for decades make sense of planned change. Concerns for uncertainty are drawn from historic correspondences, contemporary websites, and interviews relating to redevelopment of Elephant and Castle. I find that attempts to comprehensively masterplan Elephant and Castle, a central London neighbourhood and strategic transportation interchange, have overshadowed the lives of people living and working there for over a century. Few areas of London have been the focus of so many attempts to systematically demolish, plan, and reconstruct. Almost every decade, and with each successive government, architecturally drawn plans proposing new buildings, parks, and roadways have been published with ambitious timelines for when transformations can be achieved. But despite the frequency of masterplans that have been set out to redevelop the area large-scale transformations at Elephant and Castle have progressed in disjointed and uncertain ways. Zeiderman et al. recognizes that "[urban] uncertainty crosses multiple levels and scales" (ibid.: 298). Rather than a feature of lives lived in the market and housing of places such as Elephant and Castle, it is a condition that politicians, planners, and developers also need to navigate. However, knowledge of how these changes will unfold is unevenly shared, being denied people living and working in the masterplanned area thus undermining their ability to settle. Elephant and Castle is a place of false starts in masterplanning processes, planning interrupted during economic declines, contracts renegotiated as developers seek more favourable terms for redevelopment, and masterplans redrawn as land ownerships have been contested. These repetitive cycles of unfinished masterplanning have transformed landscape relations, resulting in uncertainty for people planning the future of their lives and businesses at Elephant and Castle.

The notion of incompleteness that I discuss in this chapter is founded on landscapes as open-ended relations between people and places that are perceived in different ways. I have previously argued (Wall, 2017, 2020) that understanding and exploring techniques of working with incomplete and open-ended relations is necessary if landscape is accepted as a process. This is not to ignore what Zeiderman et al. describe as "the precarious conditions in which many city dwellers reside" (2015: 299), but rather to understand what is at stake during times of uncertainty as core to understanding the piecemeal unfolding of masterplanning processes. I have also proposed that new material and representational approaches must be imagined if incremental, collective practices can challenge claims of single authored masterplans and give voice to individuals and narratives that are frequently excluded. The incomplete knowledge from which mapping and masterplanning is constructed is of particular interest as it points to the open-ended nature of both practices. Zygmunt Bauman describes a new type of uncertainty–"not knowing the ends"[1] that is concerned with a lack of knowledge about the future and the difficulty in making decisions (2000: 61). I find at Elephant and Castle that uncertainties about the regeneration unsettled many families and businesses before any formal process of evicting tenants and traders occurred.[2] When decisions were made to remain or leave Elephant and Castle they were done so anxiously, with incomplete knowledge about how futures would unfold. Bauman describes: "The state of unfinishedness, incompleteness and underdetermination is full of risk and anxiety; but its opposite brings no unadulterated pleasure either, since it forecloses what freedom needs to stay open" (ibid.: 62). The incomplete, as Zeiderman et al. describe of uncertainty, offers moments and sites of possibility. In addition to the material benefits that can be gained by property owners, developers, and local government, open-ended processes, gaps in local knowledge, and uncertainty in progress provide opportunities for some voices to be heard. The question of whether the concerns of local residents, market traders, and established businesses are considered during urban redevelopments is explored in this chapter.

Practices of landscape citizenships, as I explore in this chapter, are formed between notions of rights and belonging as they relate to land—the rights to belong to a place and opportunities to consider that place as one's own. These are not relations with nation states—as Matless describes a "landscaped citizenship" (1997: 147) where "landscape becomes occasion for national citizenship" (ibid.: 141). Instead, I consider tangible rights, such as the promise of a 'Right to Return' for residents of Elephant and Castle's Heygate Estate—rights initially offered to residents to live in the new housing provided through the masterplanning process. I also discuss rights as associated with what Shelly Egoz describes as "rootedness in place" (2016: 165), as well as broader rights to participate in urban change. Building on Lefebvre's advocation of rights in *The Right to the City* (1996 [1968]), Harvey (2008) describes the right to the city as "… a right to change ourselves by changing the city". The opportunity to effectively engage in and identify with the remaking of neighbourhoods, streets, and parks comes to define citizenships of landscape.

Rather than state mandated processes of nationality evidenced through passports and certificates of birth, or relations with land through property ownership, I argue landscape citizenships to be an important relational framework situated through lives grounded in physical places as well as promises to belong. Landscape citizenships are lived, material, and perceived relations with land that situate belonging and longing, certainty of place, and opportunities for change. Furthermore, at Elephant and Castle, landscape citizenships hold in tension notions of land belonging to individuals, communities, and organizations with the sense that people also belong to specific landscapes. While such definitions may suggest the usefulness of masterplanning as contributing to these futures I discuss in this chapter the unsettling practices of masterplanning that have been experienced by people living and working in Elephant and Castle. The uncertainty caused by changing plans, limited knowledge of planning process, and lack of opportunity to inform the trajectory of lives and landscapes disrupts relations of belonging that situate many people in places considered to be home.

In the first section of the chapter, I describe the results of fieldwork that explores the uncertainties caused by repeated masterplanning of Elephant and Castle and the difficulties that this has caused for people living and working in the area. In the second section, I explain how during multiple workshops we created incomplete maps as a means to situate participants in their surroundings while at the same time producing collective landscape knowledge. As I bring these areas of research together at the end of the chapter, I discuss challenges for residents and traders to continue to belong and to identify with Elephant and Castle and the significant advantages for developers that result from uncertainties of urban change.

Development uncertainties

Elephant and Castle Regeneration is an ongoing process of masterplanning the Elephant and Castle neighbourhood in South London, a complex of buildings, open spaces and infrastructures that include the former Heygate Estate (now demolished), transport interchanges, and the Elephant and Castle Shopping Centre (closed, awaiting demolition), around which wrapped the Elephant and Castle Market (also closed, awaiting demolition). The redevelopment has been led, since 2007, by a partnering of Southwark Council (the local authority) with Lend Lease (a private developer). Through their purchase of the shopping centre and market in 2013 developers Delancey, and their joint venture partner APG, gained an important role in the masterplan. The involvement of private interests with urban redevelopment projects has been encouraged by the British Conservative government since the 1980s, an approach advanced under New Labour (after 1997). At Elephant and Castle this has involved a £1.5bn regeneration plan that proposes to develop the 55-acre Elephant and Castle Opportunity Area, a designation formed by the Greater London Authority (GLA) in 2004. The redevelopment encompasses an area of London already demolished and rebuilt between the 1950s and 1970s, including a housing estate of 1200 homes, Europe's first indoor shopping centre, and a system of roads and roundabouts. Although plans for renovating the shopping centre began one year after the previous redevelopment was complete in 1974, it was not until the 1990s that Elephant and Castle became the intense focus of redevelopment through comprehensive masterplanning.

Despite smaller scale improvements to Elephant and Castle in the 1990s, Southwark Council recognized that without monetizing the value of the land on which the Heygate Estate stood improvements to any infrastructures would be limited. They therefore partnered with Lend Lease, formally signing the agreement in 2010 and submitting planning applications in 2012. Southwark Council and Lend Lease have consistently pointed to the failures of the post-war development, the conditions of the buildings, and the problems of the modern road system. However, rumours have persisted that Southwark's objective was to reverse the area's demographic from 75% social housing and 25% market rate to 75% market rate with only 25% affordable (Urban 75, 2013). A local resident and employee of the local council explained:

> I don't know how true ... there was a kind of suggestion that ... Elephant and Castle had at the time about 75% social housing, which was too much, way too concentrated, and it needed to be turned into a more mixed community ... and magically everyone's lives would improve.
>
> *(Interview, 2012)*

With the demolition of the Heygate Estate and the resultant displacement of thousands of former residents across the borough and South East England the supposed objectives of the redevelopment could be realized (See Figures 11.1 and 11.2).

The masterplanned reconstruction became instrumental in reconstituting forms of citizenship—as Matless recognized in 1940s England, a period when Elephant and Castle was previously being planned for reconstruction: "these visions of a reconstructed town and country were also visions of citizenship" (1996: 425). Through the process of financial viability assessments Lend Lease persuaded Southwark Council that they would not be able to provide the required number of affordable or social rented homes (Flynn, 2016). The 'decanting'[3] of residents from the Heygate Estate has been followed by corresponding displacements from the shopping centre. In 2013, the shopping centre was described by Councillor Fiona Colley, cabinet member for regeneration at Southwark, as "the last piece in the jigsaw for the regeneration" of the Elephant and Castle area (Southwark Council, 2013). The latest ambitions for comprehensive development across Elephant and Castle include the demolition of the shopping centre and the market to make way for a new 'town centre', with more housing, office space, and retail units. The shopping centre and the market was closed on 24 September 2020, with many traders unable to relocate within the redevelopment area (35% Campaign, 2020).

The Elephant and Castle Market is the public space on which my fieldwork focused, a small sunken plaza at the heart of the regeneration. The market was a space of conversations over games of checkers in the corner of the food court, market structures installed each day and customary exchanges as goods were bought and sold. Along with the Elephant and Castle Shopping Centre, the market was at the centre of what is considered by the development partners as the 'core area' within the Opportunity Area. Although these assets are not owned by either partner, Southwark Council or Lend Lease, they were included in the 2012 Supplementary Planning Document that sets out the larger regeneration plan. The market and shopping centre have been central to the council's ambitions to "coordinate growth" at Elephant and Castle (Southwark Council, 2012: 2). As one of the final areas approved for redevelopment, Ben Campkin explains that "more recent public debates ... have centred most prominently on the shopping

FIGURE 11.1   Displacement of tenants from the Heygate Estate to other neighbourhoods across London (map by Alexis Liu, 2020, produced from research by Southwark Notes & 35% Campaign, 2014)

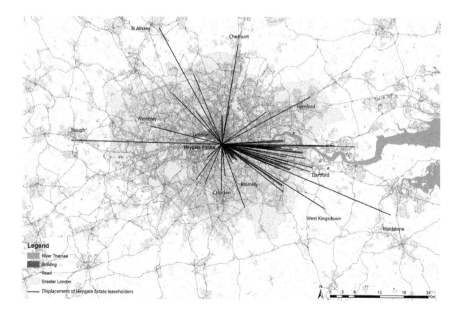

FIGURE 11.2   Displacement of leaseholders from the Heygate Estate to other neighbourhoods across the Southeast of England (map by Alexis Liu, 2020 produced from research by Southwark Notes & 35% Campaign, 2014)

centre" (2013: 68). After the Heygate Estate was 'decanted' of its 3000 residents and demolished (2011–2014) the shopping centre became the focus of further planning efforts and the plight of the shopkeepers and traders became more evident (35% Campaign, 2020). Reflecting a 'Right to Return' that never materialized for most of the Heygate Estate residents, as the closure and demolition of the shopping centre approached, almost half of the traders had nowhere to relocate. As a former shopping centre manager described of previous plans for regeneration: "I would be very surprised if any of them stayed, to be honest, just because I don't know how many businesses could wait three years from moving out, to start their business again" (Interview, 2012).

From historic and current planning documents, I found that, despite claims of comprehensive masterplanning, transformations of Elephant and Castle had unfolded in piecemeal ways. For the *Elephant and Castle Regeneration* to be realized the masterplan has been repeatedly redrawn, and agreements between the development partners have been renegotiated. Legal tools established by central government and the roles of Southwark Council and the GLA as planning controllers have been necessary to enable the development. At times when development strategies have stalled, particularly since the financial crisis (2008), the agreements signed to establish the private-public partnership remained outside of public scrutiny. The powerful position of private developers such as Lend Lease, contrasted with that of the local authority that found itself in a poor negotiating position. Having already committed to the development and unable to finance the reconstruction itself, the council needed to offer favourable terms to the developers. As a local resident describes,

> One of the main problems with Elephant and Castle is that as a regeneration process it has been going on so long. It is now so politicized, that every administration needs to be the one that on their watch they get a spade in the ground.
>
> *(Interview 2012)*

The masterplanning of Elephant and Castle progressed with a high degree of uncertainty. The stuttering progress of the masterplanning process was impacted by changes in national and local governments, the global economic collapse, and the ability of development partners to fund the reconstruction. In 2007, before many of these events even unfolded, Councillor Nick Stanton, leader of Southwark Council, explained:

> We're very conscious that we've been talking about the need to regenerate the Elephant and Castle for a long time, and it's paramount that we get on and show people that we're serious about actually delivering.
>
> *(London SE1 Community Website, 2007)*

The uncertainty provided opportunity for the developers and disadvantages to traders, residents, and some landowners. Uncertainties were exacerbated by threats by Southwark Council to facilitate the compulsorily purchase of residential and commercial properties. Businesses were hesitant to remain in the area as long leases in the shopping centre were run down so that the owners could redevelop when their plans were approved. Uncertainty caused by masterplanning resulted in market stalls ceasing trade, businesses closing down, investment in improvements postponed, and former residents never returning. Relations between work practices, daily lives and the landscape of Elephant and Castle were reconstituted through the uncertainty of planning processes. For market stall holders, some of whom had been trading from the market for almost fifteen years one of the project officers questioned whether the former traders were "the right operators" to take up stalls in a newly designed market space (Interview, 2012). Even the market managers, who had successfully managed markets all over London were left unable to confirm their commitment to the area: "We would absolutely be willing to invest in really trying to build the market up but you have got to have some kind of certainty" (Interview, 2012). The manager goes on to describe the situation for the traders who "don't want to invest in a brand-new stall with all the trimmings if they are going to be told they have to get off next week". This nervousness to invest compounded the deteriorated appearance of the area supporting the development agendas of the council and the developers: as Campkin describes of Elephant and Castle: "demolition discourses have long contributed to its [the shopping centre's] blight" (2013: 71).

The uncertainty caused by the repeated redesign and reconstruction of the area over many decades has enabled large organizations and corporations to appropriate land and displace activities. Zeiderman et al. recognize that,

> Historically sedimented conditions of uncertainty lead to strategic attempts by distinctly located and unequally positioned urban actors to stabilize select elements of the urban milieu, which enables certain outcomes and constrains others while generating more uncertainties along the way.
>
> *(2015: 285)*

The uncertainty for people living and working in Elephant and Castle was not being addressed through the masterplanning process but rather exacerbated. Furthermore, the uncertainty that led to businesses relocating before the shopping centre was closed and families leaving the Heygate Estate in advance of being evicted facilitated the ambitions of developers and local authority.

> These uncertainties can never entirely be overcome, nor is overcoming uncertainty always the objective, since uncertainty can be advantageous for some and disadvantageous for others, a problem at one scale and a solution at another.
>
> *(Zeiderman et al., 2015: 285)*

The remodelling of transport infrastructures provided a consistent narrative and rationale for the demolition of buildings and the building over of public spaces—with both periods of masterplanned change (1950s–1970s and since 2007) resulting in the transfer of large swathes of land between powerful landowners, developers, and government. Southwark Council and Lend Lease employed planning and legal mechanisms to acquire properties that impeded their developments: the means provided in Section 106 of the Town and Country Planning Act 1990 and the Planning and Compulsory Purchase Act 2004 were used by the council to favour their development partner, Lend Lease. This includes Southwark Council removing tenants from the Heygate Estate, then transferring the cleared land to Lend Lease, and then threatening to acquire and demolish the shopping centre so that Lend Lease could complete the regeneration. In April 2020, Southwark Council stepped in using further Compulsory Purchase Orders to acquire the shopping centre with the aim of removing remaining traders whose continued presence caused uncertainty for the private developers, Delancey. The redevelopment process facilitated by Southwark Council has consistently ignored the potential of local initiatives to improve the area and they have dismissed the value of renovating old buildings, resulting in the eviction of residents, the closure of businesses, and the demolition of public spaces.

The regeneration of Elephant and Castle has faced significant resistance from residents, business organizations, and community groups. As one of the early redevelopment projects initiated by a local authority in England and that recognized the demolition of social housing was one of few mechanisms for releasing the value of the land in their boroughs, Elephant and Castle has drawn significant criticism. From *The 35% Campaign* to the *Elephant Amenity Network* or *Heygate Was Home* to *Latin Elephant*, the missteps of the local authority and the strategic appropriations by large private developers have been carefully documented. Protests have been held, sit-ins organized, and alternative masterplans proposed—but despite the visible injustices that have been experienced by many residents and traders over the last three decades the displacement of families and businesses and the complete transformation of landscapes has continued. Opportunities that have arisen through the uncertainties of the regeneration process have been seized by private developers and facilitated by the local authority. Smaller scale interests, of traders, employees, leaseholders, and residents, have in contrast experienced the difficulties of such large and uncertain urban transformations.

Incomplete mapping

In 2013, while I was undertaking fieldwork at Elephant and Castle, I co-organized a workshop, *Cartographies and Itineraries*—a workshop for the *NYLON* conference (London School of Economics, 2013). Bringing together my research into Elephant and Castle Market with ethnographic research of street markets in Naples undertaken by my colleague Antonia Dawes, we aimed with *Itineraries and Cartographies* to enable participants to explore Elephant and Castle with the aid of drawn and written itineraries. Rather than producing a guide that participants could follow we were interested to facilitate short encounters between people and places around

Elephant and Castle by providing selected information and brief introductions. We produced three contrasting mapped itineraries—*Walking the Elephant*, *Buying the Elephant*, and *Selling the Elephant*. Each contained combinations of historical and contemporary information, intentionally left some parts empty, and omitted certain knowledge. The structure of the maps opened up areas for participants to add to, write notes, and make comments. Participants were asked to use the maps to explore the area with the intention that the incomplete maps would reveal questions as they tried to read historical accounts within transformed environments or as they got lost navigating streets without knowledge of their names. It was a short workshop, but the conversations in developing the itineraries raised questions of mapping that I have continued to explore in more recent *Incomplete Cartographies* workshops (University of Greenwich 2016; TU Wien 2017). Together they have revealed potential practices of belonging and making citizenships that are central to this chapter. The questions raised by the gaps in the brief itineraries and incomplete maps required participants in the workshops to first situate themselves within their surroundings, second, engage with other people whose lives are part of these places, and third, formulate collective landscape narratives of places through further marking the maps.

Making maps is an additive process of constructing and editing. Lines are drawn in additive ways in order to produce maps: mapmakers must also decide what knowledge is omitted and accept what is difficult to represent. A balance is found by the cartographer between adding sufficient information to communicate what is necessary for the purpose of the map, while leaving out information that is less useful. This process makes maps and processes of mapping focused on specific means—to navigate, to demarcate, to control, to own. While the maps we produced for the workshops were more focused on what Cosgrove describes as "… processes of mapping rather than with maps as finished objects" (1999: 1), they also framed specific relations with land. We needed to make choices on what to include and what to exclude as we produced the mapped itineraries. Through the more recent *Incomplete Cartographies* workshops, where many people were asked to contribute their knowledge to the maps, I have identified specific formations of collective knowledge and individual identity through mapping. By beginning with a framework that was sufficiently open, participants were able to record their observations as well as the accounts or other participants. The partial nature of the information provided in the initial mapped itineraries created ambiguities that required participants to carefully observe and talk with people.

Such mapping can provide a means to describe worlds as they change, to make sense of uncertain surroundings, and to recognize capacities to inform future plans. It requires landscapes to be closely observed and carefully considered as maps are remade, for participants to situate their knowledge and actions in relation to other people and the land. Opportunities to contribute to the authorship of maps can also allow people to reconsider identities of citizenship. In his essay, 'The Uses of Cartographic Literacy: Mapping, Survey and Citizenship in Twentieth-Century Britain', David Matless explores maps as popular documents, "… concerning how

FIGURE 11.3 Three Cartographies and Itineraries for Elephant and Castle, folded (Dawes and Wall, 2013)

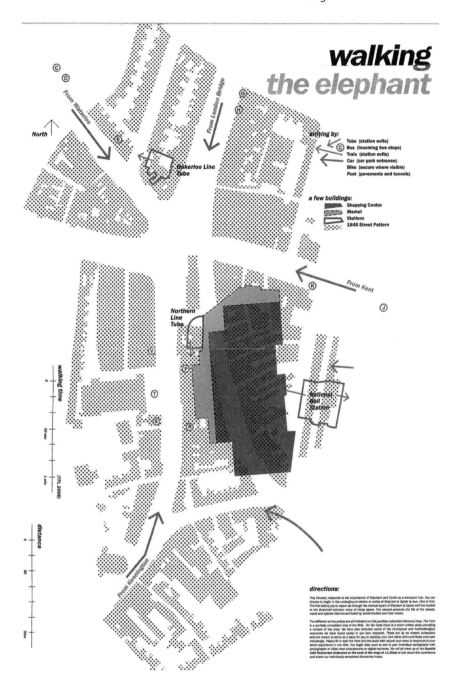

FIGURE 11.4   Walking the Elephant, Cartographies and Itineraries for Elephant and Castle (Dawes and Wall, 2013)

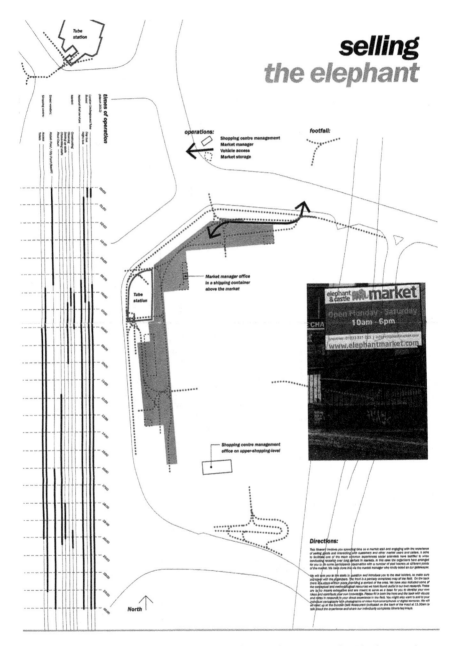

FIGURE 11.5 Selling the Elephant, Cartographies and Itineraries for Elephant and Castle (Dawes and Wall, 2013)

the map should be used, who is able to use it, what forms of knowledge it should register, what kinds of citizenship it should cultivate" (1999: 194). Linking practices of survey with those of mapping, Matless reflects on the work of regional survey in the mid-twentieth century. He describes how "Regional survey sought to cultivate citizenship through a new form of local knowledge, stressing observation over book learning…" (ibid.: 207). While it reflects a more paternalistic approach than the one followed in our workshops—more concerned with "making of local citizens" (ibid.) than providing opportunities to construct relations with their surroundings—it points to the potential of mapping in the making of citizenships and the relationship of knowledge to mapping. Matless recognizes that in England in the 1930s and 1940s, "citizenship was to emerge from careful observation through eyes and map" (1997: 145). Ståle Angen Rye and Nanang Indra Kurniawan also recognize the possibility of mapping "as acts of citizenship" (2017: 156). From their research in Indonesia, they find that the making of Indigenous citizenships involves extended periods of time, across a range of scales, and participation of multiple actors. This making of citizenships reflects some of the approaches to *Incomplete Cartographies* where incremental stages of mapping, questions of territory, and collective knowledge are core concerns.

Matless (1999) describes the priority that proponents of regional survey gave to observation over local narratives—traditions of folk-lore that were considered to distract from local knowledge. However, he also recalls the significance of abstraction: "… the map as a document is at once concrete and abstract" (ibid.: 198). Reflecting on the requirements of Patrick Geddes "… that a synthesising vision of the region demanded literal as well as conceptual overview", Matless describes a "model of thought" (ibid.) close to notions of landscape. In addition to observation, recording, drawing, and editing required during processes of mapping, Matless points to the significance of how landscapes are perceived and constructed conceptually (ibid.). The maps produced in the *Cartographies and Itineraries* and *Incomplete Cartographies* workshops reflect the need to include more than just that observed, but to make note of conversations with market traders, accounts of shop keepers, and narratives told by older residents. These processes of mapping combine a range of methods of data collection and analysis, all of them, however, bringing to the foreground the social constructedness of landscapes. Mapping becomes a way of making sense of and situating collective worlds, and through partially constructed maps participants can situate themselves as well as adding to the map to share their local knowledge.

The intentional incomplete nature of the workshop maps provided opportunities to produce knowledge and to explore potential futures. The uncertainty inherent in the map's incompleteness advanced the capacity of the map as a tool to understand and make sense of relations with the land. The information omitted was not obscured in order to increase uncertainty or rewrite histories of the neighbourhood, instead careful construction of the incomplete maps provided a framework that enabled participants to discover and make sense of places. Rye and Kurniawan describe "… the importance of understanding participatory mapping as a process that involves the continuous creation and redefinition of knowledge" (2017: 156). They continue: "It is not so much a matter of becoming proper citizens of the state but of the continuous construction of citizenship" (ibid.). The mappings produced during the workshops, therefore, can be seen as processes of forming landscape citizenships that affirm political, material, representational relationships. The processes of navigating and engaging with places through incomplete maps and itineraries points to the potential for mapping as a tool for situating, forming identities, and senses of belonging.

Breaking, making, belonging

The formation of citizenships is mediated by the agency afforded to individuals and communities to participate in mapping and planning practices. If landscape citizenships are forged through associations that are practiced in relation to the land, then we can recognize that such citizenships are limited as participation is restricted. The degree and nature of inclusiveness in practices of mapping and processes of masterplanning come to define the scope of relations that individuals and communities can forge with people, places, and futures. In contrast to the open nature of the mapping workshops, the redevelopment of Elephant and Castle has been defined by uncertain progress with limited access to decision making. Relations that residents of the Heygate Estate and traders at the Elephant and Castle Shopping Centre have with the area have been undermined as both sites were acquired for demolition. Although the Heygate Estate and the shopping centre were planned as different phases of the regeneration process both redevelopments have been highly facilitated by Southwark Council and both have resulted in significant displacement of families and businesses. The lack of involvement in decision making processes has left residents and businesses without the right to determine the future of their lives and landscapes. If the formation of landscape citizenships is established through prolonged interactions between people and land we can conclude that the uncertainty created by the regeneration process has denied the ongoing citizenship of many shopkeepers, market traders, pensioners, and families of Elephant and Castle.

I also recognize concerns for belonging as residents and commercial traders are denied knowledge of how future changes will unfold. If both mapping and citizenships are dependent on the continuous formation of knowledge the degree to which residents and traders are able to define, disseminate, access, or are denied knowledge comes to inform senses of belonging. The ability of local councillors and private developers to rewrite histories of Elephant and Castle while making arguments for comprehensive redevelopment denies long-term residents and traders their own narratives. Furthermore, the continuous changes to the regeneration process and the lack of information shared with local communities at Elephant and Castle undermines associations between people and their surroundings. As residents and businesses have been displaced from Elephant and Castle relations of belonging have been redefined. *Heygate Was Home*, a website that traces narratives associated with regenerating the Heygate Estate, highlights the longing that residents to the housing estate have for their demolished neighbourhood, friends now scattered across the borough, and the resultant loss. It also reveals how residents were misled by the information provided by the local authority during the redevelopment process and shows the impact of narratives of politicians and filmmakers of a deteriorating estate troubled by crime. One resident explains: "This whole scheme has been a shambolic act of deception" (Heygate Was Home, n.d.). Exacerbating the uncertainty caused by a lack of knowledge about the regeneration, the histories of the area rewritten by the developers and local authority have denied residents the agency of their own past.

Conclusions

What can we conclude from discussing together the uncertainty caused by prolonged processes of urban redevelopment and practices of mapping that embrace incompleteness as a means of producing collective knowledge? The similarities of large scales, top-down representations, and controlling mechanisms in traditions of masterplanning and mapping are evident. However, I have found that the opportunities and obstacles caused by uncertainty, as described by Zeiderman et al. (2015: 283) resonate differently in the case of redevelopment (*Elephant and Castle Regeneration*) than they do in the mapping workshops (*Cartographies and Itineraries* and *Incomplete Cartographies*). The uncertain progress of masterplanning at Elephant and Castle has created opportunities for the private developers while making it difficult or impossible for many smaller businesses, families, and individuals to remain. Their exclusion from informing the process of redevelopment, and the knowledge denied, contrasts with the aims of the mapping workshops that were to include otherwise excluded narratives and to enable people to relate to their surroundings. But the question of whether practices of participatory mapping, using incomplete maps in academic workshops, can produce knowledge and contribute to forms of citizenship is overshadowed by the experiences of residents and traders—people who have been denied information, lost opportunities, and had homes and businesses demolished. Jerry Flynn, a former resident of the Heygate Estate, explains that decisions for regeneration at Elephant and Castle were made in secrecy, between the council and the developers and they were determined by viability assessments "contingent … on uncertain 'facts', opinion and argument" (2016: 284). While working with incomplete maps may forge new associations for participants—and even strengthen forms of belonging for residents and traders who can further situate themselves through the process of mapping—the ruptures caused by regeneration at Elephant and Castle highlight the massive imbalances in relations of power that limit opportunities for individuals, families, and smaller businesses and generate great gains for private developers.

I conclude that belonging requires persistent practices that are unsettled by piecemeal and repeated urban changes. Through the uncertainties of masterplanning, Elephant and Castle has witnessed a gradual decline in activity in the market and shopping centre. Initially, some businesses relocated and some residents moved away in anticipation of the redevelopment and then with fewer residents visiting the shopping centre other businesses struggled to continue trading. Furthermore, with a lack of knowledge of the regeneration process traders were unable to plan, businesses postponed renovations to their shops and restaurants, and the market managers were less willing to invest in improvements to the market. The small, repeated, local actions of passing through the market on the way to the bus stop or sitting in the same seat each day in one of the shopping centre cafes reflect lived practices undermined by decades of uncertainty—uncertainty that has eased the advance of redevelopment. The inability of traders and families to maintain their daily routines disrupted their relationships with Elephant and Castle. The regeneration has interrupted the continuity for citizens to identify with their neighbourhood. Traders and residents have been denied the oppurtunity to tell their stories of Elephant and Castle, overshadowed by the council/developer regeneration narratives, and by being displaced the right to inform their own future has been circumscribed. Associations that defined the landscapes of Elephant and Castle have been reconstituted: as long-established practices of belonging have been unsettled new forms of belonging are forged as incomers buy properties in the new development.

The aim of the *Cartographies and Itineraries* and *Incomplete Cartographies* was not to establish new forms of citizenship or identities of belonging, instead to open up opportunities to build collective knowledge of places through practices of working with incomplete maps. The uncertainty of navigating Elephant and Castle, and other neighbourhoods, with maps that have partial information with often overlapping temporalities may initially unsettle. But the intentional gaps left within the incomplete maps provide moments of pause, images to clarify, and sentences to finish. The brief nature of the mapping workshops enabled participants to situate themselves and slightly increase knowledge of Elephant and Castle. Some of the shopkeepers and visitors that participants spoke with were interested to understand their relation to the places shown on the map. However, as Rye and Kurniawan highlight, participatory mapping needs to embody "continuous creation and redefinition of knowledge" and the "continuous construction of citizenship" (2017: 157). During the *Incomplete Cartographies* workshops several participants were interested in furthering their knowledge through repeated interactions with people they had previously not spoken with and developing more layered associations with their neighbourhoods. The continuity of landscape relations, how people relate with each other and their surroundings, proves an essential dimension of landscape citizenships. This highlights the importance of rights to landscape and rights to remain that were negated during the redevelopment of Elephant and Castle.

Rye and Kurniawan point to the potential of mapping as contributing to political action and claiming of rights: "Mapping is never neutral or value free. It can be used to consolidate and preserve political power … but it can also serve as a means of protest and resistance" (2017: 150). As Jerry Flynn (2016) reveals in the maps of displaced Heygate council tenants and displaced Heygate leaseholders at Elephant and Castle, maps are powerful instruments that can counter the silence or secrecy of local government and private developers. But despite the use of such mapping in struggles for belonging I find that the extreme imbalances of power, that in the case of Elephant and Castle are facilitated by Southwark Council, can appear insurmountable. The uncertainties caused by the planning processes at Elephant and Castle have exacerbated already pronounced imbalances of power—creating more opportunities for the private developers and local government and greater injustices for traders and residents. Uncertainties have been exploited and created by Lend Lease and Delancey during the masterplanning process in ways that have disempowered local residents and traders. Associations that families and businesses have established with and across Elephant and Castle over decades have been systematically dismantled. The limited opportunities for displaced tenants and leaseholders to return to Elephant and Castle after being removed from the Heygate Estate were determined through agreements between Southwark Council and Lend Lease. As the land bank at Elephant and Castle comes to belong to private developers, long-term residents, traders, and shopkeepers lose their sense of belonging.

Through disruption to their daily practices and the denial of knowledge about the future—in addition to the rewriting of past narratives—the inability of citizens to maintain relations with the Elephant and Castle in which they have grown up and lived denies continuity of citizenship. While the lack of agency of local people in decision making and the denial of knowledge about the planning process could be addressed by more inclusive, open-ended, participatory processes of mapping, the uneven distribution of power has benefited profit-oriented developers over more situated relations of belonging.

Notes

1 Bauman quoting Gerhard Schulze (1997: 49).
2 Bauman elaborates: "… of 'clearing the site' in the name of a 'new and improved' design; of 'dismantling', 'cutting out', 'phasing out', 'merging', or 'downsizing', all for the sake of a greater capacity for doing more of the same in the future—enhancing productivity or competitiveness" (2000: 151).
3 The term 'decanting' has been increasingly adopted as a way of describing the removal of tenants from local authority housing estates as they are prepared for demolition and regeneration. This includes moving tenants out of their homes voluntarily (as well as evicting them by force) and placing them into other accommodation (sometimes hundreds of miles away).

Bibliography

Bauman, Zygmunt (2000). *Liquid Modernity*. Cambridge: Polity Press.
Campkin, Ben (2013). *Remaking London*. London: IB Taurus.
Cosgrove, Denis (1999). *Mappings*. London: Reaktion Books.
Egoz, Shelly (2016). *The Right to Landscape: Contesting Landscape and Human Rights*. London: Routledge.
Flynn, Jerry (2016). 'Complete control: Developers, financial viability and regeneration at the Elephant and Castle'. *City*, 20(2), pp 278–286.
Harvey, David (2008). 'The right to the city'. *New Left Review*, 53, pp 23–40.

Heygate Was Home (n.d.). *Heygate Was Home* [Online]. Available at: heygatewashome.org (Accessed 1 October 2020).

Lefebvre, Henri (1996 [1968]). *The Right to the City*. Translated by Eleonore Kofman and Elizabeth Lebas. Malden, MA: Blackwell Publishing.

London SE1 Community Website (2007). *Lend Lease chosen as Elephant and Castle development partner* [Online]. Available at: london-se1.co.uk/news/view/2835 (Accessed 3 November 2017).

Matless, David (1996). 'Visual culture and geographical citizenship: England in the 1940s'. *Journal of Historical Geography*, 22(4), pp 424–439.

Matless, David (1997). 'Moral geographies of English landscape'. *Landscape Research*, 22(2), pp 141–155.

Matless, David (1999). 'The uses of cartographic literacy: Mapping, survey and citizenship in twentieth-century Britain'. In: Denis Cosgrove, ed. *Mappings*. London: Reaktion Books.

Rye, Ståle Angen and Nanang Indra Kurniawan (2017). 'Claiming indigenous rights through participatory mapping and the making of citizenship'. *Political Geography*, 61, pp 148–159.

Schulze, Gerhard (1997). 'From situations to subjects: Moral discourse in transition'. In: Pekka Sulkunen et al., eds. *Constructing the New Consumer Society*. New York: Macmillan, pp 38–57.

Southwark Council (2012). *Elephant and Castle Supplementary Planning Document (SPD)* [Online]. Available at: southwark.gov.uk/assets/attach/12788/EIP29-Elephant-Castle-SPD-and-OAPF-2012-.pdf (Accessed 5 February 2017).

Southwark Council (2013). *Section 106 Annual Report 2012–14* [Online]. Available at: southwark.gov.uk/assets/attach/1600/S106_Annual_Report_2012_14.pdf (Accessed 5 February 2017).

Southwark Notes (2014). *Heygate Displacement Maps* [Online]. Available at: southwarknotes. wordpress.com/heygate-estate/heygate-dispacement-maps/ (Accessed 30 October 2020)

Urban 75 (2013). *Elephant and Castle regeneration outrage: 2,535 new homes and just 79 social rented units* [Online]. Available at: urban75.org/blog/elephant-and-castle-regeneration-outrage-2535-new-homes-and-just-79-social-rented-units/ (Accessed 19 January 2014).

Wall, Ed (2017). 'Post-landscape *or* the potential of other relations with the land'. In: Ed Wall and Tim Waterman, eds. *Landscape and Agency: Critical Essays*. London: Routledge.

Wall, Ed (2020). *The Landscapists: Redefining Landscape Relations*. Oxford: Wiley.

Zeiderman, Austin, et al. (2015). 'Uncertainty and urban life'. *Public Culture*, 27(2), pp 281–384.

35% Campaign (2020). *The harsh reality of relocation for shopping centre traders* [Online]. Available at: 35percent.org/2020-04-23-reality-for-traders-elephant-castle-shoppping-centre/ (Accessed 30 October 2020).

12

LEGACIES OF VIOLENCE

Citizenship and sovereignty on contested lands

Danika Cooper

What follows has been written on contested ground. I write from my office at the University of California, Berkeley, on the ancestral, traditional, and contemporary lands of the Chochenyo Ohlone Nation. The Morrill Act of 1862 stole nearly 11 million acres from 250 Indigenous nations and communities for the establishment of 52 agricultural colleges across the United States. Of this land, the University of California acquired 148,636 acres—lands taken from over 120 tribal nations across California and for which the University paid $0 but from which it continues to benefit in perpetuity (Lee and Ahtone, 2020). This history highlights the ongoing relationship between higher education, Indigenous struggles for survival and sovereignty, and the dispossession of their lands, past and present.

It is my ongoing responsibility to reflect on and address the violence and dispossession that underlie the creation and growth of the United States. In 2012, President Barack Obama declared: "Unless you are one of the first Americans, a Native American, we are all descended from folks who came from somewhere else ... For just as we remain a nation of laws, we have to remain a nation of immigrants" (Obama, 2012). As a child of immigrants, I know first-hand the value of welcoming immigration, but all non-Indigenous occupiers of this land are required to learn about dispossession as a part of occupying Indigenous lands; to acknowledge that the violence that occurred in the creation of the United States was not inevitable but rather strategic and intentional (Saunt, 2020; Blackhawk, 2008); and to realize that until we adequately address the question of reparations as a landed interrogation such systems and their violent legacies persist.

On 9 July 2020, the United States Supreme Court made a landmark decision to alter the map of Oklahoma by honouring the boundaries of the Muscogee (Creek)

Nation's reservation lands. The decision represents a critical victory for Indigenous struggles for land and sovereignty as it is one of the few instances wherein the United States government has been held responsible in adhering to and fulfilling treaty obligations. Ian Gershengorn, a lawyer who argued on behalf of the Muscogee Nation, said of the ruling, "Congress persuaded the Creek Nation to walk the Trail of Tears with promises of a reservation—and the Court today correctly recognized that this reservation endures" (KickingWoman, 2020). This case, *McGirt v Oklahoma* (2020) foregrounds the continued relevance and urgency to revisit historical land

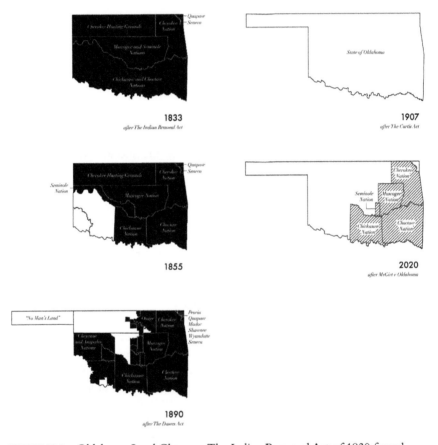

FIGURE 12.1 Oklahoma Land Changes: The Indian Removal Act of 1830 forced Indigenous nations in the east to move to new lands west of the Mississippi River, known as "Indian Territory." As the United States expanded further west, Indian Territory became valuable for white settlement and lands that had been promised to Indigenous nations were renegotiated to satisfy American expansionist desires. By 1907, Indian Territory had been altogether seized and replaced by State of Oklahoma. In the summer of 2020, the United States Supreme Court, the highest judicial institution in the land, declared that the United States must honour the established boundaries of the reservations. This case sets a judicial precedent for redressing broken treaties and obligations (by Danika Cooper, 2020)

legislation in an effort to confront its legacy as a tool for ongoing violence against Indigenous nations and peoples; to reconcile its inconsistencies and discrepancies; and to demonstrate how contested possessions of land in the United States continue to intersect with American citizenship and Indigenous sovereignty.

The political and economic power of the United States is both materially and ideologically built upon its landholdings. Its dominion over land has always been an act of violence against Indigenous peoples; a violence that is measured, mapped, legislated, commodified, and mobilized through its land laws. Examining these laws through an anti-colonial lens disrupts dominant narratives that neutralize the history and production of the American landscape. Indeed, the power of these laws lies in their apparent neutrality by normalizing propertied systems and erasing landed claims for Indigenous sovereignty. Reading the original texts of these land laws exposes how parcelling, allocating, selling, and regulating land have all been part of a larger strategic and cumulative effort to brutally erode Indigenous peoples' rights as both members of their own sovereign nations and as citizens of the United States. In this context, land laws entwine American citizenship with histories of stolen land and systems of private property.

When the Indian Citizenship Act was passed in 1924, nearly 150 years after the founding of the United States, it granted American citizenship to all Indigenous peoples (*Indian Citizenship Act*, 1924). Though citizenship is often understood as the means by which individuals are protected from violence and afforded rights by the state, for the Indigenous peoples of the United States, citizenship has been the very mechanism that permits violence because American citizenship continues long-term processes of land dispossession and forced relocations of Indigenous people. Federal land laws, those that exist at the level of the nation-state, underwrite and complicate conceptions of Indigenous sovereignty and citizenship, and position the United States as a land that has been both imagined and physically transformed through an ongoing campaign of violence—politically sanctioned, racially motivated, protracted, and enduring. This chapter argues that the formation of nations relies on access to a land base; that the establishment of the United States as a nation was grounded in the dispossession of Indigenous lands; and that throughout American history, the policies that have governed those stolen lands have been violent instruments to shape the citizenship of Indigenous peoples to their own nations and to the United States.

The role of land in the formation of nations

Land and nation are indivisibly linked—land is the physical embodiment of the ideological and conceptual project of nationhood, and generally, a nation's influence heavily relies on its land base. This power arrangement is true for both the United States and many of the Indigenous nations its creation dispossessed. However, in the case of the United States, its nationhood relies upon orchestrated and strategic violence to expel all those who have impeded the American accumulation and control over land. This colonial and hegemonic conception of nationhood eliminates any other competing formulations—even those that had long preceded the founding of the United States. Locating the relationship between sovereignty and land as an aspect of global human rights, the United Nations Subcommittee on Racism, Racial Discrimination, Apartheid and Decolonization published a report in 1977 in which they traced the incontestable connection between a society's land and its culture, economics, and politics:

> A distinction must be made between property in its economic and legal aspects and property considered as a social institution. The territorial question of American Indian peoples in the United States is fundamentally an economic question, that is, as the source of livelihood, but also involves the survival of human societies, and is, therefore, a question of human rights, and a nationalities question. A people cannot continue as a people without a land base, an economic base, and political independence, as distinguished from a religious group or an ethnic minority of fundamentally the same historical character as the majority society.
>
> *(quoted in Churchill & LaDuke, 1992: 243)*

Arguing that human rights should extend even beyond the United Nations' framing, Shelley Egoz, Jala Makhzoumi, and Gloria Pungetti argue that all people have a 'right to landscape' on which "the tangible needs for survival and the intangible, spiritual, emotional and psychological needs that are quintessential to the human experience" overlap with physical land (Egoz, Makhzoumi and Pungetti, 2011: 5). United States federal laws which legalize the seizure of land, however, violently alienate people from the landscape to which they are bound or related (Olwig, 2005: 20).

Further, Abenaki scholar Lisa Brooks argues that Indigenous concepts of nationhood fundamentally differ from those of the settler-colonial framework: she writes that nationalism in the Indigenous context

> relies on the multifaceted, lived experience of families who gather in particular places [...] that draws on theoretical and epistemological models that arise from [...] the many, varied, complex, and changing modes in which native nations have operated on the ground, in particular places, over a wide expanse of time.
>
> *(2006: 244)*

The influence of land in consecrating Indigenous nationhood and structuring other aspects of society far exceeds that of the western context: legal scholar Nell Jessup Newton notes that "tribal people [have] different concepts of the relationship of people to land than those encapsulated in the concept of property in Euro-American cultures" (1999: 261). In many Indigenous cultures, relationships to land are at the centre of identity creation and meaningfully contribute to their socio-cultural, spiritual, and political systems (Hiller and Carlson, 2018; Simpson, 2017; McAdam, 2015; Byrd, 2011; Little Bear, 2000; Cajete, 1994). Tewa scholar Gregory Cajete explicitly articulates this idea, writing that:

Every cultural group established their relations to [their place] over time. Whether that place is in the desert, a mountain valley, or along a seashore, it is in the context of natural community, and through that understanding they established an educational process that was practical, ultimately ecological, and spiritual. In this way they sought and found their life.

(1994: 113)

Importantly, the exact relationship to land varies with each individual nation, tribe, and community and their layered, complex circumstances and temporalities; thus, presenting a single theory of the interconnections between land and culture would be both difficult and reductive (McCoy, 2002: 429). For instance, some Indigenous nations were historically nomadic while others were sedentary; some cultivated their lands for sustenance, others hunted; some were deeply communal with other nations and tribes, while others were insular.[1] Today, vast differences remain between how contemporary Indigenous groups envision and engage with land, due to the continuing evolution of ancestral traditions and external forces that produced new and involuntary relationships to land. Perhaps most emblematic of those forces is the national reservation system, which has ensured that some Indigenous groups now occupy reservation lands on or adjacent to their ancestral lands, while others have access to reservations far from ancestral homes with little or no connection to those lands. Still others have been denied access to any lands altogether.

Indigenous lands, both ancestral homelands and assigned reservation parcels, are sites of prolonged Indigenous occupation as well as the generator of cultural, economic, and physical sustenance. These territories hold religious and spiritual significance, while also representing a "defined, legally protected territory within [which] tribes exercise their inherent sovereign right to self-government" (McCoy, 2002: 442). Mohawk activist and scholar Patricia Monture-Angus defines sovereignty as "a question of identity (both individual and collective) more than it is a question of an individualized property right. Identity […] requires a relationship with territory (and not a relationship based on control of that territory)" (2000: 36). Monture-Angus's explicit reference to property points to a key difference between Indigenous relationships to land and that of the settler colonial paradigm, which has largely relied on the accumulation of and control over land.

Conflicting ontologies of nationhood present one of the most fundamental differences between settler colonialism and the Indigenous nations they encountered. This ontological difference is complicated by Patrick Wolfe's assertion that "land is necessary for life," and that "contests for land can be—indeed, often are—contests for life" (2006: 387). Following Wolfe's logic, then, the fight to occupy, govern, and control land forms the core of who has the right to live. In settler America, those with the right to land and the right to life are white settler citizens to the exclusion of everyone else. Thus, it might be understood that it is this assumed right to life for some that has repeatedly justified and normalized the dispossession of Indigenous lands. Despite global acknowledgements that land contributes to a society's culture, identity, and nationhood, the American federal government has long employed

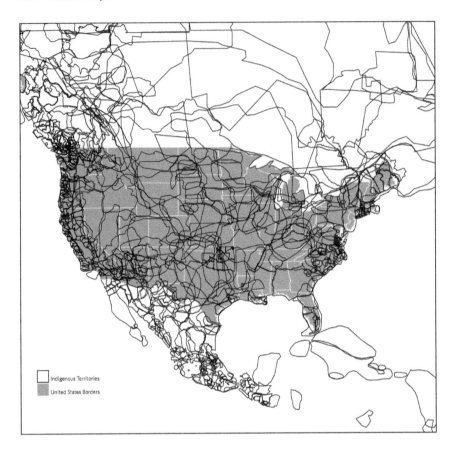

FIGURE 12.2   Indigenous Territories: Before the establishment of the United States of America, the North American landmass was occupied by Indigenous nations who engaged with the land in myriad of ways. Some were nomadic, moving with weather patterns or trade routes, while other were sedentary; some were hunters, while others gathered their food; some established massive infrastructure to control water while others adapted to seasonal hydrological flux. This map shows the overlapping and changing boundaries of Indigenous nations and demonstrates the incompatibility of the geopolitical boundaries of the United States with Indigenous use and occupation of land (by Danika Cooper, 2020, based on data from native-land.ca (Accessed 2020))

racialized policies for the ownership, regulation, and management of land as a political instrument, dismantling Indigenous sovereignty and weakening tribal culture, a topic more explicitly explored in the following sections.

From sovereign to savage: racial difference as tool to erode sovereignty

The American legal framework has been employed for the assertion of American political power by legalizing its control over and ownership of land. In such assertions for power, American expansionist agendas explicitly required Indigenous racialization, dispossession, and disenfranchisement. In 1778, politician David Ramsey wrote, "The remotest depths of our national frontiers will be transformed into agricultural fields, fabulous mines, and cities[...] Those very spots that are now howled over by savage beasts and more savage men" would be the necessary resources for fulfilling the aspirations of an American empire (quoted in Worster, 2001: 114). Here, Ramsey employs the term "savage men" as a political tool to reinforce portrayals of Indigenous peoples as immoral, uncivilized, and in need of Christian redemption; such representations justified ecocidal and genocidal policies that promoted the growth of economic and political systems (Berkhofer, 1979: 113).

In the period immediately following the American Revolution (1765–1783), the Continental Congress needed to establish itself as a legitimate democratic republic.[2] The new nation's legitimacy depended on recognition by Indigenous nations, many of which had already been formally recognized as sovereign nations by various European crowns. Thus, relationships between Indigenous nations and

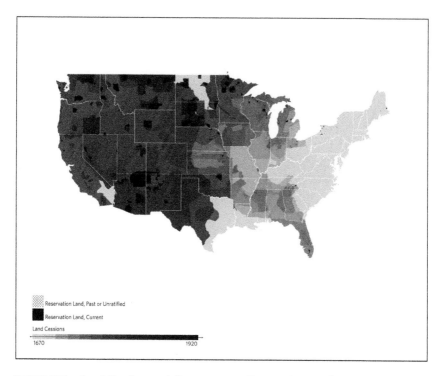

FIGURE 12.3 Land Cessions and Contemporary Reservations in the United States: Tracking the establishment of the United States through time reveals that the nation's accumulation of land meant the dispossession of land from Indigenous nations and communities. Federal land policies and land cessions legitimized this dispossession by upending Indigenous occupation and dividing land into private parcels for its white citizenry. The federal reservation system, which set aside lands for Indigenous peoples, physically and symbolically isolated Indigenous peoples from the white settlers who began to "legally" occupy Indigenous lands (by Danika Cooper, 2020)

FIGURE 12.4 Legislation Collage: Federal land laws legitimated and legalized the dispossession and disenfranchisement of Indigenous peoples. Centering legal texts as cultural representations of values that guide action highlights how the legislative system has long been employed to launch violent, coordinated, and protracted attacks against Indigenous sovereignty and nation-building. In particular, American citizenship—continually constituted through constructions of landscape and property—was intentionally mobilized to dispossess Indigenous peoples from their lands by eroding their claims to nationhood and sovereignty (by Danika Cooper, 2018)

Homestead Act
1862

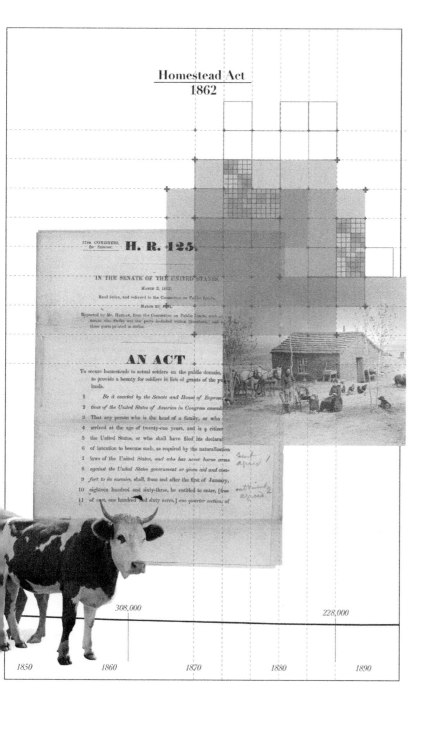

31TH CONGRESS, 2D SESSION. **H. R. 125.**

IN THE SENATE OF THE UNITED STATES.

MARCH 3, 1862.

Read twice, and referred to the Committee on Public Lands.

MARCH 25, 1862.

Reported by Mr. HARLAN, from the Committee on Public Lands, with amendments, viz: Strike out the parts included within [brackets], and insert those parts printed in *italics*.

AN ACT

To secure homesteads to actual settlers on the public domain, and to provide a bounty for soldiers in lieu of grants of the public lands.

1   *Be it enacted by the Senate and House of Represen-*
2   *tatives of the United States of America in Congress assembled,*
3   That any person who is the head of a family, or who has
4   arrived at the age of twenty-one years, and is a citizen of
5   the United States, or who shall have filed his declaration
6   of intention to become such, as required by the naturalization
7   laws of the United States, *and who has never borne arms*
8   *against the United States government or given aid and com-*
9   *fort to its enemies,* shall, from and after the first of January,
10   eighteen hundred and sixty-three, be entitled to enter, [free
11   of cost,] one hundred and sixty acres, one quarter section of

308,000   228,000

1850   1860   1870   1880   1890

Dawes Act
1886

INDIAN LAND FOR SALE

GET A HOME
OF
YOUR OWN

❈

EASY PAYMENTS

PERFECT TITLE
❈
POSSESSION
WITHIN
THIRTY DAYS

FINE LANDS IN THE WEST
IRRIGATED   GRAZING   AGRICULTURAL
IRRIGABLE   DRY FARMING

In 1910 the Department of the Interior Sold Under Sealed Bids Allotted Indian Land as Follows:

Location.	Acres.	Average Price per Acre.	Location.	Acres.	Average Price per Acre.
Colorado	5,211.21	$7.27	Oklahoma	34,664.00	$19.14
Idaho	17,013.00	24.85	Oregon	1,020.00	15.43
Kansas	1,684.50	33.45	South Dakota	120,445.00	16.53
Montana	11,034.00	9.86	Washington	4,879.00	41.37
Nebraska	5,641.00	36.65	Wisconsin	1,069.00	17.00
North Dakota	22,610.70	9.93	Wyoming	865.00	20.64

FOR THE YEAR 1911 IT IS ESTIMATED THAT 350,000 ACRES WILL BE OFFERED FOR SALE

For information as to the character of the land write for booklet, "INDIAN LANDS FOR SALE," to the Superintendent U. S. Indian School at any one of the following places:

CALIFORNIA:
Hoopa.
COLORADO:
Ignacio.
IDAHO:
Lapwai.
KANSAS:
Horton.
Nadeau.

MINNESOTA:
Onigum.
MONTANA:
Crow Agency.
NEBRASKA:
Macy.
Santee.
Winnebago.

NORTH DAKOTA:
Fort Totten.
Fort Yates.
OKLAHOMA:
Anadarko.
Cantonment.
Colony.
Darlington.
Muskogee.
Pawnee.

OKLAHOMA—Con.
Sac and Fox Agency.
Shawnee.
Wyandotte.
OREGON
Klamath Agency.
Pendleton.
Roseburg.
Siletz.

SOUTH DAKOTA:
Cheyenne Agency.
Crow Creek.
Greenwood.
Lower Brule.
Pine Ridge.
Rosebud.
Sisseton.

WASHINGTON:
Fort Simcoe.
Fort Spokane.
Tekoa.
Tulalip.
WISCONSIN:
Oneida.

244,437

| 1900 | 1910 | 1920 | 1930 | 1940 |

FIGURE 12.4 Cont.

Indian Citizenship Act
1924

1950   1960   1970   1980

the United States were governments on equal footing. Between 1778 and 1868, the United States Senate ratified 367 treaties with Indigenous nations and communities (Spirling, 2012: 86), also establishing additional legal arrangements, such as executive agreements, federal statutory obligations, international agreements, unratified treaties, and other instruments of foreign affairs (Corntassel and Witmer, 2008: 7; Robbins, 1992: 89–90). By engaging in formal treaties, the United States recognized the sovereign status of the Indigenous governments. As late as 1903, federal authorities were still negotiating with Indigenous leaders as the heads of separate nation-states. Indigenous peoples held, and continue to hold, a clear and binding legal entitlement to conduct themselves as sovereign nations. The phrase "Indigenous nation" is therefore neither rhetorical, symbolic, nor metaphorical; instead, it denotes the right to act authoritatively as an autonomous state, occupying a distinct physical space and territory (McCoy, 2002: 443).

But despite extensive engagements with Indigenous nations and communities as part of its international relations, the United States held that it was entitled to exercise political power over them. In 1818, Secretary of War John C. Calhoun wrote

> By a proper combination of force and persuasion, of punishment and rewards, [Indigenous peoples] ought to be brought within the pales of law and civilization. Left to themselves, they will never reach that desirable condition … When sufficiently advanced in civilization, [Indigenous peoples] would be permitted to participate in such civil and political rights as the respective States.
>
> *(Jaimes, 1992: 124)*

Historical discrepancies in the status of US–Indigenous nations' relations allowed the United States government to dictate when Indigenous groups were regarded as sovereign entities and when they were not. These inconsistencies were frequently exploited to produce and strengthen favourable conditions for the United States especially in relation to its accumulation of land.

Assaulting sovereignty through seizing land: systems of private property

Nearly all of the treaties that the United States entered into with Indigenous nations revolved around acquiring lands, but it was through federal land laws—policies that did not require the consent of Indigenous nations—that the United States exercised its power most assertively. In the first century of the United States' existence, federal land laws worked to dispossess Indigenous peoples of their lands, establishing private property as the organizing principle of the nation, racially segregating it, and growing the agrarian economy for the benefit of its white settler citizens. Through a coordinated series of laws, beginning with the Land Ordinance Act in 1785, the American federal government has enacted policies that deploy colonial conceptions of space and racially motivated socio-economic agendas: a process that political theorist Robert Nichols calls "recursive dispossession", wherein the consequences of past dispossessions are repeatedly enacted, looping back onto themselves, with each iteration building from and strengthening the original act of dispossession (2019: 9).

The colonial conception of land as a commodity to be parceled and sold as private property has led to persistent social and economic inequality. Systematizing access to land through a logic of private property ensures that some people (i.e., landholders) have the legal right to reap the economic, social, and political benefits of ownership, while those without private property are fully excluded from such benefits. Indigenous and enslaved peoples have been actively excluded from participating in this system through the theft of Indigenous lands, the agrarian economy and its resultant labour practices, the implementation of racist federal land and citizenship policies, and the enactment of the national reservation system. Critical race theorist Cheryl Harris argues that connections between property and race are the backbone of an agrarian economy because they "establish and maintain racial and economic subordination" (1993: 1716). This subordination of one group of people under another creates a sociopolitical system that legal scholar Brenna Bhandar calls the "racial regimes of ownership" in which

> scientific techniques of measurement and quantification, economic visions of land and life rooted in logics of abstraction, culturally inscribed notions of white European superiority, and philosophical concepts of the proper person who possessed the capacity to appropriate [...] worked in conjunction to produce laws of property and racial subjects.
>
> *(2018: 6)*

Bhandar further argues that the United States' accumulation of capital through its land base has always relied upon racial difference because "blackness and Indigeneity came to signify a lesser value not only in relation to white European settlers but with respect to relations of ownership" (2018: 106).

Connections between private property, capitalism, and race were legalized through land legislation and socially enacted through citizenship policies: these connections have produced vast disparities in social and economic systems, the legacies of which remain violent and present today. In the United States, a constitutive relationship exists between state power, the dispossession of land from Indigenous peoples, and transformation of that land into property, to be exploited for the accumulation of wealth through enslaved labour forces. This relationship is most clearly visible through the parallel development of federal land policies and citizenship requirements.

Laws and consequences

Over the course of 200 years, the United States established, solidified, and legalized its dominion over land and people. Positioning federal land policy as a buttress— holding up and reinforcing settler colonialism—fortifies the relationship between landscape, nation-making, sovereignty, and citizenship as "a continually unfolding process", the effects of which are still being expressed and exposed today (Bhandar,

2018: 24). Examining individual land laws both as standalone policies with particular strategies and specific consequences as well as part of a larger constellation of strategic and systemic approach to land, citizenship, and Indigenous sovereignty reveals how the federal legislation that governs the landscape is deeply embedded within violent agendas for the accumulation of wealth and power.

Land Ordinance Act of 1785

> *The surveyors, as they are respectively qualified, shall proceed to divide the said territory into townships of six miles square, by lines running due north and south, and others crossing these at right angles, as near as may be, unless where the boundaries of the late Indian purchases may render the same impracticable, and then they shall depart from this rule no farther than such particular circumstances may require.*
>
> *(Land Ordinance of 1785)*

The Land Ordinance Act of 1785 defined how lands acquired after the Revolutionary War would be measured, divided, and allocated among its new citizens. The Treaty of Paris (1783) officially ended the war, not only establishing America as "free sovereign and Independent States", but also giving over all lands that had previously been controlled by the British Empire. The Land Ordinance Act, and subsequent land policies, helped solidify pervasive cultural attitudes and political agendas of the time that explicitly linked private property to an American national identity. In correspondence with a fellow politician, John Adams wrote in 1776:

> The balance of power in a society, accompanies the balance of property in land. The only possible way, then, of preserving the balance of power on the side of equal liberty and public virtue, is to make the acquisition of land easy to every member of society; to make a division of land into small quantities, so that the multitude may be possessed of landed estates. If the multitude is possessed of the balance of real estate, the multitude will take care of the liberty, virtue, and interest of the multitude in all acts of government.
>
> *(Adams, 1776)*

This national identity translated, for Adams and many other politicians of the post-Revolutionary era, into the desire to survey, enclose, and divide the land into distributable parcels. These spatial processes were imagined as attendant to a democratic distribution of power that would prevent oligarchical claims that could lead to tyranny. Thus, the creation and distribution of private property became an essential component of the early goals of the new nation, regardless of who had previously or currently occupied those lands.

During Britain's colonial occupation, many Indigenous nations had adamantly refused the concept of land ownership, and their defiance continued when lands were transferred to the young nation. However, through both coercion and violent means, many Indigenous nations were forced to surrender their lands to the federal government. By 1830, over 700 million acres of land had been transferred to the United States through land cessions.[3] Surveyors parcelled these territories into individual townships of 36 square miles and further subdivided them into one-square-mile sections, which were then sold to settler families. Dependent on revenue generated from land sales in the creation of an agrarian-based republic, the United States government gave settlers legal claim to land and restricted the rights of the Indigenous peoples who had previously occupied them. Widespread transformations of the continent into distributable parcels of private property were thus contingent upon continued and indisputable access to the land.

Indian Removal Act, 1830

> *That it shall and may be lawful for the President to exchange any or all of such districts, so to be laid off and described, with any tribe or nation of Indians now residing within the limits of any of the states or territories, and with which the United States have existing treaties, for the whole or any part or portion of the territory claimed and occupied by such tribe or nation, within the bounds of any one or more of the states or territories, where the land claimed and occupied by Indians, is owned by the United States, or the United States are bound to the state within which it lies to extinguish the Indian claim thereto.*
>
> (Indian Removal Act, 1830)

Under the Indian Removal Act, Indigenous groups were either involuntarily relocated onto reservation lands set aside for their use or were forced to live on very small portions of their ancestral lands. All other ancestral lands outside of the small parcels reserved for their occupation were either given or sold to white settlers, or were retained for federal use. President Andrew Jackson, known for his vicious policies designed to eradicate, weaken, and disperse Indigenous populations, proclaimed during his 1830 *State of the Union* address to Congress that the Indian Removal Act would physically segregate the Indigenous peoples from white settlers and that under the supervision of the government, they would eventually "cast off their savage habits, and become an interesting, civilized and Christian community" (Jackson, 1830). Throughout his tenure as president, Jackson continued to advocate for removal with increasingly violent language, proclaiming to Congress in December 1833:

> That those tribes can not exist surrounded by our settlements and in continual contact with our citizens is certain. They have neither the intelligence, the industry, the moral habits, nor the desire of improvement which are essential to any favorable change in their condition. Established in the midst of another and a superior race, and without appreciating the causes of their inferiority or seeking to control them, they must necessarily yield to the force of circumstances and ere long disappear.
>
> (Jackson, 1833)

Jackson's reference to "improvement" is significant, as the term is used in two ways: first, to prove that Indigenous peoples lack a desire for self-improvement and are therefore socially inferior; second, to justify their isolation from white settler citizens. Many conceptions for the future of the United States at this moment relied upon the enclosing of land to transform it into productive grounds for agrarian growth and wealth. That the Indigenous nations did not share this approach to land indicated to Jackson and many others their lack of "intelligence, industry, [and] moral habits", rather than as a marker of ideological and ontological differences of what land meant and how land and culture were connected. Anthropologist Karen Blu makes the case that the federal government's reservation system came directly out of this ontological difference in these two conceptions of land; she argues that the reservations were a "product of the struggle between Indians and Whites for primacy of place—for the right to inhabit and control place and space" (Sarmiento and Hitchner, 2017: 229).

Beyond specifying who had rights to specific lands, the reservation system served the federal government's political and cultural agendas in two important ways. First, it physically and symbolically separated Indigenous people from white settlers, allowing white settlers the opportunity to "legally" occupy Indigenous land without having to commingle with its rightful owners; second, it legislated "a long process of decoupling land from established social bonds and reconstructing it as a commodity like any other" (Harris, 2010: 263). Shiri Pasternak outlines the relationship between social ordering and land, emphasizing that reservations were "driven mainly by attempts by church and state to carve 'individuals' from impenetrable social blocks of tribal life" (2015: 182). This approach upholds Anne Bonds and Joshua Inwood's assertion that the "logics of extermination" are central to the success of white settler colonialism as "the building of new settlements necessitates the eradication of Indigenous populations, the seizure and privatization of their lands, and the exploitation of marginalized peoples in a system of capitalism established by and reinforced through racism" (2016: 716). Initiated and legalized by the reservation system, subsequent legislation continued to promote social violence against Indigenous nations and segregation from white colonial settlers.

Homestead Act, 1862 | Desert Land Act, 1877

> *That any person who is the head of a family, or who has arrived at the age of twenty-one years, and is a citizen of the United States [...] shall, from and after the first January, eighteen hundred and sixty-three, be entitled to enter one quarter section or a less quantity of unappropriated public lands.*
>
> (Homestead Act, 1862)

> *That it shall be lawful for any citizen of the United States, or any person of requisite age "who may be entitled to become a citizen, and who has filed his declaration to become such" and upon payment of twenty five cents per acre—to file a declaration*

under oath with the register and the receiver of the land district in which any desert land is situated, that he intends to reclaim a tract of desert land not exceeding one section, by conducting water upon the same.

(Desert Land Act of 1877)

Aimed to fulfil desires for an American agrarian empire, the ratification of the Homestead Act incentivized the western migration of its population by providing any adult citizen with 160 acres of public land to cultivate, creating "a class of prosperous small farmers whose own prosperity fed the economic development of the nation" (White, 1991: 143). After 15 years, the Desert Land Act increased the allotment size to 640 acres of desert land and required settlers to reclaim, irrigate, and cultivate the lands. Underpinning these policies was the widely held belief that the western lands were *terra nullius* ("nobody's land"); in positioning Western lands as "empty" or "without 'civilized' society," the US government justified its "imposition of a new economic and spatial order on 'new territory', either by erasing the pre-capitalist Indigenous settlement or confining it to particular areas" (Kain and Baigent, 1992: 329). Referring to these lands as *terra nullius* reinforced a wilful blindness to the ways in which the land sustained daily practices and livelihoods of Indigenous populations (Pasternak, 2015: 182). By the late 19th century, land laws had successfully partitioned the American West both spatially and ideologically into three kinds of land uses: white settler agrarian communities, reservations for displaced Indigenous nations, and otherwise unoccupied lands.

General Allotment Act (Dawes Severalty Act), 1887

That in all cases where any tribe or band of Indians has been, or shall hereafter be, located upon any reservation created for their use, either by treaty stipulation or by virtue of an act of Congress or executive order setting apart the same for their use, the President of the United States be, and he hereby is, authorized, whenever in his opinion any reservation or any part thereof of such Indians is advantageous for agricultural and grazing purposes, to cause said reservation, or any part thereof, to be surveyed, or resurveyed if necessary, and to allot the lands in said reservation in severalty to any Indian located thereon …

… And every Indian born within the territorial limits of the United States to whom allotments shall have been made under the provisions of this act, or under any law or treaty, and every Indian born within the territorial limits of the United States who has voluntarily taken up, within said limits, his residence separate and apart from any tribe of Indians therein, and has adopted the habits of civilized life, is hereby declared to be a citizen of the United States, and is entitled to all the rights, privileges, and immunities of such citizens, whether said Indian has been or not, by birth or otherwise, a member of any tribe of Indians within the territorial limits of the United States without in any manner affecting the right of any such Indian to tribal or other property …

(General Allotment Act, 1887)

In a culminating effort to erode and erase the tribal cultures and their traditions, the General Allotment Act (better known as the Dawes Severalty Act) used citizenship as a forceful assimilation mechanism. The Act gave the President of the United States the power to breakup reservation land into allotments of private property that could then be distributed to individuals. Any Indigenous person who agreed to the terms was given a parcel of reservation land for her own private use and granted American citizenship, an exchange that further emphasized the importance of property and the destruction of communal occupation of lands. In campaigning for allotment to Congress, Senator Dawes emphasized the relationship between private property and "civilized" society:

> [The Indians] have got as far as they can go, because they own their land in common ... There is no enterprise to make your home any better than that of your neighbor's. There is no selfishness, which is at the bottom of civilization. Until this people consent to give up their lands and divide them among their citizens so that each can own the land he cultivates, they will not make much progress.
>
> *(Witkin, 1995: 360)*

Further articulating the goals of the Dawes Act, Thomas J. Morgan, the Commissioner of Indian Affairs, argued in 1889 that

> tribal relations should be broken up, socialism destroyed, and the family and autonomy of the individual substituted. The allotment of land in severalty, the establishment of local courts and police, the development of a personal sense of independence and the universal adoption of the English language are means to this end.
>
> *(White, 1991: 115)*

Senator Dawes's and Commissioner Morgan's statements emphasize, yet again, the central role that the contested legal possession of land has played in the erosion of Indigenous culture and sovereignty. The resulting replacement of traditional modes of collective use and occupancy with a system of private property became, in the approving words of President Theodore Roosevelt, "a mighty pulverizing engine to break up the tribal mass" (Schmidt, 2011: 5). In exchange for renouncing their identity and entitlement to reservation lands, those who accepted allotments were granted American citizenship, underlining the strong and long-standing association between citizenship and private property in the United States. The loss of reservation lands—lands that were already pitiful compensation for the original dispossession of land—was further violence enacted upon Indigenous peoples through yet another seizure of land and destruction of communal tribal bonds. Ultimately, the Dawes Act was a success. In fewer than 20 years, Indigenous communities had lost almost half of their lands: in 1881, Indigenous communities held over 150 million acres of land; by 1890, they held about 100 million; in 1900, they were down to just under 78,000 acres (White, 1991: 115).

Curtis Act, 1898

> *That when the roll of citizenship of any one of said nations or tribes is fully completed as provided by law, and the survey of the lands of said nation or tribe is also completed, the commission heretofore appointed under Acts of Congress, and known as the "Dawes Commission," shall proceed to allot the exclusive use and occupancy of the surface of all the lands of said nation or tribe susceptible of allotment among the citizens thereof, as shown by said roll, giving to each, so far as possible, his fair and equal share thereof, considering the nature and fertility of the soil, location, and value of same; but all oil, coal, asphalt, from and mineral deposits in the lands of any tribe are reserved to such tribe, and no allotment of such lands shall carry the title to such oil, coal, asphalt, or mineral deposits; and all town sites shall also be reserved to the several tribes, and shall be set apart by the commission hereto-fore mentioned as incapable of allotment ...*
>
> *(Curtis Act, 1898)*

An amendment to the Dawes Act, the Curtis Act resulted in the breaking up of tribal governments and communal lands in "Indian Territory", the region set aside for the forced relocation of displaced Indigenous nations under the policy of "Indian removal". The Choctaw, Chickasaw, Muscogee (or Creek), Cherokee, and Seminole nations occupied this land and had been previously exempt from the Dawes Act due to the terms of previous treaties. The Curtis Act ignored earlier treaties in order to authorize the abolishment of tribal courts, tribal governments, and tribal land claims in "Indian Territory". Dissolving tribal authority left these territories subject to federal claims. In 1907, "Indian Territory" lands—reservations that were given to the Muscogee, Cherokee, Chickasaw, Choctaw, and Seminole nations—were combined with the unincorporated Oklahoma Territory to create the state of Oklahoma, once again dispossessing Indigenous communities of land and producing the contestation that formed the basis of the 2020 Supreme Court case *McGirt v Oklahoma* discussed earlier (French, 1999: 245).

The Dawes and Curtis Acts led to a large displacement of Indigenous populations from their former communal cultural lands to urban centres, contributing to what sociologist Laurence Armand French and others have called a "cultural genocide" (French, 1999: 245). Coined by sociologist Raphael Lemkin, the term "cultural genocide" refers to the "destruction of national pattern of the oppressed group" and "the imposition of the national pattern of the oppressor" (2008: 79). Lemkin argues that "if the culture of a group is violently undermined, the group itself disintegrates and its members must either become absorbed in other cultures [...] or succumb to personal disorganization and, perhaps, physical destruction" (Short, 2010: 835). By the end of the 19th century in the United States, federal land policies were thus political instruments that wilfully and knowingly stripped Indigenous nations of their lands, sovereignty, and cultural practices. In coordination with these land laws, the denial of citizenship of Indigenous peoples while not recognizing them as members of their own sovereign nations worked to further erode their cultural practices and political agency.

Indian Citizenship Act, 1924

Be it enacted by the Senate and House of Representatives of the United States of America in Congress assembled, That all non-citizen Indians born within the territorial limits of the United States be, and they are hereby, declared to be citizens of the United States: Provided, That the granting of such citizenship shall not in any manner impair or otherwise affect the right of any Indian to tribal or other property.

(Indian Citizenship Act, 1924)

In 1924, the Indian Citizenship Act imposed United States citizenship on all Indigenous peoples, whether they desired it or not. By unilaterally extending American citizenship without collaboration with Indigenous peoples or nations, it has been considered by many as "repressive emancipation", a term used to define the process of "liberating" a people from conditions that they themselves do not deem unjust or oppressive (Witkin, 1995: 355). It might also be noted that some state governments still denied Indigenous peoples the rights afforded by citizenship, such as voting and holding office. The state of Arizona, for example, denied Indigenous peoples voting privileges for nearly two decades after the Indian Citizenship Act was passed (Deloria and Lytle, 1983: 222).

Under the Citizenship Act, the connections between land, economics, and citizenship were again strengthened. In the 1920s, it had become apparent to land speculators, federal agents, and settlers that much of the lands apportioned under the Dawes and Curtis Acts—initially believed to be "worthless"—were in fact rich in mineral deposits and natural resources. Coal, copper, gold, uranium, oil, and natural gas were all to be found on lands either held by Indigenous individuals or underneath reservation lands. It was therefore imperative, from an economic standpoint, that the federal government and its corporate interests gain access to these resources (Jaimes, 1992: 127). Now that all members of Indigenous groups were American citizens, and no longer representatives of their original nations, resource and land negotiations could be conducted with more lenient restrictions that were often more advantageous for the federal government than for the Indigenous nations who had previously occupied the land.

Indian Reorganization Act, 1934

Be it enacted by the Senate and House of Representatives of the United States of America in Congress assembled, That hereafter no land of any Indian reservation, created or set apart by treaty or agreement with Indians, Act of Congress, Executive order, purchase, or otherwise, shall be allotted in severalty to any Indian ...

... Any Indian tribe, or tribes, residing on the same reservation, shall have the right to organize for its common welfare, and may adopt an appropriate constitution and bylaws, which shall become effective when ratified by a majority vote of the adult members of the tribe, or of the adult Indians residing on such reservation, as the case may be, at a special election authorized and called by the Secretary of the Interior under such rules and regulations as he may prescribe ...

(Indian Reorganization Act of 1934)

In 1934, Congress passed the Indian Reorganization Act (IRA), which had initially been intended to reverse assimilationist policies of the past, restore Indigenous sovereignty and self-governance, and reinstate Indigenous authority over the management of their lands and resources. In spite of its potential to radically transform the federal government's approach to Indigenous nations, in practice the act actively eroded Indigenous sovereignty by mandating that tribal governance be made through "constitutions" or "charters" that were federally approved and regulated by the Bureau of Indian Affairs (a federal organization) rather than through governance structures and institutions established by each nation. Under the IRA, all decisions were subject to the approval of the Secretary of the Interior, further eroding the authority of Indigenous nations to affirm their own cultural practices, sacred histories, citizenship requirements, judicial systems, and governmental bodies that had previously provided the basis for Indigenous nationhood. These new "governments" were charged with increased responsibilities, including economic planning. They oversaw mineral lease negotiations, contracted with external corporate agencies, and negotiated long-term agricultural and ranching leasing, water rights, and land transfers, all of which needed the direct approval of the Bureau of Indian Affairs.

House Concurrent Resolution 108, 1953

> *Whereas it is the policy of Congress, as rapidly as possible, to make the Indians within the territorial limits of the United States subject to the same laws and entitled to the same privileges and responsibilities as are applicable to other citizens of the United States, to end their status as wards of the United States, and to grant them all of the rights and prerogatives pertaining to American citizenship; and Whereas the Indians within the territorial limits of the United States should assume their full responsibilities as American citizens …*
>
> (House Concurrent Resolution 108, 1953).

The House Concurrent Resolution 108 unilaterally ended all federal responsibility to Indigenous nations and communities, ushering in the Era of Indian Termination, whereby the federal government used American citizenship as a justification for the denial of Indigenous sovereignty and recognition, reservation land entitlements, and civil rights. The resolution gave Congress power to re-evaluate existing treaties between Indigenous nations and the federal government and to break up, seize, and sell reservation lands. When questioned about whether treaties would be abrogated, Senator Watkins, a major proponent of the resolution, emphatically stated:

> We have arrived at the point where we do not recognize now within the confines of the United States any foreign nations. You now have become citizens of the one nation. Ordinarily the United States does not enter into treaties […] between any of its citizens and the Federal Government […] So it is doubtful now that from here on treaties are going to be recognized where the Indians themselves have gone to the point where they have accepted

citizenship in the United States and have taken advantage of its opportunities. So that question of treaties, I think, is going to largely disappear.

(Witkin, 1995: 381)

Subsequent policies worked to further uphold the termination agenda and to use coercive citizenship to nullify sovereignty and treaty rights. One year later, Public Law 280 placed any reservation that had not already been terminated under state jurisdiction, directly limiting the jurisdictional and legislative power of existing reservation governing structures. Then, in 1956, Public Law 959 shrunk the allocation of federal funds to reservations and almost immediately resulted in reduced education and health services, increased unemployment rates, and reduced overall reservation *per capita* income (Robbins, 1992: 99). Under this law, anyone willing to "voluntarily" relocate from the reservation to a federally approved relocation centre would be provided with federal funding to subsidize their relocation costs and was offered professional training. The results of termination were devastating: from 1953 to 1958, 109 Indigenous nations or portions of nations were dissolved, and their reservation lands taken back by the federal government, resulting in nearly 35,000 Indigenous people who "migrated" from reservation lands to urban centres (Robbins, 1992: 99). Throughout the 1950s, federal termination policies both encouraged Indigenous assimilation through citizenship and undermined Indigenous rights to sovereignty and protection of homelands by disassembling and destabilizing their status as nations.

Though the termination agenda was eventually abandoned, the consequences of its policies have remained potent largely because no clear alternative policy emerged. On the one hand, Congress did not restore recognition to all Indigenous nations that were denied, nor did it return lands that were lost. On the other hand, Congress extended federal recognition to Indigenous nations that it had previously denied (Walch, 1983: 1202–1203). Nevertheless, those nations that were recognized in the post-termination era were still often denied basic aspects of sovereignty by Congress, which ultimately held the power to determine which nations to restore and which to leave terminated.

Indian Civil Rights Act, 1968

The consent of the United States is hereby given to any State not having jurisdiction over civil causes of action between Indians or to which Indians are parties which arise in the areas of Indian country situated within such State to assume, with the consent of the tribe occupying the particular Indian country or part thereof which would be affected by such assumption, such measure of jurisdiction over any or all such civil causes of action arising within such Indian country or any part thereof as may be determined by such State to the same extent that such State has jurisdiction over other civil causes of action, and those civil laws of such State that are of general application to private persons or private property shall have the same force and effect within such Indian country or part thereof as they have elsewhere within that State

(Civil Rights Act of 1968)

The Indian Civil Rights Act, passed as a rider to the Civil Rights Act of 1968, forbade states from assuming jurisdiction over lands occupied by Indigenous nations and communities without their permission and provided for retrocession, or the return of civil and criminal jurisdiction from the states back to the Indigenous nations themselves. The Indian Civil Rights Act sought to undo the consequences of Public Law 280 by reinstating tribal sovereignty and the protection of individual rights into federal policy. This law, however, only afforded these rights to those Indigenous nations and communities that had already been federally recognized and which had their own reservation lands.

Today the federal government has denied over 500 Indigenous nations and communities who seek recognition (List of unrecognized tribes in the United States, 2020). The Office of Federal Acknowledgement (OFA), part of the Bureau of Indian Affairs, validates the "authenticity" of tribal identity through 'anthropological, genealogical and historical research methods,' despite having a political and economic stake in its outcome (US Department of Interior, Indian Affairs, 2017). The evaluation criteria require the petitioning tribe to prove, among other things, that it "has been identified as an American Indian entity on a substantially continuous basis since 1900" and that it "has maintained political influence or authority over its members as an autonomous entity from 1900 until the present" (*Procedures for Federal Acknowledgement of Indian Tribes,* 2015). Angela A. Gonzalez and Timothy Q. Evans call into question this "objective" process:

> This process, granted authority and made enforceable through federal law and regulation, has been used to assault the right to be Indian in much the same way that it has been used to assault American Indian rights to lands and resources. And just as the federal government made legible and therefore controllable lands and resources, so, too, has it exercised its authoritarian power by imposing what constitutes the qualities and characteristics of an Indian tribe.
>
> *(2013: 58–59)*

Through "acknowledgement", the federal government imposes arbitrarily determined standards and timelines upon Indigenous identity and, by doing so, regards Indigeneity as static and non-evolving. Because acknowledgement is both controlled by the federal government and is also the primary tool by which Indigenous nations and communities affirm their existence as distinct political entities within the American system, it is another part of the systemic attempt to strip Indigenous nations of their autonomy and sovereignty. Further, lack of acknowledgement directly limits the rights and legislative processes of these groups as sovereign entities, renders their claims to homelands mute in the American system, and prevents them from becoming eligible to receive federal services. This process of acknowledgement is yet another demonstration that land, federal law, and property are entwined and used to politically justify and enact ongoing violence against Indigenous populations, despite their status as American citizens.

Counter-Narrating History for an Anti-Colonial Future

American citizenship—embedded within the laws of the United States and also constituted through constructions of landscape and property—was intentionally mobilized to dispossess Indigenous peoples of their lands and to erode their claims to nationhood and sovereignty. Assembling and critically analyzing the federal laws that have worked to legalize the dispossession and disenfranchisement of Indigenous peoples allows for a counter-narrative to our conventional landscape histories and exposes their insidious, long-term consequences both on the physical landscape and in the lived experiences of Indigenous peoples who inhabit it.

These past injustices remain deeply inscribed into the landscape's present conditions; addressing the past and proposing an anti-colonial, anti-racist future requires giving voice to counter-narratives—those stories that have either been altogether left out of the dominant historic record or wiped clean of their violent foundations. These counter-narratives situate the production of the American landscape in direct dialogue with settler colonization, cultural genocide, and racial hierarchies. Not meaningfully incorporating these histories into the dominant historical record risks replicating and reproducing colonial relations and dynamics (Hiller and Carlson, 2018: 64; Smith, 2012).

Positioning legal texts as socio-historical artefacts reveals the United States legislative system as a mechanism to launch violent and coordinated attacks against Indigenous sovereignty and nation-building. This kind of critical reflection on American history is a reminder that the future is not predetermined nor fixed, but is instead constantly constructed and negotiated through its legal, economic, and socio-political systems. As Unangax̂ scholar Eve Tuck and settler scholar K. Wayne Yang have asserted, building an anti-colonial, anti-racist future is not a metaphor (2012); rather, it is a deliberate socio-political process that requires action toward reparative justice. Chickasaw scholar Jodi Byrd asserts that remediation efforts which do not adequately upend and address the root cause of injustice signals a failure "to grapple with the fact that such discourses further re-inscribe the original colonial injury" (2011: xxiii). Following from Byrd, modes for reparative justice must necessarily address private property as the primary apparatus in generating economic growth and political power in the United States and which was built upon the dispossession of Indigenous lands (Rifkin, 2017; Simpson, 2017; Corlett, 2003). A more just future thus depends upon instating new land and citizenship policies that affirm Indigenous sovereignty and ignite a structural shift towards redressing these past injustices.

Policies, in chronological order

> *Land Ordinance of 1785* (1785). **HR 9A–C3.1.**
> *Indian Removal Act* (1830). *Stat.* **4**: 411.
> *Treaty with Creeks* (1833). *Stat.* **7**: 417.
> *Homestead Act* (1862). *Stat.* **12**: 392.

Treaty with the Creek Indians (1866). *Stat.* **14**: 785.
Desert Land Act of 1877 (1877). *Stat.* **19**: 377.
General Allotment Act (1887). *Stat.* **24**: 388-391.
Curtis Act (1898). *Stat.* **30**: 495.
Indian Citizenship Act (1924). *Stat.* [Online] **43**: 253.
Indian Reorganization Act of 1934 (1934). *Stat.* **48**: 984.
House Concurrent Resolution 108 (1953). *Stat.* **67**: B132.
Civil Rights Act (1968). *Stat.* **82**: 77-81.
Procedures for Federal Acknowledgement of Indian Tribes (2015). *USC* [Online] **25**.

Notes

1 The high variability of these cultural practices and relationships to land is evidenced even within the same approximate geographic region: the O'odham and Ancestral Puebloans are examples of nations that have shared regional geographies, and yet, are dramatically different when evaluating how land figures into their cultural and daily practices. The O'odham people, who historically inhabited the Sonoran Desert in current day Arizona and northern Mexico were semi-nomadic, following seasonal water flows. In the summer months, they practiced *ak-chin* farming, sowing seeds at the mouths of washes after the first rains of the season and harvesting the resulting tepary beans, acorns, and gourds a few months later (di Cintio, 2012: 15). In the winter months, O'odham men made a ceremonial pilgrimage to the Gulf of Mexico, nearly 100 miles from their summer villages, to mine sea salt used as both a healing mineral and as a good to be traded with other regional nations and peoples (Demby, 2018). Contrastingly, the Ancestral Puebloans who historically occupied Chaco Canyon, a geography at the intersection of present-day Arizona, Utah, Colorado, and New Mexico, were a sedentary society known for their sophisticated urban settlements. They carved these into the canyon walls with multistory, masonry architecture and connected their settlements through an intricate regional road network (Cordell, 1984: 103). Despite the dry climate, Puebloans devised sophisticated irrigation systems of dams, canals, and terraces that carefully collected and managed runoff water off of the surrounding cliffs and watered fields of maize, beans, and squash (Frazier, 2005: 98–99).

2 Originally, the Continental Congress was a convention of delegates from the British American colonies. After declaring independence from Great Britain in 1776, the Continental Congress acted as the main governing structure for the United States until 1789 when the American Constitution was signed, which ushered in the current American political system.

3 The figure of 700 million acres of land was calculated using historical land cession data from 1776 to 1830. Data is available at: data.fs.usda.gov/geodata/edw/edw_resources/meta/S_USA.TribalCededLandsTable.xml. Also referenced was the "Invasion of America" research project led by Claudio Saunt. Available at: arcgis.com/apps/webappviewer/index.html?id=eb6ca76e008543a89349ff2517db47e6.

Bibliography

Adams, John (1776). 'From John Adams to James Sullivan, 26 May 1776' [Online]. Available at: founders.archives.gov/documents/Adams/06-04-02-0091 (Accessed 4 November 2020).

Berkhofer, Robert F (1979). *The White Man's Indian: Images of the American Indian, from Columbus to the Present*. New York: Vintage.

Bhandar, Brenna (2018). *Colonial Lives of Property: Law, Land, and Racial Regimes of Ownership*. Durham, NC: Duke University Press.

Blackhawk, Ned (2008). *Violence over the Land: Indians and Empires in the Early American West*. Cambridge, MA: Harvard University Press.

Bonds, Anne and Joshua Inwood (2016). 'Beyond white privilege: Geographies of white supremacy and settler colonialism'. *Progress in Human Geography*, 40(6), pp 715–733.

Brooks, Lisa (2006). 'At the gathering place'. In: Jace Weaver, Craig S. Womack, and Robert Warrior, eds. *American Indian Literary Nationalism*. Albuquerque, NM: University of New Mexico Press, pp 225–252.

Byrd, Jodi A (2011). *The Transit of Empire: Indigenous Critiques of Colonialism*. Minneapolis, MN: University of Minnesota Press.

Cajete, Gregory A (1994). *Look to the Mountain: An Ecology of Indigenous Education*. Durango, CO: Kivakí Press.

Churchill, Ward and Winona LaDuke (1992). 'The political economy of radioactive colonialism'. In: M. Annette Jaimes, ed. *The State of Native America: Genocide, Colonization, and Resistance*. Boston: South End Press, pp 241–266.

Cordell, Linda S (1984). *Prehistory of the Southwest*. Orlando, FL: Academic Press.

Corlett, J Angelo (2003). *Race, Racism, and Reparations*. Ithaca, NY: Cornell University Press.

Corntassel, Jeff and Richard C Witmer (2008). *Forced Federalism: Contemporary Challenges to Indigenous Nationhood*. American Indian law and policy series. 3. Norman, OK: University of Oklahoma Press.

Deloria, Vine and Clifford M Lytle (1983). *American Indians, American Justice*. Austin, TX: University of Texas Press.

Demby, Samantha (2018). *Keeping the Salt in the Earth* [Online]. Available at: nacla.org/blog/2018/07/05/keeping-salt-earth (Accessed 14 April 2020).

di Cintio, Marcello (2012). Farming the monsoon: A return to traditional Tohono O'odham Foods. *Gastronomica: The Journal of Food and Culture* 12(2), pp 14–17.

Egoz, Shelley, Jala Makhzoumi, and Gloria Pungetti (2011). 'The right to landscape: An introduction'. In: Shelley Egoz, Jala Makhzoumi, and Gloria Pungetti, eds. *The Right to Landscape: Contesting Landscape and Human Rights*. Farnham: Routledge, pp 1–22.

Frazier, Kendrick (2005) *People of Chaco: A Canyon and its Culture*. New York: W.W. Norton & Company.

French, Laurence Armand (1999). 'Native American reparations: Five hundred years and counting'. In: Roy Brooks, ed. *When Sorry Isn't Enough: The Controversy over Apologies and Reparations for Human Injustice*. New York: New York University Press, pp 241–247.

Gonzalez, Angela A and Timothy Q Evans (2013). 'The Imposition of Law: The Federal Acknowledgment Process and the Legal De/Construction of Tribal Identity'. In: Amy E Den Ouden and Jean M O'Brien, eds. *Recognition, Sovereignty Struggles, and Indigenous Rights in the United States*. Chapel Hill, NC: University of North Carolina Press, pp 37–64.

Gowen, Annie (2020). 'Supreme Court says nearly half of Oklahoma is an Indian reservation. What's next?' *Washington Post* [Online]. Available at: washingtonpost.com/national/supreme-court-says-nearly-half-of-oklahoma-is-an-indian-reservation-whats-next/2020/07/10/8c2aba02-c2e7-11ea-b4f6-cb39cd8940fb_story.html (Accessed 25 July 2020).

Harris, Cheryl I (1993). 'Whiteness as Property'. *Harvard Law Review*, 106(8), pp 1707–1791.

Harris, Douglas C (2010). 'Book Review of The Law of the Land: The Advent of the Torrens System in Canada, by Greg Taylor (Toronto: Osgoode Society for Canadian Legal History, 2008)'. *University of British Columbia Law Review*, 43, pp 259–267.

Hiller, Chris and Elizabeth Carlson (2018). 'These Are Indigenous Lands: Foregrounding Settler Colonialism and Indigenous Sovereignty as Primary Contexts for Canadian Environmental Social Work'. *Canadian Social Work Review/Revue Canadienne de Service Social*, 35(1), pp 45–70.

Jackson, Andrew (1830). *President Andrew Jackson's Message to Congress 'On Indian Removal' (1830)* [Online]. Available at: ourdocuments.gov/doc.php?flash=true&doc=25&page=t ranscript (Accessed 5 October 2018).

Jackson, Andrew (1833). *President Andrew Jackson: Fifth Annual Message (1833)* [Online]. Available at: presidency.ucsb.edu/ws/index.php?pid=29475 (Accessed 5 October 2018).

Jaimes, M Annette (1992). 'Federal Indian Identification Policy: A Usurpation of Indigenous Sovereignty in North America'. In: M Annette Jaimes, ed. *The State of Native America: Genocide, Colonization, and Resistance*. Boston: South End Press, pp 123–137.

Kain, Roger JP and Elizabeth Baigent (1992). *The Cadastral Map in the Service of the State: A History of Property Mapping*. Chicago: University of Chicago Press.

KickingWoman, Kolby (2020). 'Muscogee (Creek) Nation's 1866 boundaries upheld by Supreme Court'. *High Country News* [Online]. Available from: hcn.org/articles/justice-muscogee-creek-nations-1866-boundaries-upheld-by-supreme-court (Accessed 13 July 2020).

King, Tiffany Lethabo (2019). *The Black Shoals: Offshore Formations of Black and Native Studies*. Durham, NC: Duke University Press.

Lee, Robert and Tristan Ahtone (2020). 'Land-grab universities'. *High Country News* [Online]. Available at: hcn.org/issues/52.4/indigenous-affairs-education-land-grab-universities (Accessed 10 April 2020).

Lemkin, Raphael (2008). *Axis Rule in Occupied Europe: Laws of Occupation, Analysis of Government, Proposals for Redress*. Clark, NJ: The Lawbook Exchange.

List of unrecognized tribes in the United States (2020). *Wikipedia* [Online]. Available from: en.wikipedia.org/w/index.php?title=List_of_unrecognized_tribes_in_the_United_States&oldid=953375422 (Accessed 28 April 2020).

Little Bear, Leroy (2000). 'Jagged Worldviews Colliding'. In: Marie Battiste, ed. *Reclaiming Indigenous Voice and Vision*. Vancouver, BC: University of British Columbia Press, pp 77–85.

McAdam (Saysewahum), Sylvia (2015). *Nationhood Interrupted: Revitalizing Nêhiyaw Legal Systems*. Saskatoon, SK: Purich Publishing.

McCoy, Padraic I (2002). 'The Land Must Hold the People: Native Modes of Territoriality and Contemporary Tribal Justifications for Placing Land into Trust through 25 C.F.R. Part 151'. *American Indian Law Review*, 27(2), pp 421–502.

Monture-Angus, Patricia (2000). *Journeying Forward: Dreaming First Nations' Independence*. Halifax, NS: Fernwood.

Newton, Nell Jessup (1999). 'Indian Claims for Reparations, Compensation, and Restitution in the United States Legal System'. In: Roy L Brooks, ed. *When Sorry Isn't Enough: The Controversy over Apologies and Reparations for Human Injustice*. New York: New York University Press, pp 261–269.

Nichols, Robert (2019). *Theft Is Property!: Dispossession and Critical Theory*. Durham, NC: Duke University Press.

Obama, Barack (2012). 'Remarks by the President at Naturalization Ceremony' [Online]. Available at: obamawhitehouse.archives.gov/the-press-office/2012/07/04/remarks-president-naturalization-ceremony (Accessed 25 April 2020).

Olwig, Kenneth R (2005). 'Representation and alienation in the political land-scape'. *Cultural Geographies*, 12(1), pp 19–40.

Pasternak, Shiri (2015). 'How Capitalism Will Save Colonialism: The Privatization of Reserve Lands in Canada'. *Antipode*, 47(1), pp 179–196.

Rifkin, Mark (2017). *Beyond Settler Time: Temporal Sovereignty and Indigenous Self-Determination*. Durham, NC: Duke University Press.

Rifkin, Mark (2019). *Fictions of Land and Flesh: Blackness, Indigeneity, Speculation*. Durham, NC: Duke University Press.

Robbins, Rebecca L (1992). 'Self-Determination and Subordination: The Past, Present, and Future of American Indian Governance'. In: Annette M. Jaimes, ed. *The State of Native America: Genocide, Colonization, and Resistance*. Boston: South End Press, pp 87–121.

Sarmiento, Fausto and Sarah Hitchner (2017). *Indigeneity and the Sacred: Indigenous Revival and the Conservation of Sacred Natural Sites in the Americas*. New York: Berghahn Books.

Saunt, Claudio (2020). *Unworthy Republic: The Dispossession of Native Americans and the Road to Indian Territory*. New York: WW Norton.

Schmidt, Ryan W (2011). 'American Indian Identity and Blood Quantum in the 21st Century: A Critical Review'. *Journal of Anthropology*, 2011, pp 1–9.

Short, Damien (2010). 'Cultural genocide and indigenous peoples: a sociological approach'. *International Journal of Human Rights*, 14(6), pp 833–848.

Simpson, Leanne Betasamosake (2017). *As We Have Always Done: Indigenous Freedom through Radical Resistance*, 3rd ed. Minneapolis, MN: University of Minnesota Press.

Smith, Linda Tuhiwai (2012). *Decolonizing Methodologies: Research and Indigenous Peoples*, 2nd ed. London: Zed Books.

Spirling, Arthur (2012). 'US Treaty Making with American Indians: Institutional Change and Relative Power, 1784–1911'. *American Journal of Political Science* 56(1), pp 84–97.

Tuck, Eve and K Wayne Yang (2012). 'Decolonization is not a metaphor'. *Decoloniztion: Indigeneity, Education & Society*, 1(1), pp 1–40.

US Department of Interior, Indian Affairs (2017). *Office of Federal Acknowledgment (OFA)* [Online]. Available at: bia.gov/as-ia/ofa (Accessed 15 October 2018).

Walch, Michael C (1983). 'Terminating the Indian Termination Policy'. *Stanford Law Review*, 35(6), pp 1181–1215.

White, Richard (1991). *'It's Your Misfortune and None of My Own': A History of the American West*. Norman, OK: University of Oklahoma Press.

Witkin, Alexandra (1995). 'To Silence a Drum: The Imposition of United States Citizenship on Native Peoples'. *Historical Reflections/Réflexions Historiques*, 21(2), pp 353–383.

Wolfe, Patrick (2006). 'Settler colonialism and the elimination of the native'. *Journal of Genocide Research*, 8(4), pp 387–409.

Worster, Donald (2001). *A River Running West: The Life of John Wesley Powell*. New York: Oxford University Press.

13

THE COMMON LINE PROJECT

Lines, landscapes, and digital citizenships

Paula Crutchlow, John Drever, Chris Hunt, Pete Jiadong Qiang, Volkhardt Müller, Steven Palmer, and John Wylie

Prologue

On 29 February 2020, at Harwes Farm in Lancashire, England, a group of local mothers, caretakers, and young children called *Mums2mums* worked with us to create a first set of tree models to be planted, digitally, upon *The Common Line*. This Line is the longest possible straight line that can be traced across the landmass of Britain without crossing any seas or tidal waters. In the morning, we gathered small branches and cuttings from the trees and hedges immediately around the farm. Then we went inside and began to construct the tree models. The branches and clippings were mounted upon small wooden boards and their bases fixed and moulded with clay dug from a seam near the farm. After that, people were free to spend time painting and decorating the trees as they wished (see Figure 13.1). As the model trees took shape, we discussed the Line and the fractal aspect of trees—each part an iteration of the whole, like the shape of the coastline—and the project generated more questions and issues than answers.

Once the tree models were complete to the satisfaction of their creator, we used our mobile phones to 3D scan them. This is an imperfect process, but one that can be done using several different readily available apps. The resulting 3D digital tree images on our phones were then ready to be 'planted' upon *The Common Line*.

The Line runs adjacent to Harwes Farm, but with time and the still-wintry, muddy moorland conditions in mind, we opted to drive around to the most easily accessible nearby spot *en masse* in a minibus. How strange in retrospect to think of the carefree way in which we crammed onto the bus, just a few short weeks before the coronavirus pandemic brought our fieldwork—including a planned return to Harwes—to an abrupt halt. We were a group of around 15–16 people in total, led by our host and the owner of Harwes Farm, Gillian Taylor, and by our partner Paul Hartley from *In-Situ Arts*, a Pendle based organization who facilitate art making

FIGURE 13.1 At Harwes Farm (photo by John Wylie)

alongside the local communities they are embedded in. On arrival at our chosen site, it was apparent that in the immediate vicinity *The Common Line* ran through three field and property boundaries in rapid succession. These questions of property, ownership, and access emerge sharply almost everywhere along the Line, we have found. We were still able to gather upon the Line here, though, using flags to mark its immediate course and presence. Then we demonstrated to the group how, using augmented reality (AR) software, we can use our mobile phones to plant the 3D digital trees into the live and living landscape.[1] The wind was whistling ahead of another band of incoming sleet, there were small children running free across the site, our tech was glitching—and it felt like a significant moment.

Introduction

We wanted to begin this chapter with this story from our work on *The Common Line*, because we would like to emphasize from the start that this project is primarily about encounter, dialogue and process. We choose to emphasize these aspects of *The Common Line* here in part because our experience has been that people can sometimes, if firstly presented with an overview image of the Line (see Figure 13.2), perceive this as primarily an abstract, conceptual and cartographic project. We sense this sometimes in reactions and questions from academic and artistic audiences, for example. Later in this chapter we will have more to say about linearity, and

FIGURE 13.2   *The Common Line*—full map view (credit: *The Common Line*)

the relations between lines and landscapes, but our work is firstly about meeting and talking with people who live and work on or close to the Line. We arrive as 'strangers with an idea', as one of us has precisely put it, reliant on the hospitality and local knowledges of those we meet. And 'we' in this context, the authors of this chapter, are a diverse group of artists, technologists and academics from different backgrounds The story of how we assembled ourselves is in itself part of the story of *The Common Line*, which we will detail further below. Our work and our approach is thus multi-faceted and hybrid, drawing together academic interests in the interfaces of landscape, ecology, and technology with more direct concerns around land ownership and land access in Britain, explored collaboratively through situated practices of socially engaged and participatory arts.

When we encounter and speak with people in and around the Line, whether by chance or design, we suggest that the conversations and discussions that ensue are in themselves a form of critical and creative landscape practice. Our work could also be described as a form of digital landscape phenomenology, or an exploration of the 'situated digital' (Catlow and Waterman, 2015); because it is about people moving through and locating themselves within landscapes, most commonly using their phones, in a process that combines—and frequently disrupts—bodies, topographies and technologies. It is in these ways that we believe our work speaks to the linked concerns of landscape and citizenship that this volume addresses.

The Common Line is the longest possible straight line that can be drawn across Britain, without crossing any sea or tidal waters, and we were making tree models at Harwes Farm because the ultimate aim for the Line is that it will be a linear forest: a participatory, digital land art work manifested through the planting of trees of all kinds along its entire length—actual trees and imagined trees, native and non-, biological and digital.

The full, interactive version of the Line can be viewed via our website: thecommonline.uk/. We would encourage readers of this chapter to spend a few minutes browsing and investigating it while reading. In this chapter, we will reflect upon how our ongoing work on *The Common Line* may prompt fresh consideration of issues of landscape, digital interaction, and commons in the United Kingdom today. We will begin by outlining the genesis of the Line and its initial framing and determination; we will also introduce our current approach and working practices, as a diversely skilled group of researchers, artists and technologists concerned to imagine and enact the Line not only as a concept or image but as a lively space of encounter between different humans and non-humans. It is above all in this sense that we believe *The Common Line* can serve to host and stimulate new dialogues and experiences in relation to landscapes, identities, and commonality in Britain.

The chapter will then move on to discuss *The Common Line* regarding landscape specifically. This will involve some thinking about the relations between landscapes and lines, and, in this context, about orientations (Ahmed, 2006). In this section, we will reflect firstly on the relations between lines and questions of power and representation in landscape, and then secondly on how lines can help to understand landscapes in terms of bodies, movements, and alignments.

In a last substantive section, we consider questions of citizenship, commonality, and the digital in the context of the Line. One predicate of this book is a normative argument that citizenship and landscape should necessarily be considered together. We think it is straightforward in some ways to understand our work on *The Common Line* as an exploration—and even affirmation—of such an argument. Virtually every encounter with the Line that we have facilitated to date has involved conversations about location, environment, time, and identity. If this has not often explicitly involved debates about 'the right[s] to landscape' (Makzhoumi, Egoz, and Pungetti, 2011), then issues of property, access, local knowledge and claims to belonging have often been prominent in the ways that people have talked when they have encountered *The Common Line*. Our work here connects with established traditions concerned with elucidating relations between landscape and identity in the UK (Darby, 2000; Matless, 1998). But our approach to and understanding of citizenship in this context is notably influenced by the digital elements of *The Common Line*. People need to use their smartphone or another's, frequently with location services enabled, to access and participate in the Line. We seek to minimize and highlight issues of corporate surveillance and datafication that this raises by using transparent and minimal data collection, and open source, public domain softwares and geodata as much as possible. More widely, this raises questions regarding digital citizenship, and here we would endorse and follow Isin and Ruppert's (2020) arguments, in noting that while 'the digital' often escapes the frame of the nation-state in which citizenship is most frequently understood, everyday digital technologies such as smartphones may still offer spaces for active citizenship and the emergence of new forms of political subjects. In the final substantive section of this chapter, we will reflect upon our desires for forms of digitally enabled citizenship that elude both national modes of belonging *and* the seductions of a free-floating yet perpetually captured digital imaginary proffered by today's data-hungry global corporations.

Our conclusions at this time will necessarily be provisional but at the end we will draw again upon our more direct experiences on *The Common Line* to reflect upon how issues and practices of landscape intersect, in frequently challenging ways, with issues of belonging.

The Common Line: genesis, determination, and evolution

The concept of *The Common Line* was created in Spring 2017 by artist Volkhardt Müller. German-born but resident in the city of Exeter in the southwest of England for several years, Müller's practice has long involved explorations of and intervention into the politics of landscape, particularly in Devon, the largely rural county in which Exeter sits. Devon's landscape is often viewed and framed through a classical English rural and pastoral aesthetic, and Müller's work—individually and in collaboration with other members of the artist group *Blind Ditch*—explores and contests this framing, for example through staged *tableaux vivant*, contrasts of urban and rural, conversations with landowners and farmers, and investigations of trees as landscape icons and beacons.[2]

From its first articulation, the Line was already envisioned as a line of planted trees across Britain, with each planting enabled though human labour, stewardship, and care. The necessity of wide and diverse public participation needed to negotiate such a project was equally evident, and in many ways such participation and engagement is *the* key objective of our work. But the Line was also from the very outset a set of questions. Where exactly is *The Common Line* and how might its course be plausibly determined? How practicable, if at all, might the idea of a linear forest across Britain be? How could the Line be brought into presence, encountered, and accessed by different kinds of audiences? What kinds of individual and collective agencies and productive flourishings could be experienced by human and more-than-human actors through the Line?

In the light of such questions, it was apparent from our early conversations that *The Common Line* would necessarily have to take shape via a computational, cartographic and more widely techno-cultural assembly of screens, maps, plots, calculations, activities, and performances. These issues came further to the fore when a funding opportunity regarding 'digital immersion' in places and landscapes arose, and the current core project team crystallized in response. Alongside Volkhardt Müller, other members of the Blind Ditch group involved in work on the Line include performance maker and cultural geographer Paula Crutchlow, sound artist and acoustic ecologist John Drever, and producer and participatory arts practitioner Cat Radford.[3] Creative technologist Christopher Hunt, trading under the name of *Controlled Frenzy*, facilitates the digital development and realization of *The Common Line* through augmented and virtual reality technologies, working in collaboration with architectural designer and researcher Pete Jiadong Qiang. Public Art curator and campaigner Alex Murdin has contributed to the reach and visioning of the work. Lastly, academics John Wylie and Steven Palmer bring expertise in, respectively, cultural geographies of landscape, and remote sensing and digital mapping.

The determination of the exact trajectory of *The Common Line* across mainland Britain involved firstly acknowledging some broad parameters and constraints. The curvature of the earth and the topographical variability of its surface are two such factors. Equally, the in-practice infinite length and fractal nature of Britain's coastline, alongside the daily and monthly fluctuations of the tides, presents a challenge to calculation. To pragmatically address these questions, Steven Palmer downloaded a polyline shapefile digital representation of the UK Coastline from Natural Earth™, a free public domain cartographic dataset. A python script, using the Arcpy module™, was used to convert this coastline shapefile—a Britain with approximately 4000 sides—into a series of nodes or points, spaced 10 metre apart. To ensure that the calculated Line did not cross any sea tidal waters in its course, restrictions were imposed at estuarine points—for example, at Teddington Locks in west London, the tidal limit of the River Thames. The python script then systematically calculated the distance between every two nodes around the perimeter of the UK coastline, and identified the pair that are separated by the longest distance without crossing the coastline boundary. The resulting 880 km of 'Common Line' is the longest possible line linking two points on the coastline of mainland Britain. We are confident that this determination is robust and would be replicated by anyone using similar cartographic and algorithmic tools.

The Common Line thus commences upon a coastal promontory in north-west Scotland, near the hamlet of Mellon Udrigle. Tracking roughly SSE from there, it crosses the Scottish Highlands, and runs through the cities of Stirling and subsequently Carlisle on the English side of the border. Further south, its midpoint lies in the upland Howgill Fells in Yorkshire. From there, the Line passes through the Peak District and later through the centre of Milton Keynes. Crossing the Thames at Teddington Locks—the river's tidal limit—*The Common Line* reaches the south coast of England at Seaford, west of Eastbourne. Our most current visual rendition of the Line articulates it as a line of 44,000 plot points, spaced at 20 metre intervals across Britain. Each of these plots are potential locations for the planting of the trees, living and virtual, through which the proposition of *The Common Line* will be ultimately realized.

Defining *The Common Line* as a linear forest allows us to conceptualize individual trees within a broader ecosystem, raising discussion on the functional, structural and compositional qualities of new tree cover, as well as the ideological, institutional, legal, and operational basis of forestation in the UK more broadly. In every location, attention will need to be paid to soil and climate conditions, in order to determine the most suitable tree species for plantation. There will also of course be sites, chiefly roads and residential areas, in which digital plantation of 3D AR tree models will be the optimal course of action. Twenty metres is the maximum distance trees can be spaced to constitute a 'forest' when fully grown.[4] This 20 metre spacing also addresses the current challenge of GPS 'drift'; the continual movement of a precise, digitally identified geographic location that happens due to the changes in time and location data used to calculate its position. Constructed and processed by multiple satellites in relation to positionings of individual mobile devices, GPS data are affected by a meshwork of weather, mobilities, obstructions, and infrastructure failings. The challenge of GPS drift limits how closely the individual digital identities that are needed for attaching data to each site can be spaced.

To properly introduce *The Common Line*, it is necessary to outline these kinds of details, decisions, and calculations. This runs the risk, however, of presenting the project in primarily technical and technocratic terms. Perhaps the most important point to stress is that the Line is *not* simply a line or set of points on a map. It is a marking, performance, and experiencing of landscape through the integration of living and computational ecologies. The Line thus creates socio-technical-material spaces of encounter wherein citizens can interact with each other and debate the circumstances of landscape, identity, and environment as they are mediated and negotiated in Britain today.

Our practice and ethos in this work is thus profoundly dialogical in orientation. Drawing upon the proposition of art as a medium for critical engagement with social realities developed in discussions of dialogical aesthetics and socially engaged art practice (see Kester, 2013, for example), we seek to bring *The Common Line* to life precisely through encounters and conversations with publics on the Line and both in proximity to and distance from the Line, in symposiums, presentations, and online fora. While our work has not been explicitly shaped by posthuman agendas, *The Common Line* is thus clearly a hybrid and multiple entity, or ensemble, composed of bodies and systems-in-relation—of plants, animals, people, and machines, performances, sounds, voices, and digital augmentations.

At the time of writing we are very much mid-stream with this work. The majority of our initial investigations into *The Common Line* occurred in the city of Carlisle (transected by the Line) and the wider surrounding Cumbrian countryside in the far north of England. In this first phase, we also located and visited the mid-point of *The Common Line* on the Howgill Fells in Yorkshire and staged there a first plantation of a digitally drawn tree. We have conversed with and interviewed a range of landscape stakeholders and ecological experts, arriving at a practical understanding of the challenges of planting and caring for tree species We have subsequently carried out further site investigations and iterations of our approach at locations on the Line including Milton Keynes and Hounslow in west London, and most recently the north Lancashire moorlands, as discussed at the start of this chapter. Through this iterative, site-based work, we are creating a prototype experience of the Line, using smartphone-based technologies to guide public participants and volunteers through their own process of discovery and interaction.

We are experimenting with different forms of mapping, digital display, soundings and orientation, to enable alignment and understanding, and we are designing 3D renditions of trees, tree-shapes and tree related words and sounds that can be digitally located upon the Line using augmented reality technologies.

Throughout these activities we have been guided above all by a principle that digitally mediated and material landscapes cannot, and indeed should not, be conceived of as oppositional or indeed distinct from one another. There is also no critical or artistic insight to be gained from supposing a practical or ontological separation of the digital and material (see Kinsley, 2014 for a discussion). Our ongoing use of digital platforms and frameworks does raise issues of course. The commercial and proprietorial nature of many digital platforms and systems raises ethical questions over their incorporation and use. Yet, while it is in one sense undeniable that digital technologies such as smartphones are every bit as much ideological framing devices as picturesque landscapes, we also insist upon a non-dualistic and performative credo in our work. In other words, there is no pristine, pre-technological landscape that awaits discovery *via* digital devices, and equally there is no immaterial realm of data that might be draped upon or layered onto a physical landscape. We inhabit world's always already digital material. *The Common Line* must therefore be viewed as an attempt to understand and abide within landscapes that are digital, technological, cultural, and material from the outset.

Lines, landscapes, and *The Common Line*

As we reflect further upon our work on *The Common Line*, we ponder how the Line might intersect with wider thinking about landscape as a spatial trope. This involves locating the Line within wider cultural, political, and scientific narratives of landscape, linearity, and visuality. In doing so, we acknowledge the ways in which our work must therefore continually negotiate its inheritance and re-purposing of some classic landscape motifs. We also explore how other recent work on lines and landscapes might speak to *The Common Line*.

Can a line be a landscape? Whatever wide differences there may be across the varied branches of landscape thinking and practice, one common theme is an assumption that a landscape is an areal rather than simply a linear phenomenon. This holds true across many of the ways in which landscape has been most persuasively defined and understood. The European Landscape Convention (ELC), for example, foregrounds the areal aspect of landscape from the outset: "'Landscape' means an area, as perceived by people, whose character is the result of the action and interaction of natural and/or human factors" (Council of Europe, 2000: 2).

Here, the areal nature of landscape is taken to be self-evident to the point where it does not require further definition or analysis. Even if the vagueness of the term 'area' runs the risk of tautology within the ELC definition, the appeal is to a common-sensical and widely shared assumption that landscape infers three-dimensional spatial extension: a landscape has width, depth, and variable elevation. A landscape may therefore *contain* lines, and lines of various kinds (roads, rivers, power lines) may *run across* the landscape and be integral to its meaning, experience, and functioning–but a sense persists that a 'landscape' may not be reduced simply to these lines; instead, it will possess a quality of totality which transcends them.

At the same time, lines, and the diverse infrastructures that accrete around them, can dominate landscapes and the lives of those who inhabit them. Border lines, at once barriers and gateways, are likely the supreme expression of how linear tracings can decisively shape landscapes. In this context, one of our abiding concerns since we began to work on *The Common Line* has been the manner in which its visual and cartographic representation echoes a sense of the line—and of straight lines especially—as an instrument of sovereign or imperial command and control. On one level this is a concern about a potentially naïve subscription on our part to an abstract and mathematical understanding of space and spatial relations. If you zoom in on our current map of *The Common Line*, it disaggregates into a multitude of points (our potential tree-planting sites, see Figure 13.3 below).

At this scale, the Line can seem far removed from the lived experience of landscape; it also has a visibly arbitrary quality, cutting its way across mixed land uses. It is precisely this quality, in which a form of visual rationality combines with elevated indifference to local circumstances, that is frequently found within colonial

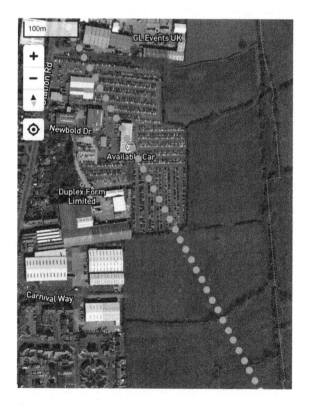

FIGURE 13.3   *The Common Line*—detail (credit: *The Common Line*)

cartographies and vertical power regimes, both historic and present-day (see Ryan, 1996, for an important early discussion; see also Adey et al., 2013; Cosgrove, 2003). *The Common Line* thus risks alignment with the aloof and distanced landscape visions associated with the planner, architect, or surveyor, and our work is thus characterized by an ongoing effort to negotiate myriad relations between map and territory.

If the straight line traced across a map in a bird's-eye view is one dominant motif in the relationship between landscapes and linearities, a clear correlate is of course the equally straight line of linear perspective—the underpinning geometric scaffolding of classical landscape art. The relationship between these two rectilinearities in the histories of Western visions of earth and landscape is a recurrent investigative theme—notably in the work of Denis Cosgrove and Kenneth Olwig, for example. As Olwig (2011: 408) writes:

> Perspectival representation of the landscapes of the Earth emerges when the top-down vertical projection of a standard map is tilted so that a horizontal view of a scene of the Earth emerges. When the angle of cartographic projection is changed, and one looks out into the space of a linear perspectival drawing, an illusion of depth is created … Perspective thus creates the illusion of a both unifying and infinite singular space.

The straight line of perspective thus carries similar connotations to that of the vertical map view—it implies a landscape commanded by the eye of the beholder, and it correlates visual landscape with certain elite subjectivities: the painter, the landowner, the traveller, the military commander. Once again *The Common Line* finds itself necessarily contextualized within deeper critical histories of visual and artistic landscape practice. In a different register, however, Olwig's remarks are of special interest for us, because they pinpoint the exact visual translation that we would like people, using their mobile phones to access *The Common Line,* to be able to experience. In other words, our initial ambition was for a smartphone experience akin to that found on Google Earth™, where subscribers can toggle between map view and 'street view'. Having first used the map view to navigate to within touching distance, people on the ground would be able to see *The Common Line* live, so to speak, through their phone screens, in precise relation to the perspective afforded by their own bodies, stretching off into the distances along its computed trajectory. For now, however, this lies somewhat beyond the limits of technical possibility. Ironically, the visual technologies of the screen space—technologies in which our *Common Line* work is inevitably immersed—still glitch and falter at the precise moment of live presence and witness.

If the straight lines of map and perspective are perhaps the most obvious connections between landscapes and lines, a very different set of associations emerges via Tim Ingold's work, and in particular his book *Lines: a brief history* (2007). For Ingold, perhaps more than any other recent author, a landscape does not *contain* lines—it is *composed* of lines. The lines that compose the landscape are in no way straight, he argues; nor, however, are they merely a random tangle. Instead they are *lifelines*, tracks of human and non-human movement, of energies, actions, and desires. To be alive is to move and, Ingold (2007: 76) argues, "to be instantiated in the world as a line of travel" and thus to sojourn through landscapes as a continual 'wayfarer'. Here, wayfaring,

> Is the most fundamental mode by which living beings, both human and non-human, inhabit the earth … the inhabitant is … one who participates from within in the very process of the world's continual coming into being, and who, in laying a trail of life, contributes to its weave and texture.
>
> *(ibid.: 81)*

As these claims indicate, Ingold's account of landscape in terms of lifelines combines a classically phenomenological focus upon embodied human movement and sensation with a more vitalist emphasis upon wider living and elemental forces: the lifelines of weather patterns, shifting shorelines, animal migrations, organic growth, and decay. If this seems an oddly pre-modern vision of life and landscape, then Ingold's response would be that much of modern life has indeed barricaded us within worlds composed of hard surfaces and straight lines. Wayfaring, as a mode of continual line-making, *is* beholden to creative actions and capacities, whether these be basket-weaving, kayak-making, or song-writing (three of Ingold's examples), and is thus in the broadest sense a 'technological' making-practice. But there is also here a strong claim that things have gone awry in contemporary Western and urban societies characterized by assumptions regarding strict differences between thinking and doing, mind and body, culture and nature. Lifelines have become straight lines. The final chapter of *Lines* is entitled "how the line became straight", and here Ingold notes how the elaboration of increasingly rectilinear and rationalized landscapes in modernity is accomplished via a contrast with a 'nature' for whom straight lines are supposedly 'abhorrent'. Beyond noting the clearly gendered dimensions of this distinction, for Ingold it signifies a general evacuation of life into an abstract, universalized, and geometric spatiality:

> The line of wayfaring, accomplished through the practices of dwelling and the circuitous movements they entail is *topian*; the straight line of modernity, driven by a grand narrative of progressive advance, is *utopian*.
>
> *(ibid.: 167)*

The Common Line is nothing if not utopian in its aspirations to enroll many thousands of people over many years in a tree plantation project some 880 km long. Of course, by 'utopian', Ingold—and Olwig, upon whose work he is drawing—primarily mean 'placeless'. Their concern regards a universalizing, expanding realm of 'nowhere', an abstract space encroaching upon and even annihilating a purportedly more original, embodied space of the lived. In its calculation, determination, and execution, *The Common Line* is wholly beholden to this kind of abstract spatiality. In its indifferent straightness, *The Common Line* could seem to exemplify a thoughtless, placeless modernity. But we would argue our project is both topian and utopian: *The Common Line* offers the possibility of a landscape that is simultaneously 'nowhere' *and* 'now-here'. Every one of its sites, every plot-point, is envisaged as a new line of growth and becoming, a tree plantation that, with carefully steward-ship, could establish a lifelong connection between individual biographies and local ecologies. In this specific focus upon each point as a lifeline extending into the future, *The Common Line* seeks to nurture and set in motion multiple lived tempor-alities, both human and non-human.

More widely, the prompt for us from Ingold's work is not to wishfully invest in some pre-modern or arcadian vision of how lives and landscapes intertwine. Nor is it a case of investigating how phenomenological approaches might encounter and combine with digital and computational spaces. As already discussed, *The Common Line* is instead predicated upon a vision of lines and landscapes that are always already lived, digital, and non-human at once. This is to acknowledge that there is no body or landscape that is 'pre-technological'. And equally to acknowledge that the 'abstract' (the map, the plan, the diagram) is not a category that is alien to the 'lived'. Again, to proceed in this way (as De Certeau, 1984, for example, does for strategic purposes) is to risk an ontological separation of thought and action, digital and physical. Instead we would follow McCormack (2012: 719) in arguing for "the necessity of abstraction for any effort to think through the processual materiality of lived space".

A further significant context for considering lines and landscape comes from Sara Ahmed's (2006) work on orientation. Alignment and orientation are key aes-thetic and technical considerations for our work on *The Common Line*. As noted above, we cannot yet toggle between a bird's-eye and ground-perspective viewpoint on screen. What this means in practice is that, whilst people can navigate to par-ticular points on the Line, it remains very difficult for them to gain an 'establishing shot', so to speak—an evocative and quickly graspable sense of the Line's trajectory at ground level. To put this another way, it is challenging to align yourself in situ with *The Common Line*, and thus to connect with the proposition that in the UK we need to re-think our relations with landscapes both local and afar to find new commonalities across regions and ecologies.

Ahmed's (2006) work is important in this context, because of the way in which she explores and reveals the always political and culture dimensions of phenom-enological concepts of alignment and orientation (something that tends to remain in the background in Ingold's work). Her chief concern in *Queer Phenomenology* is with racialized and sexualized orientations—a focus which allows her to show how the landscapes we inhabit are aligned and oriented in ways that establish straightness and whiteness as directional and spatial norms (ibid.). For us, then, Ahmed offers a language to communicate with: by travelling to *The Common Line* and aligning yourself with it, you are able to re-orient yourself to the needs of both local ecol-ogies and distant strangers. As Ahmed states,

Following a line is not disinterested: to follow a line takes time, energy and resources, which means that the "line" one takes does not stay apart from the line of one's life, as the very shape of how one moves through time and space.

(2006: 17)

The Common Line may well be a straight line, with all the associations and difficulties it presents, yet in its very straightness, and in the lived experience of attempted alignment, it is also a line of deviation, dislocation, and re-alignment.

Citizenship: commoning digital systems?

Our work on *The Common Line* is nested within questions of citizenship and commonality in the wider context of our now routine inhabitation of digital worlds. Viewed broadly, the premise of citizenship is an ability and capacity to speak and act on things that matter to us. Participation in civic matters, however, requires legal status of identity and belonging, combined with knowledge of custom and conduct, and perception governed through configurations of material and institutional technologies. Ideals of citizenship are also increasingly enrolled into matters of concern positioned beyond, in opposition, or in addition to those governed by the nation state—for example, urgent global issues of climate change or the racialized disparities highlighted by Black Lives Matter. These rights-based calls-to-action harness the performative power of digitally networked cultures to increase the force of their imaginaries, with the forms of participation and politicized subjectivities they engender being shaped by the socio-technological structures they are embedded in (Isin and Ruppert, 2020).

The naming of *The Common Line* purposefully evokes the historical and contemporary resources, rights, and social processes that have been variously categorized as being a source of public wellbeing or common-wealth. Based on an understanding that some things belong to all of us, and not just some of us, commons are understood as freely available cultural and natural resources. In this sense, what is held in common is as much shared language, sociality, and collective intelligence as clean water and air, mountains, seas, and forests. As argued by Elinor Ostrom and Massimo De Angelis (in Arvidsson, 2019), commons are also a social institution for collective action: a framework, set of relations, and lived practices for *making* something valuable as much as valuing it. Where capitalist systems metabolize what is common into a resource commodity regulated by the interests of states and corporations, the establishment and governance of commons are constituted and regulated by the communities and social systems who own them, who negotiate and impose their own collective sanctions for misuse. Whilst the practice of commoning is often set in radical and romantic opposition to capitalism, they are not necessarily exclusive to one another, and indeed each supports the survival of the other (Arvidsson, 2019).

Linking the new commons movement to global ideals of green governance, commons activist and green policy strategist David Bollier (2014) describes it as "a social system for the long-term stewardship of resources that preserves shared values and community identity"; something we both inherit and continue to create together through a negotiation that expresses who we are. This demands those participating to live in closer connection with, and communicate transparently about, what might be considered common in order to understand and negotiate how to manage those things together for the benefit of all over time—in this sense echoing the arguments of Egoz et al. (2018) regarding 'landscape democracy'. The activity of commoning is therefore a social vision and ideological outlook that holds the potential to stand in resistance to a dominant hegemony of neoliberalism, capitalist regimes of accumulation, and the environmentally and socially damaging short-term thinking that they promote. It involves inclusive, situated, and embedded ways of evolving practices, lifestyles, and attitudes that work towards this governance. Happening both outside of and within the capitalist production process itself, these practices are necessarily underpinned with understanding of policy, strategy, and law as much as knowledge and skills of how to inspire and coordinate social cooperation and collaboration (Arvidsson, 2019; Bollier, 2014).

The initial frameworks for *The Common Line* have been imagined and developed by the project team as a digital land artwork that is simultaneously an aesthetic medium for discussion with social reality (Kester, 2013) and a social infrastructure for collective action. These frameworks are reliant on existing digital infrastructures and biological ecosystems as much as human participation to transform the abstract proposition of the Line into an activity of commoning. As much as the topological and physical imposition of *The Common Line* might evoke colonial comparisons, the digital networks that support the project's participatory infrastructure need to also be understood as potentially colonizing and extractive architectures (Mejias, 2013). Digital access is geographically, economically, and socially uneven, and coercive modes of protocol-based governance are coded directly into digital architectures. Digital platforms, or socio-technical assemblages with the smart phone at their centre, directly shape the form and effects of participation by managing, organizing, and mining social relations through datafication. Many kinds of governance and management are being transformed by this dynamic and totalizing approach to data collection at an individual level, the analysis and valuing of which is transforming the production and expression of subjectivities and the places and spaces we inhabit (Couldry and Powell, 2014; Van Dijk, 2014).

The extent of our current resources means that, by necessity, we must produce the digital architecture of *The Common Line* using open-source public domain, free to use, or low cost proprietary softwares. Therefore, as much as navigating the patchwork of private, corporate, and public land the Line crosses, we also need to come to grips with how to work within and draw attention to these processes in ways that support, rather than hinder, the project's aims. The creation of cooperative platforms is one way of working within the capitalist framework towards new, less extractive, and more just modes of organization. Other interventions might be citizen-produced data sets and alternative analytics that make different choices about what is measured and valued, and using that data to make alternative economies of information. Additionally, whilst the production and sharing of content contributes directly to digital economies, the symbolic, expressive, and discursive potential of digital media can still be harnessed through the representation, narration, and negotiation of alternative worlds and viewpoints (Couldry and Powell, 2014).

At the time of writing, we are creating a digital platform centred on a smart phone app that will facilitate a set of digital and landscape interactions to be experienced while on the Line. The app will allow publics to locate and encounter the Line as a series of potential tree planting sites that they can help assess the viability of, thereby creating an open data set of site conditions. At the site of encounter, an augmented reality (AR) experience will render a visible and audible virtual tree line, situating *The Common Line* as a three-dimensional layer through smartphones. This will also be a participant-curated experience of AR tree models and sounds, chosen by publics from an open database of digital assets generated through participatory activities, such as those we trialled in Pendle.

Acting as a proxy for the living trees, the AR trees aim to establish community through a disjuncture of landscape experience for individual participants. This draws attention to the forms of participation that we already experience, instigating reflective distance and making space for new realizations and politicized subjectivities. In this way, participation in *The Common Line* works towards the creation of a common experience for those who participate—one that hopefully leads to the connections and empathies that support the action of planting living trees (Tyżlik-Carver, 2014). Ultimately, therefore, the app will offer publics means to plant and care for their own living or AR tree somewhere along the Line, designed and managed through investigatory and participatory processes. In other words, we aim for the Line to be responsive to public input—and this endeavour holds inherent tensions that challenge our ambitions for commoning from the outset. We must remain alert to the tension between an advocacy of commoning at the macro-level, and the potential re-inscription of privatized modes of ownership at individual tree sites. Whilst the terms of use will be inscribed into the overarching digital frameworks, sanctions for misuse will need to be negotiated through lived experience and practice in common with the Line as it evolves in situ.

Digital artist and activist Marc Garrett (2018) proposes that, to build places, spaces, and identities that are independent from the "overpowering infrastructures and systems" and that facilitate neoliberalism's hegemony, we must use our artistic agencies as means of unlocking these proprietorial systems. Agency to do this, whether individual or collective as Garrett suggests, is based in a combination of artistic intention, direct intervention in processes of production and consumption, and the emergence of new sets of values and practices held between art, academia and technology (ibid.). It is through this form of open, creative, discursive, and socially engaged work that *The Common Line* attempts to common the digital systems within which it is embedded.

Conclusion

On 10 July 2018, several members of *The Common Line* team walked to the Line's midpoint. Serendipitously, the midpoint is found close to Cautley Spout, the highest waterfall in England. A notable scenic spot and the epitome of a romantic northern English landscape, the waterfall is set high and wild on the Howgill Fells in Yorkshire. The trip up to the waterfall, while clearly marked, was a stiff hike on a warm and sunny day. Our destination midpoint was in a valley some 20–30 minutes further northwest from there—a further trudge across unmarked boggy moorland.

At the waterfall we sought to mark out the Line's trajectory with flags designed by Volkhardt Müller (see Figure 13.4).

Our activities attracted the attention of several groups of hikers passing through. People were sometimes intrigued, sometimes nonplussed by our explanations of what we were doing and by our descriptions of the Line. But Volkhardt encountered a rather harsher reaction from one passing pair. The moorland we stood upon is a denuded landscape—like much of upland Britain, its current condition and appearance is largely the result of overgrazing by sheep. This, however, did not prevent strong opinions being expressed, to the effect that a straight line of trees would represent an 'unnatural' blight on the landscape. The conversation concluded, in the context of this wide-open space, with references to there being 'too many people' in Britain.[5] Despite this, we continued on foot, undeterred, to the midpoint of the Line. There, we then planted our first prototype AR tree—a Douglas Fir, hand-drawn with Google Tilt Brush™. To any observers we would have seemed a group merely checking our mobile phones, but what we were seeing through them was the potential for a transformed landscape.

In introducing *The Common Line* and the ambitions and complexities of our work regarding it, we have reflected upon the challenges of both determining and even beginning to research the Line. We have discussed the ways in which the project might be contextualized within, and draw insights from, wider thinking about lines and landscapes. And we have argued that our participatory work on the Line involves a critical and creative reappraisal of citizenship and commonality in the context of digital imaginaries and experiences.

We remain hopeful, utopian even, in our visions for how *The Common Line* might unfold as a work of public participation, eventually able to sustain itself autonomously. But our less-positive encounter at Cautley Spout serves as a reminder of how fraught the relationships between landscapes and senses of belonging and identity can be, particularly in a time where renewed racial and exclusionary senses of nationalism and national entitlement limit our capacities for imagining more diverse landscape futures.

On one level, landscape can be understood as a visual myth of not belonging. This argument is put forward concisely by the art historian Robin Kelsey (2008), who believes landscape is a technique for setting the world at a distance from us, but only so that we can deny our involvement and our belonging. On the one hand, we claim that we do not belong to the world, while on the other hand we act as if

FIGURE 13.4   At the mid-point of the Line, Cautley Spout (photo by John Wylie)

it belongs to us as our property (ibid.). In this way, landscape is a visual form which maintains a "fantasy of not belonging to the totality of life of a terrestrial expanse" (ibid.: 204).

If this line of thinking sets us on a path towards legitimizing environmental degradation through distancing, then the opposite route is perhaps even less promising. Investments in and longings for landscape are often characterized by forms of 'homeland thinking' (Wylie, 2016), which tend to handcuff people and land together—as if 'a' people and 'a' landscape somehow naturally belong together. This frequently leads towards the exclusionary narratives of ethnic nationalism. Without denying the sustenance, succour, and satisfaction that people may find in the landscapes they think of as 'theirs' by virtue of inhabitation, it remains very challenging for us to reckon with or endorse this line of thought.

The Common Line moves through and moves on, even as it embeds and locates itself in multiple grounds. It is both topian and utopian. The Line cuts across a myth of not-belonging by encountering landscape as always everywhere embedded and dialogical. Equally it cuts across a myth of belonging by displacing, by moving always in the direction of others and differences by exposing claims to exclusive communion with landscape as always contingent.

Acknowledgements

The Common Line is a partnership project between Blind Ditch, Controlled Frenzy, and University of Exeter. The project has received funding from the AHRC-EPSRC 'Next Generation of Immersive Experiences' Programme, 2017–18, Grant Ref: AH/R009171/1; from the University of Exeter EPSRC Impact Accelerator Fund (2020); and the Invention, Creativity, and Experience fund at the University of Goldsmiths (2020). We have been generously supported through partnership activities with a range of associations and institutions, including: Rural Recreation, University of Cumbria Institute for the Arts, the National Forestry School University of Cumbria, the Woodland Trust, In-Situ Arts, Harwes Farm CIC and Mums2Mums group in Pendle. The project has been further developed through conversation with participants at talks, workshops, and conferences including: *Bath Spa University Environmental Humanities public lecture* (2019), *Tracing the Groundwork*, Milton Keynes Art Centre (2018), *Approaching Estate* hosted by the Central Saint Martins 'Sensing Site' research group at Furtherfield Gallery and Commons, Finsbury, London (2019) and *Walking's New Movements Conference*, University of Plymouth (2019). Particular thanks to the organizers of the *Landscape Citizenships Symposium* (2018) and editors of this volume, Tim Waterman, Ed Wall, and Jane Wolff, for their engagement, support, and comments in the writing of this chapter.

Notes

1 AR software uses a smartphone's built-in knowledge of its position to place and anchor 3D digital objects within the space visible through the phone camera. The user is then able to walk around these objects and examine them from different perspectives on their screen.
2 The following project and artist websites indicate the range and key concerns of Muller's artistic work: toposexeter.uk/; volkhardtmueller.com/; blindditch.org/.
3 The Blind Ditch group formed in 1999, based around the Dartington College of Arts in Devon, and have worked episodically on creative projects ever since. The group are named

after a hidden landscape feature common in Devon that is used to enliven waterlogged or 'clammy' soil—and are also important biological ecosystems. Their website states:

> We are passionate about creative process as a way of provoking public conversation and debate. We want to challenge and entertain audiences and participants as thinking citizens, in surprising and empowering ways. If there were such a thing as experimental, risk-taking and accessible art in whatever form, we hope that we make it ... We run reasonably democratic, interdisciplinary art making processes that invite all kinds of collaborators—other artists, experts and publics—to make things happen with us. We respect genre but we don't aspire to it.
>
> *(blindditch.org/about/)*

4 Personal communication with Dr. Andrew Weatherall, National Forestry School, University of Cumbria, UK. Most attempts to scientifically define and delimit a forest commonly focus upon overall acreage and tree height rather than density per se (Chazdon et al., 2016).

5 Müller's own account of this encounter can be found on our website: thecommonline. uk/2018/07/20/hallucinations-of-the-sublime/.

Bibliography

Adey, Peter, Mark Whitehead, and Alison Williams, eds. (2013). *From above: the politics and practice of the view from the skies.* London: Hurst & Company.

Ahmed, Sara (2006). *Queer phenomenology: objects, orientations, others.* Durham, NC: Duke University Press.

Arvidsson, Adam (2019). 'Capitalism and the commons'. *Theory, Culture & Society*, 37(2), pp 3–30.

Bollier, David (2014). *Green Governance 5: the commons as a growing global movement* [Online]. P2P Foundation. Available at: blog.p2pfoundation.net/green-governance-5-the-commons-as-a-growing-global-movement/2014/06/28 (Accessed 6 May 2020).

Catlow, Ruth and Tim Waterman (2015). *Situating the Digital Commons* [Online]. Available at: furtherfield.org/situating-the-digital-commons-a-conversation-between-ruth-catlow-and-tim-waterman/ (Accessed 7 January 2020).

Chazdon, Robin L, et al. (2016). 'When is a forest a forest? Forest concepts and definitions in the era of forest and landscape restoration'. *Ambio*, 45(5), pp 538–550.

Cosgrove, Denis (2003). *Apollo's eye: a cartographic genealogy of the earth in the western imagination.* Baltimore, MD: Johns Hopkins University Press.

Couldry, Nick and Alison Powell (2014). 'Big Data from the bottom up'. *Big Data & Society*, 1(2), pp 1–5.

Council of Europe (2000). *European landscape convention.* Florence, Italy: Council of Europe. Available at: coe.int/en/web/conventions/full-list/-/conventions/rms/09000016800 80621 (Accessed 8 May 2020).

Darby, Wendy (2000). *Landscape and identity: geographies of nation and class in England.* Oxford: Berg.

De Certeau, Michel (1984). *The practice of everyday life.* Berkeley, CA: University of California Press.

Egoz, Shelley, Jala Makhzoumi, and Gloria Pungetti (2011). *The right to landscape: contesting landscape and human rights.* Farnham: Ashgate Publishing.

Egoz, Shelley, Karsten Jørgensen and Deni Ruggeri, eds. (2018). *Defining landscape democracy: a path to spatial justice.* London: Edward Elgar.

Garrett, Marc (2018). 'Unlocking proprietorial systems for artistic practice'. *Research Values*, 7(1). Available at: aprja.net/article/view/115068 (Accessed 14 July 2020).

Ingold, Tim (2007). *Lines: a brief history*. London: Routledge.

Isin, Engen and Evelyn Ruppert (2020). *Being digital citizens*. London: Rowman & Littlefield.

Kelsey, Robin (2008). 'Landscape as not belonging'. In Rachel DeLue and James Elkins, eds. *Landscape Theory*. London: Routledge, pp 203–213.

Kester, Grant (2013). *Conversation pieces: community and communication in modern art*, 2nd ed. Berkeley, CA: University of California Press.

Kinsley, Samuel (2014). 'The matter of "virtual" geographies.' *Progress in Human Geography*, 38(3), pp 364–384.

Matless, David (1998). *Landscape and Englishness*. London: Reaktion Books.

McCormack, Derek (2012). 'Geography and abstraction: towards an affirmative critique'. *Progress in Human Geography*, 36(6), pp 715–734.

Mejias, Ulises Ali (2013). *Off the network: disrupting the digital world*. Minneapolis, MN: University of Minnesota Press.

Olwig, Kenneth (2011). 'The Earth is Not a Globe: landscape versus the 'Globalist' Agenda'. *Landscape Research*, 36(4), pp 401–415.

Ryan, Simon (1996). *The cartographic eye: How explorers saw Australia*. Cambridge: Cambridge University Press.

Tyżlik-Carver, Magda (2014). 'Towards an Aesthetics of common/s: beyond participation and its post' [Online]. *New Criticals*. Available at: newcriticals.com/towards-aesthetics-of-commons-beyond-participation-and-its-post/print (Accessed 14 July 2020).

van Dijk, Jose (2014). 'Datafication, dataism and dataveillance: Big Data between scientific paradigm and ideology'. *Surveillance & Society*, 12(2), pp 197–208.

Weston, Burns and David Bollier (2011). *Regenerating the Human Right to a Clean and Healthy Environment in the Commons Renaissance* [Online]. Common Law Project. Available at: commonslawproject.org/sites/default/files/Regenerating%20Essay%2C%20Part%20I. pdf (Accessed 5 June 2020).

Wylie, John (2016). 'A landscape cannot be a homeland'. *Landscape Research*, 41(4), pp 408–416.

14

WONDERING THROUGH THE LOOKING GLASS, AND BACK OUT OF THE 'BOX'?

A meta-epilogue

Kenneth R. Olwig

> "How queer it seems," Alice said to herself, "to be going messages for a rabbit!"
>
> *Lewis Carroll (1866: 43)*

It is my thesis that most, perhaps all, of the chapters in this wonderful collection are wrestling at some level with the question of whether to try to work outside or within—and thereafter subvert—the Euclidean/cartographic/perspectival 'BOX'. I will return to this thesis, but first it is necessary to illuminate this somewhat black BOX. The 'queer' nature of this BOX has been most provocatively explored in the writings of the mathematician and photographer Charles Lutwidge Dodgson aka Lewis Carroll, the author of *Alice's Adventures in Wonderland* and its sequel, *Through the Looking Glass, and What Alice Found There* (Carroll, 2015 [1871]: 40). Dodgson loved Euclidean geometry, but was distressed by the absurdity of projective geometry's distortions to it and, by extension, to the substantive material world (Bayley, 2009; Dodgson, 1879).

The 'BOX'

Figure 14.1, by Carroll's chosen illustrator Sir John Tenniel, shows a scene from *Through the Looking Glass*, in which the trace of the BOX becomes visible. Carroll and Tenniel satirize here the absurdity of landscape as a chessboard upon which anthropomorphic animals, plants, and distorted people are boxed into the geometries of life as in a chess game. In Wonderland, those who do not play the game are judged to be queer. From Alice's point of view, however, it is this chessboard landscape, and its citizenry closeted up in its boxed space, that is queer, prompting her to exclaim at the sight of Tenniel's landscape: "It's a great huge game of chess that's being played—all over the world—if this is the world at all, you know" (Carroll,

FIGURE 14.1 Illustration of landscape as chessboard by John Tenniel from Lewis Carroll's 1871 *Through the Looking Glass, and What Alice Found There* (credit and copyright: John Tenniel)

2015 [1871]: 40). Since much of the world has come to look something like this scene, with land divided into boxed properties, Alice's question suggests that perhaps we have all collectively gone through the looking glass, and are wandering around the Wonderland world's landscape?

"In the beginning was the word" in our logocentric universe and images are thus secondary illustrations to the word, but it is to be remembered that Dodgson was a geometer and photographer. In geometry, diagrams play a key role, and images, of course, are the focus of photography, suggesting that the images in the Alice books were more than mere illustrations. Tenniel, furthermore, was the most influential political cartoonist of his time, known for his sometimes radical engagement with political and social reform as satirically imaged in his drawings. Carroll carefully integrated text and image so that they formed an emblematic unity. Figure 14.1 is a perceptive, somewhat surreal, cross between the modern map's Euclidean geometry, with its boxed graticule made up of the lines of latitude and longitude, and an illusory perspectival landscape scene. This illusion is created when the top-down, vertical projection of the map is tilted (as one can easily do today on Google Earth) toward a more *oblique* angle. The origin of the modern map is linked to the iconic ancient Greco-Roman astronomer, geometer, astrologer, cartographer and Platonist Claudio Ptolemy (c. 100–c. 170 AD). It emerged as the standard modern map when it was rediscovered—in what later became understood as the Renaissance. Indeed it arguably contributed significantly to the making of this Renaissance and with it its geometrically planned cities and architecture. It also thereby came to help shape what became the ideology of a polarizing, dichotomizing modernism which defined the immediate past as a stagnant, traditional, and backward 'Medieval' interval age placed between the fall of the Roman Empire and its rebirth in the Renaissance. It was this Renaissance that then became, in modernist ideology, the ur-foundation of subsequent stages of development, each seeking to destroy the previous stage. In the process devlopment supposedly 'progressed' from emplaced 'tradition' into the open universal space of modernity with its endless prospects lying just beyond a (receding) horizon. But is empire and its attendant colonialism really to be emulated? And does society in fact develop in progressive stages?[1]

The invention of the perspectival technique, by which the places of the world are subsumed into geometric space, is most famously identified with the legendary Renaissance architect Filippo Brunelleschi (1377–1446). Standing outside Florence Cathedral, the iconic dome of which he designed, he took a mirror image of a perspectival drawing he had constructed of the adjacent Baptistry, and by looking at the image of it in a mirror through a hole in the drawing, he created a wondrous parallel 3D illusion of the actual Baptistry alongside. The purpose of the hole was to fix and focus the singular eye in the optimal location for achieving the illusion that one has virtually entered into a parallel perspectival space behind the mirror's surface—where the virtual image seen is located. If the eye is improperly located, the objects viewed will look distorted in size and shape. Brunelleschi's dome worked much the same way. Inside the cathedral, with your eye looking up from directly under the hole of its eyelike oculus, you will achieve a correct perspectival illusion of looking into the concentric hierarchy of the Christian heavens and the Ptolemaic cosmos as represented on a fresco.

Alice enters Wonderland through a mirror and a hole—looking downward instead of upward, as in Brunelleschi's dome—but also thereby into the space of a Ptolemaic globe. She enters Wonderland by going straight down a well-like rabbit hole penetrating a vast distance toward the centre of the earth and beyond, into an upside-down realm that Alice hesitantly terms "the antipathies". Alice asks, as she falls: "I wonder what Latitude or Longitude I've got to" (Carroll, 1866: 5). She is, in effect, falling down the space of an infinite series of globes within globes. Ptolemy, whose globe she is falling through, envisioned the cosmos as a series of spheres within spheres, with the earth at its centre, so Alice is falling toward the centre of the Ptolemaic cosmos.

The Euclidean space of the map as boxed chessboard

The Ptolemaic map is made by plotting locational points at the coordinate intersection of the lines of latitude and longitude of the map's rectangular graticule. Though the Euclidean lines and their points of intersection are visible on the map, this is deceptive because it is in fact impossible to see them. As Euclid puts it: "A point is that which has no part", by which he means that a point has no width, length, or breadth, but is an indivisible location. The dot on a map, no matter how small, is therefore (literally) a perversion of Euclid because the Euclidean point has no tangible size whatever, and it is thus not even invisible because it can only exist in the mind—get the *point?* The same is true of the line, which is "breadthless length", and this means that border lines in the Euclidean space of a map will be inherently edgy lines of conflict because they only make sense in the wondrous, non-existent, realm of Euclidean space. Their breadthlessness also means that there will always be room for a new striation in the necessarily substanceless, and thereby homogenous, space that lies between them. This is also true when such breadthless lines enclose a box, so there will be space for an infinite number of boxes within a box.

The perspectival wonderspace of scenic landscape

Tenniel's image is ingenious because it brings out how perspectival depiction does not, first and foremost, r*epresent* an actual pre-existing landscape, but rather expresses an underlying Euclidean geometric structure, the agency of which has predetermined and shaped this landscape. The image thereby reveals both the imagining of a prospective future landscape, planned using an invisible geometry, and the visible outcome of this planning. The surveyor, the planner and the cadas- tral map thus come first, and then the physical landscape is made to conform to their abstractions. In the case of Figure 14.1, the result is the homogenizing chess- board space of uniform crops divided by straight rows of trees and bushes and crossed by rows of straightened waterways. Such a quadratic landscape facilitates both property taxation and the efficient mechanized intensive mass production of uniform and marketable crops, which, in turn, facilitates the taxable flow of capital. There are, as noted by several contributors to this book, other flows that are key to the landscape's life. In this chessboard landscape, however, the elemental flow of water has been channelled above ground through straightened waterways that once meandered generating fertile and water-cleansing meadows and, below ground, by invisible rows of drains. That this boxed-in landscape is antithetical to the landscape of elemental flows is becoming apparent with climate change because increasing flows of water cannot be handled by the drains, and once straightened streams are now re-meandering themselves, bursting through their channels and flooding fields and settlements (Olwig, 1996, 2016).

Whereas the lines of Euclidean space, and its inherent 'digital', striated character, remain apparent in the map's striated graticule, the lines in perspectival drawing are normally not visible, having become an invisible, behind-the-scenes structure, the geometric laws of which determine the shape and appearance of the visible material world. Perspective thus creates the chimera of a smooth, unstriated, seamless geo- metric space in which one's gaze can move out to the point of infinity on the horizon. The fact of the matter, however, is that this is an illusion and one's eye is actually moving from one striated scale to the next, and the lines of perspective are not structuring the material world—they are distorting it. This is illustrated by my addition of two Alices to Figure 14.2. The two identically sized figures are placed near the actual top and bottom of the flat quadratic page, but the one appears to be smaller and in the foreground whereas the other appears to be larger and in the

FIGURE 14.2 Illustration of landscape as chessboard by John Tenniel with two Alices inserted into the landscape. Note that she is actually the same size in both the perspectival 'foreground' and 'background', though she looks bigger in the 'background' (credit and copyright: John Tenniel)

distant background. The opposite would be true, of course, in the world outside the looking glass, where figures appear smaller when seen at a distance. Alice, to enter Wonderland's queer world, must thus shrink in size and shape her organic body to its geometric structure. Carroll appears to suggest that this is akin to eating or drinking (or smoking) certain hallucinatory substances, indicating that Wonderland has a narcotic effect.

The constitutional perspective of the state

As the German philosopher Ernst Cassirer points out in his book, *The Myth of the State*, many who wished to establish a centralized state in the Renaissance and the Enlightenment sought a "Euclidean method of politics" involving "axioms and postulates that are incontrovertible and infallible" (1946: 166). This "myth" provided the rationale for the modern state. It came to fruition with the 1648 Peace of Westphalia that effectively ended the Thirty Years' War and is often seen as establishing the birth of the territorial state as a cartographically bounded spatial body of territory. It not only fostered the spatial enclosure of states within an absolute homogenous cartographic space, it also made possible the subdivision of the state into sub-dividable spatial blocks.[2] The new spatial regime of what became the nation state thereby fostered the enclosure of estates, farms, and urban areas

FIGURE 14.3 Illustration by John Tenniel from *Alice in Wonderland* where Alice finds herself to be out of scale and boxed (credit and copyright: John Tenniel)

into individually owned properties that bestowed citizenship in the state upon its owners, legitimating the state as an expression of the 'natural' incontrovertible axioms and postulates of its underlying geometry. The absolute space of Euclidean politics likewise justified the top-down figure of the cosmic 'Sun King' as the absolute (literal and figurative) *ruler*—who thinks globally and *rules* locally. It is for this reason that space has come to be thought of as being more cosmopolitan than place. This idea seems logical from a Euclidean geometric perspective because it conflates location within the coordinates of space with place.

The substanceless Euclidean point of location is subsidiary to the global framework of the map—maps are flattened globes—but the equation of location and place in fact involves an etymological distortion. *Place* derives not from the Latin *locus*, but from the Greek *plateia*, as do words like plaza, piazza, and marketplace. In Greece, the *plateia* was (and still is) where things like representative legal and governmental assemblies (and lots more) take place and make place. It is not a locus determined by geometry, but a place where people assemble. From where, well before the Renaissance, such places were foundational to the polities then known as 'landscapes'. To this day, countries such as England are governed by representative assemblies ranging from those adjudicating the use of customary legal rights of commoners in particular communities, to parliamentary and judicial national assemblies which seek to assemble a compatible common law governing the rights of the entire country, and to international bodies that work with transnational legal issues. Societies governed this way are not insular, because common use rights do not govern bounded private spaces but the common resources needed for given activities—the herding of grazing animals (including tame geese) thus can involve a moving flow over vast distances, as can the Medieval wandering of pilgrims, craftsmen, students, and minstrels. It wasn't unenclosed place that bound people to a narrow location, but the enclosed, boxed-in space of the state with its unforgivable trespassing of private property.

One of the places where this boxing process was most manifest was the 'Terra Firma' of the Veneto, colonized, enclosed, and drained by the 17th century merchants of Venice and iconically architected and landscaped by Andrea Palladio (1508–1580). This ideal scenic 'Palladian' landscape, that subsequently spread to Britain, was celebrated by the young Denis Cosgrove (1993) as an ideal of modern spatial progress which seemed to justify the alienation of the customary rights to the land of the native agrarian population that it implicated (though Cosgrove (2003) later altered his thinking). The Palladian landscape ideal spread to the United States, particularly under the stewardship of the principle writer of the country's constitution, Thomas Jefferson. He not only designed a Palladian-style villa (*Monticello*) for himself, and pioneered the preservation of designated iconic natural park landscapes as a "natural" parallel to the Palladian; he also began the cartographic process by which the western United States was largely divided into rectangular states, that were again subdivided into rectangular counties, townships, private properties, and eventually individual voting booths. This space became the foundation of individualized citizenship in 'Jeffersonian democracy'.

The process of cartographically boxing-in nations was ideologically naturalized in countries with natural coastal borders like Britain and the United States, with its God-given Manifest Destiny to connect "sea to shining sea". The establishment of a unified German state was challenged, however, both because its core area was made up of diverse politically and religiously fragmented territories, and because much of Europe's culturally German population was dispersed in colonies that effectively governed much of Eastern Europe. Mapping thus played a key role in the process of German political unification and cultural (and racial) homogenization. Carroll wrote at the time when Germany, led by Prussia, was nearing the culmination of the internal unification process. He satirized the German *mappa*-philia in a book, *Sylvie and Bruno*, in which the pair meet a German named "Mein Herr" and the following dialogue ensues:

> "That's another thing we've learned from your Nation," said Mein Herr, "map-making. But we've carried it much further than you. What do you consider the largest map that would be really useful?"
> "About six inches to the mile."

"Only six inches" exclaimed Mein Herr. "We very soon got to six yards to the mile. Then we tried a hundred yards to the mile. And then came the grandest idea of all! We actually made a map of the country, on the scale of a mile to the mile!"

"Have you used it much?" I enquired.

"It has never been spread out, yet," said Mein Herr: "the farmers objected: they said it would cover the whole country, and shut out the sunlight! So we now use the country itself, as its own map, and I assure you it does nearly as well".

(Carroll, 1893: 169–170)

One of the ways the Germans covered the land with the space of the map was by promoting the kind of spatially and vegetatively enclosed landscape seen in Figure 14.1. This was done both within what became Germany proper and, particularly, in the area of Germany's eastern territorial expansion, notably in Poland where the open landscape was seen to be a sign of racial inferiority. This, in turn, justified the expropriation of the land and the cleansing of its 'slavish' people. Germany has been roundly criticized for this plan, but this is perversely 'unfair' given that the Germans were greatly inspired by the often highly celebrated American cultivating and civilizing of the 'wild west'. Comparatively little attention, by contrast, has been paid to the consequent American virtual genocide of the continent's native population.

Out of the 'BOX'

The framework I have developed above provides an important context for knitting the diverse chapters in the book together and linking them to ongoing discourses concerning the ideological role of the modernist definition of landscape as scenic space, and the role of architecture, the state, and liberal economics in the design and planning of the modernist landscape. All the chapters are wrestling in various ways with the question of working outside or within—but subverting—the Euclidean/cartographic/perspectival BOX.

Boxed in and out

It is not surprising that it is the authors of the two chapters focusing on the situation of the native population of the USA and Canada who have most clearly approached landscape citizenship from outside the BOX. Danika Cooper's 'Legacies of Violence: Citizenship and Sovereignty on Contested Lands' is a valuable, historical study of how European settler colonists were made citizens of the BOX while the native population was simultaneously deprived of their landscape citizenship by the BOX—and their population genocidally 'shrunk' to fit the boxed-in reservations reserved for them. James Bird, Ange Loft, and Jane Wolff's 'Landscape Citizenships: A Conversation among Treaty People' takes a more personal and emic approach to many of the same issues broached by Cooper. It brings out, among other things, the important role of language as an alternative to space and scenery as a route to landscape identity and citizenship.

Jala Makhzoumi's socially engaged chapter, 'Beirut's Public Realm and the Discourse of Landscape Citizenships', is also concerned with the way native populations are boxed out of their customary common realm by those with the capital and power to spatially enclose and colonize it as private property. She does this through an analysis of "the perception and valuation of landscape as 'nature', and of landscape as a common communally-shared space that underlies the discourse of activists protesting the Dalieh of Raouche and the Horsh Pine Forest". 'Unearthing Citizenships in Waste Landscapes' by Catherine De Almeida takes an innovative look at another kind of commons—that of waste. She is mindful of the historical meaning of the commons, pointing out that "'waste' and 'common' were used to characterize notions of use, not site ecologies". A commons was (and is) thus not a spatially bound property; rather it was defined by the resources to which commoners had rights for use (e.g. pasture, wood, berries, etc.), and therefore by where this use could take place. The term 'waste' was applied negatively to commons by those 18th and 19th century 'liberal' economists who wished to enclose and privatize them, thus alienating the commoners from their landscape. Enclosure thereby paved the way both for today's agribusiness and the practice of dumping stuff on what was deemed to be wasteland, but which today is being reclaimed for people's recreation.

Borders and lines

The commons are defined by what takes and makes place, and not by their geometric border lines. The same is also true of the common borderlands of seacoasts, as seen in Anna S. Antonova's thoughtful 'Narrating Landscape Citizenship on the Coast: Conflicting Views from the Bulgarian Black Sea and Yorkshire North Sea Shores'. The nature of the seashore landscape as a source of citizenship thus differs considerably, depending upon whether it is experienced as bordering an archipelagic society connected by the open sea or as bordering a land-based society. In his chapter, 'Situating Landscape Citizenships: Borders, Margins, Hybridity, and the Uncanny', Joern Langhorst, like Antonova, focuses on borderlands, the role of water, and the multiplicity of landscape citizenships. He also has a sharp eye for the border as a common liminal zone, the meaning of which is constituted through its crossing. His borderland, as a transitional space that gives entry to the uncanny, has much in common with Carroll's rabbit hole and looking glass as a route to insight in a land of wonder—a Germanic word that can mean both miracle and strange. Yigit Turan, on the other hand, in 'Superkilen: Coloniality, Citizenship, and Border Politics', problematizes not so much the line of an external border, but rather the internal, wedge-like borderlines that box people in and out. Copenhagen's linear Superkilen park (super wedge park), as she dissects it, is the frame of a gentrifying looking glass, reflecting the wondrous designs of real-estate speculators.[3]

In Maria Gabriella Trovato's insightful 'Spatial Inequalities and Marginalization: Displaced Syrians in the Bekaa Valley, Lebanon', the refugee camp functions as a kind of liminal question mark to the BOX. Citizenship, also that of refugees, thereby comes to be seen as being "rooted in social and economic rights that are grounded within the material and ecological landscape of the city and the countryside", not by being boxed in by the spatial borders of the state. As she puts it: "Since landscape is place, nature, and culture-specific, the idea transcends nation-state boundaries and, as such, can be understood as a universal theoretical concept similar to the way in which human rights are perceived". Natural disaster areas, like refugee camps, likewise can facilitate reflection on the normal abnormality of the BOX, as in the case of Mary M. Nelan's 'The Constructed Identity of Disaster Aid Workers and Their Place in the Affected Community'. Nelan's reflection is stimulated by the experience of being an engaged disaster aid worker; this has led to the realization that the landscape with which one identifies as a member of a community of disaster aid workers is basically of a social character and "does not have to be fixed in one place". For this reason, "citizenship to that landscape does not have to be rooted in a single event" as experienced in a particular location.

Subverting the BOX from within

The BOX, like it or not, has become firmly entrenched as a 'reality' in the straightened and "straight" mainstream of society and on the drawing boards of architects and planners. This suggests that another, more practicable, way of breaking out of the box is to subvert it from within, which is what Ed Wall seeks to do in 'Working with uncertainties: living with masterplanning at Elephant and Castle'. In the chapter, Wall reports on work he and his colleagues have done with residents. By punching holes into the space of the maps upon which the master planners have repeatedly erased their communities, the residents were able to reconstruct their own landscapes of belonging, and thereby gain an ability to engage in South London's notoriously uncertain planning policy. With pieces like the Shard, Elephant and Castle Pub, and the Cheshire Cat-like disappearing 'modern' yet out-of-date shopping mall, this landscape indeed recalls a modern Wonderland chessboard.

Tim Waterman and Eglé Pačkauskaité's thought-provoking chapter, 'Avuncular Architectures: Queer Futurity and Life Economies', might also be seen as subverting, or queering, the BOX from within. But this time in terms of both straight society's linear space and modernism's progressive linear notion of time, which relegates the past to the dead. The chapter's primary 'avuncular' inspiration is the eponymous *Mon Oncle* of Jacques Tati's 1958 film classic, whose nephew literally "lives" in a modernist box. For the authors, the film's loving, humorous treatment of modernist absurdity exemplifies "a 'hanging together' in a modernity which allows for the 'simultaneity of the nonsimultaneous', the 'synchronicity of the non-synchronous', and the queer athwartness which cuts across it transversally", thus violating the closeted space and time of the BOX.

Questions raised by the 'BOX'

Finally, there is the approach to the straight line in the chapter by Paula Crutchlow et al., entitled 'The Common Line Project: Lines, Landscapes, and Digital Citizenships'. It takes its literal point of departure from the practice of a conceptual art project to implant, if only in the digital imagination, a linear row of trees across Britain. The chapter, with its combination of art and material landscape, provokes a line of reasoning and a host of questions that I think are relevant to both the BOX and landscape citizenship. In this way, the BOX becomes something to think with. Thinking outside the BOX, or seeking to subvert the BOX from within, is also a way to provoke, image, and imagine alternatives to it. What would a map look like, and do, if it left the BOX? Is it possible to represent and conceptualize place in terms of the analogue and the zero, rather than the digit, the line, and the void?[4] Are there alternative cartographies that can map a topian world of flows, rather than freeze the earth in a global BOX?[5]

Notes

1 This meta-epilogue reflects my personal inspiration from the reading this book in relation both to my interpretation of Carroll's writing and to my own previous thinking and reading. I have expanded on many of the topics broached here mainly in two books with ample references to other literature: Olwig, 2019, 2002. I will not reference myself otherwise unless the topic in question is not covered by these works.

2 On the transition in meaning from "estate" as a form of social status, as represented for example in a parliament (e.g. the nobility, the commoners, etc.), to a landed property, and the parallel transition of the notion of a state as a form of government, with a given social character, to an area of territory, or, the transition from landscape as meaning a form of polity to a kind of space, see Olwig, 2002: 18–20, 50.

3 The design showpiece for Realdania, a key property developer mentioned in the chapter, is a boxy building that looks a bit like the top deck of a cargo ship (constructed on the site of a beloved green children's playground) and is appropriately named "Blox": blox.dk/english/.

4 For an approach to these ideas see Olwig, 2019: 76–87; 2006.

5 Portolan and fractal maps are explored as alternatives in Olwig, 2019: 88–103; 2017.

Bibliography

Bayley, Melanie (2009). 'Alice's adventures in algebra: Wonderland solved'. *New Scientist* [Online]. Available at: newscientist.com/article/mg20427391-600-alices-adventures-in-algebra-wonderland-solved/ (Accessed September 2020).

Carroll, Lewis (1866). *Alice's Adventures in Wonderland*. New York: D. Appleton.

Carroll, Lewis (1893). *Sylvie and Bruno Concluded*. London: Macmillan.

Carroll, Lewis (2015 [1871]). *Through the Looking Glass, and What Alice Found There*. Adelaide: eBooks@Adelaide.

Cassirer, Ernst (1971 [1946]). *The Myth of the State*. New Haven, CT: Yale University Press.

Cosgrove, Denis (1993). *The Palladian Landscape: Geographical Change and Its Cultural Representations in Sixteenth-Century Italy*. University Park, PA: Penn State University Press.

Cosgrove, Denis (2003). 'Book Review: Landscape, Nature, and the Body Politic by Kenneth Olwig'. *Geographical Review*, 93(1), pp 136–138.

Dodgson, Charles L (1879). *Euclid and his Modern Rivals*. London: Macmillan.

Olwig, Kenneth R (1996). 'Reinventing Common Nature: Yosemite and Mt. Rushmore—A Meandering Tale of a Double Nature'. In: William Cronon, ed. *Uncommon Ground: Rethinking the Human Place in Nature*. New York: WW Norton, pp. 379–408.

Olwig, Kenneth R (2002). *Landscape, Nature and the Body Politic*. Madison, WI: University of Wisconsin Press.

Olwig, Kenneth R (2006). 'Global Ground Zero: Place, Landscape and Nothingness'. In: Anne-Marie d'Hauteserre and Theano S Terkenli, eds. *Landscapes of a New Cultural Economy of Space*. Dordrecht, NL: Kluwer, pp 171–192.

Olwig, Kenneth R (2016). 'Virtual Enclosure, Ecosystem Services, Landscape's Character and the "Rewilding" of the Commons: The "Lake District" Case'. *Landscape Research*, 41(2), pp 253–264.

Olwig, Kenneth R (2017). 'Marginalia: Siida and the Alta Petroglyphs—a Fractal Alternative to Cartographic Imperialism?'. In: Tero Mustonen, ed. *Geography from the Margins*. Snowchange Cooperative. Discussion Paper: 17, pp 10–15.

Olwig, Kenneth R (2019). *The Meanings of Landscape: Essays on Place, Space, Nature and Justice*. New York: Routledge.

INDEX

Printed and bound by CPI Group (UK) Ltd, Croydon, CR0 4YY

24/10/2024

01778305-0001